T0293449

Understanding
Game Theory

Introduction to the Analysis of
Many Agent Systems with
Competition and Cooperation

Second Edition

Understanding
Game Theory

Introduction to the Analysis of
Many Agent Systems with
Competition and Cooperation

Second Edition

Vassili N Kolokoltsov

The University of Warwick, UK
St Petersburg State University, Russia

Oleg A Malafeyev

St Petersburg State University, Russia

World Scientific

JERSEY · LONDON · SINGAPORE · BEIJING · SHANGHAI · HONG KONG · TAIPEI · CHENNAI · TOKYO

Published by

World Scientific Publishing Co. Pte. Ltd.

5 Toh Tuck Link, Singapore 596224

USA office: 27 Warren Street, Suite 401-402, Hackensack, NJ 07601

UK office: 57 Shelton Street, Covent Garden, London WC2H 9HE

Library of Congress Cataloging-in-Publication Data

Names: Kolokol'tsov, V. N. (Vasiliĭ Nikitich), author. | Malafeev, O. A. (Oleg Alekseevich), author.
Title: Understanding game theory : introduction to the analysis of many agent systems with
    competition and cooperation / Vassili N. Kolokoltsov, Oleg A Malafeyev.
Description: Second edition. | Hackensack, NJ : World Scientific, 2020. |
    Includes bibliographical references and index.
Identifiers: LCCN 2020032192 | ISBN 9789811214851 (hardcover) | ISBN 9789811214868 (ebook)
Subjects: LCSH: Game theory.
Classification: LCC QA269 .K65 2020 | DDC 519.3--dc23
LC record available at https://lccn.loc.gov/2020032192

British Library Cataloguing-in-Publication Data
A catalogue record for this book is available from the British Library.

Illustrations by A. T. Fomenko.

For any available supplementary material, please visit
https://www.worldscientific.com/worldscibooks/10.1142/11681#t=suppl

Dedicated to our daughters

ANYA, DASCHA, MARGARITA

Preface

This text is devoted to game theory and its various applications. With the aim of being accessible to the widest audience the text is clearly separated into two parts with different levels of exposition.

1. The first part (Chapters 1-8) gives an elementary but systematic exposition of the main ideas of modern game theory without any special prerequisites in mathematics (secondary school level should be quite sufficient). It requires from a reader only an inclination towards logical thinking, equations, some calculations, and an acceptance of the Greek letters ϵ (epsilon), δ (delta), σ (sigma), η (eta) and Π or π (pi). Nevertheless, not following the tradition of popular science books, our presentation is concise, with rigorous definitions, quantitative as well as qualitative results. On the other hand, due to the elementary nature of this presentation, it often stops just where really serious analysis begins, giving (as a compensation) lots of references for further developments.

2. The second part is devoted mostly to the mathematical methods of the theory. Having carefully separated from the first part of this exposition all higher mathematics, we give in Chapter 9 a concise presentation of mathematical aspects and techniques of game theory. This is supplemented by examples and exercises needed to get used to these techniques and to learn to apply them. The level of mathematics required in Chapter 9 is higher than in earlier chapters, but assumes an acquaintance with only basic notions of calculus, differential equations, linear algebra and probability. Chapter 10 presents several concrete (mostly original) game theoretic models and their analysis including game theoretic treatment of rainbow options in financial mathematics, advanced models of inspection, price war games, investment competitions, etc. The sections of this chapter are supplied with problems that can be used as a starting point for individual projects

of students. The material from Chapters 1-9 fertilized by chosen models from Chapter 10 can serve as a basis for an introductory undergraduate course in game theory for both mathematics related degrees and university degrees with a minimal mathematics background (business, economics, biology, etc.). The last Chapter 11 is meant mostly for mathematics graduate and postgraduate students (and could be of interest to researchers), as it requires in places some mathematical culture, for instance some knowledge of functional analysis and stochastic processes. This chapter is devoted to selected topics of more advanced analysis (partially reflecting the authors interests and research) including differential geometry approach (transversality and catastrophe) to stability, abstract dynamic system approach to the analysis of differential games, Bellman type equations for multi-criteria optimization, turnpikes for stochastic games, connections with recently becoming popular tropical (or idempotent) mathematics, as well as with statistical physics (interacting particles). Chapter 9 with chosen parts of the last two chapters can be used for various advanced courses on the mathematical methods of game theory.

The text is aimed primarily at undergraduate students and instructors in game theory as well as at postgraduates in mathematics, system biology and social sciences. We have tried to provide an entertaining and easy read for those wishing to get acquainted quickly with (or to refresh their knowledge of) the beauty of the basic ideas of the game theory, its wide range of applicability and some recent developments. We aimed at helping to make teaching and learning a more interesting and exciting process, supplementing a course by a variety of motivating examples, historical, cultural and general science excursions.

The authors believe that due to the elementary character of the first part, the book can be used also (i) by laymen (e.g. businessmen, politicians and everyone interested in scientific problems and methods), as it introduces rigorous quantitative methods that can be helpful for an assessment of a wide range of human interactions and provides some glimpses of relevant problems from biology, economics and psychology; and (ii) by teenagers, as it is aimed to interest them in the problems of science in general, to involve them into the process of logical thinking and to stimulate their interest in mathematics.

With game theory becoming popular, there is a variety of good textbooks on the theory. As examples of accessible and extensive general introductions one can recommend books [25] and [58]. For nonspecialists introductions to particular areas of theory and applications (well reflected in their titles) one can refer to [2], [12], [21], [24], [29], [30], [31], [32], [33], [39], [40], [42], [47], [51], [53], [56], [57], [134], [138], [142], [143], [144], [159], [178], [179] and [189]. More mathematically oriented introductions can be found in [14], [36], [43], [141], [177], [183], [155], [193] and [128]. The special character of the present text is the clear separation of elementary and more advanced material, the wide covering of the theme in combination with a concise and rigorous exposition, which at the same time is generously spiced with relevant glimpses from literature and history, and finally the discussion of some advanced topics that have not yet found their place in textbooks but have a potential to become a part of the scientific culture of the future.

Almost all chapters of the first part can be read independently (only some general notions introduced in Chapter 1 are used repeatedly). So, according to their wishes and tastes, readers can start from thinking about choosing the best president or prime minister in Chapter 6, or looking through Chapter 1 and then going to explore the biological context in Chapter 4, or touching the curious quantum world in Chapter 7, or going directly to the party games of Chapter 8. On the other hand, the material in each chapter is carefully organized in a logical order and it is advisable to read each chapter from the beginning (the exception being Chapters 10, 11, whose sections are devoted to various topics and perspectives).

The first part of this book is largely based on the Lecture Notes [88] prepared for the students of the Nottingham Trent University. Further work of the authors resulted in the Lecture Notes [99] aimed primarily at the students of St. Petersburg University. The present book is based on the best parts of [88] and [99], fully revised and updated.

To stimulate mathematical and scientific imagination and to add charm to the book we illustrate it by carefully selected artistic graphics of a world renowned mathematician and mathematics imaging artist A.T. Fomenko (to whom the authors express their deepest gratitude for allowing to use his works in this text). Though these works were originally designed to illustrate the geometric structure of the Universe, they fit nicely in the circle of ideas dealt with here. Readers who would like to see more of these graphics are referred to the album [55].

It is our pleasure to thank Geert Jan Olsder and James Webb, who read carefully large parts of the book and made comments that helped to improve it immeasurably. The first named author expresses his gratitude to M. Akian, St. Gaubert, J. Binner, L. Fletcher, B. McEneaney, L. Khodarinova, R. MacKay and V. Maslov for useful discussions and the joy of collaboration on the related topics. The second named author thanks the participants of the research seminars at the Central Economics and Mathematics Institute of the Russian Academy of Science and the Institute of Mechanics and Mathematics of Ekaterinburg, at the Institute of Mathematics of the Belarus Academy of Science and the Institute of Cybernetics of the Ukrainian Academy of Science, at the Stockholm, Sankt-Petersburg, and Moscow State Universities, where he presented his research in game theory on various occasions.

Preface to the Second Edition

For the second edition the book was essentially revised and updated. When preparing the first part of our book we aimed at selecting material reflecting the most fundamental part of the theory that we did not expect to become outdated soon. Therefore the first part remained mostly unchanged (with a couple of new insightful examples added). On the other hand, for the second part, apart from correcting typos, lots of new material was prepared. In particular, Chapter 10 was essentially enlarged by more recent developments on game-theoretic treatment of option pricing and two completely new chapters were written: Chapter 12 on games with many agents analysed via the statistical limit of infinite number of players, and Chapter 13 on quantum games. These two topics represent arguably the brightest new trends in game theory of the 21st century. Due to a strong physics input, the theory of quantum games still remains separated from the main stream of game theory, and our exposition is possibly the first appearance of quantum games in a mathematics textbook.

The authors acknowledge support from the grant RFBR 18-01-007 96.

Contents

Preface   vii

Basic ideas   1

1.   Around the prisoner's dilemma   3

    1.1   What is a two-player game?   3

    1.2   Prisoner's dilemma. Dominated strategies and Pareto optimality   7

    1.3   Prisoner's dilemma for crooks, warriors and opera lovers .   9

    1.4   Discrete duopoly models and common pool resources; public goods   12

    1.5   Common knowledge, rationality and iterated elimination of strictly dominated strategies   13

    1.6   Weak dominance; debtors and creditors   15

    1.7   Nash equilibrium   16

    1.8   Battle of the sexes and Kant's categorical imperative . . .   19

    1.9   Chicken game and the Cuban missile crisis   21

    1.10   Social dilemmas   22

    1.11   Guaranteed payoff, minimax strategy, hedge   25

    1.12   Utility function   26

    1.13   General objectives of game theory; Pascal's wager   27

2.   Auctions and networks   29

    2.1   Several players; the volunteers' dilemma   29

    2.2   An example on iterated elimination of dominated strategies   31

    2.3   Second price and increasing bid auctions   32

2.4 Escalating conflicts 34
2.5 Braess paradox 35
2.6 Wardrop equilibria and selfish routing 37

3. Wise men and businessmen   39

3.1 Wise men and their wives, imp in the bottle, unexpected
    punishment 39
3.2 King Solomon's wisdom 42
3.3 Chain store paradox; centipede game 44
3.4 Normal and extensive forms of a game; battle of the sexes
    revisited 46
3.5 Dynamic games and subgame perfection; pursuit games . 48
3.6 Fair division and the ultimatum game 52
3.7 Cooperation by threat and punishment; infinitely repeated
    games 53
3.8 Computer tournaments; the triumph of the strategy
    Tit-for-Tat 59
3.9 Logical games; limits of the sequences 60
3.10 Russian Roulette; games with incomplete information . . 63

4. Hawks and doves, lions and lambs   65

4.1 Fitness and stability in population biology (general ideas) 65
4.2 Hawk and Dove games as social dilemmas 67
4.3 Mixed strategies, probability and chance 69
4.4 The theorems of Nash and von Neumann 73
4.5 Expectation and risk; St. Petersburg game 74
4.6 Symmetric mixed strategies Nash equilibria 75
4.7 Invasion of mutants and evolutionary stable strategies . . 77
4.8 The sex ratio game 80

5. Coalitions and distribution   81

5.1 Distribution of costs and gains; the core of the game . . . 81
5.2 General principles of fair distribution 85
5.3 Utilitarianism and egalitarianism; compromise set 88
5.4 Equilibrium prices 91
5.5 Linear models and linear programming 93

6. Presidents and dictators   95

    6.1 Collective choice; problems of voting 95
    6.2 Four examples of voting rules 97
    6.3 Criteria of quality of voting rules 99
    6.4 The minority principle; dictators 102

7. At the doors of quantum games   107

    7.1 Quantum bits and Schrödinger's cat 107
    7.2 Lattices and quantum logic 110
    7.3 Rendezvous of Bob and Alice 113

8. It's party time!   115

    8.1 Combinatorial games 115
    8.2 Addition and subtraction of games, order structure 119
    8.3 Impartial games and Nim numbers 121
    8.4 Games as numbers and numbers as games 125

Armed with mathematics   **127**

9. A rapid course in mathematical game theory   129

    9.1 Three classical examples of Nash equilibria in economics . 129
    9.2 Mixed strategies for finite games 133
    9.3 Evolutionary stable strategies 141
    9.4 Replicator dynamics, Nash's fields and stability 143
    9.5 Iterative method of solving matrix games 158
    9.6 Zero-sum games and linear programming 162
    9.7 Backward induction and dynamic programming 163
    9.8 Cooperative games: Nucleus and the Shapley vector . . . 170
    9.9 Revision exercises 171
    9.10 Solutions to revision exercises 175

10. Examples of game models and methods   183

    10.1 A static model of strategic investment 184
    10.2 Variations on Cournot's theme: Territorial price building . 186
    10.3 Models of inspection 188
    10.4 A dynamic model of strategic investments 195

10.5 Game-theoretic and geometric interpretation of
    risk-neutral measures 201
10.6 Interval model for option pricing 214
10.7 Options with sub-modular payoffs 224
10.8 Nonlinear and fractional Black-Scholes equations 229

11. Elements of more advanced analysis   237

11.1 Short overview 237
11.2 Two proofs of the Nash-Gliksberg theorem on the existence
    of equilibria 238
11.3 Introduction to structural stability 242
11.4 Introduction to abstract differential games 255
11.5 Cooperative games versus zero-sum games 264
11.6 Turnpikes for stochastic games 266
11.7 Games and tropical (or idempotent) mathematics 271
11.8 The first order partial differential equations in
    multi-criteria optimization problems 279
11.9 General flows of deterministic and stochastic replicator
    dynamics 285

12. Games of many agents in statistical limit   299

12.1 Controlled mean-field interacting particle systems and their
    LLN 300
12.2 Pressure and resistance framework 305
12.3 Examples: Inspection and corruption, cyber-security,
    counterterrorism, coalition building, network growth . . . 307
12.4 Minority game, its strategic enhancement and sex ratio
    game 312
12.5 Replicator dynamics with mean-field dependent payoffs:
    Controlled hawk and dove evolution 313
12.6 Mean field games (MFG) for finite-state models 315

13. Quantum games   321

13.1 On finite-dimensional quantum mechanics 322
13.2 Measurement in quantum mechanics 331
13.3 Meyer's quantum penny-flip game 336
13.4 First sequential games: Quantum Prisoner's dilemma . . . 339
13.5 First sequential games: Quantum Battle of the Sexes . . . 342

13.6 Variations on MW protocol 347

13.7 Variations on EWL protocol 350

13.8 Quantization of games with continuous strategy spaces . . 357

13.9 Finite-dimensional quantum mechanics of open systems . 361

13.10 Elements of the general theory of quantum games 371

13.11 Further links and examples 372

Bibliography 375

Index 391

PART I

Basic ideas

Chapter 1

Around the prisoner's dilemma

1.1 What is a two-player game?

> - What's our life? - A game!
> - Who is right or happy here, mate?
> - Today it's you, tomorrow me!
> *The Queen of Spades* (opera)

During its life any being attempts to achieve goals which are important to it. However, all beings are in permanent contact with others who are trying to achieve their own goals. Thus any being should always take into consideration the interests and possible actions of other beings: sometimes attempting to outwit an opponent and sometimes forming coalitions with partners that have similar interests. This is precisely what is meant by a game: an attempt to achieve one's goal in an environment where there are other beings that may have opposite or similar, but almost never identical, goals.

Science usually tries to decompose complicated interlaced interactions into simple parts (elementary bricks or links), to analyze their workings separately, and at last to reconstruct the whole chain from these simple parts. We shall start with such an elementary brick of an interaction with conflict interests, namely with a two-player game. It is convenient and customary to give the players some names. We shall often denote our players R and C, say, Ruth and Charlie. These letters are not randomly chosen. In a geometrical, or table representation (see below) they correspond to the Row and Column players. Other common names for the players are Alice and Bob with initials being the first two letters of the Latin alphabet.

The two most natural ways to describe a game are the so-called normal form (tables) and extensive form (game trees). Though the latter usually yields the most adequate description, the former is simpler and is well suited to the analysis of static games. In this Chapter we shall use exclusively the normal (or tabular) form.

A game of two players, Ruth and Charlie, *in the normal form* is specified by the set S_R of possible strategies of R, the set S_C of possible strategies of C, and by two payoff functions Π_R and Π_C. The notion of the strategy is capacious. A strategy can be described by some action, or by a sequence of actions, or more generally by a type of behavior. In any case, a strategy must define the action of a player in any situation that can arise in the process of the interaction of the players according to the rules of the game. The payoff functions specifies two real numbers $\Pi_R(s_R, s_C)$ and $\Pi_C(s_R, s_C)$ for any pairs of strategies (s_R, s_C), where s_R is from S_R and s_C from S_C. These two numbers describe the payoffs to Ruth and Charlie when they apply their strategies s_R and s_C. The payoffs can be negative, which means, of course, that in this case the player rather loses than wins.

When the number of possible strategies is not large, one can conveniently describe such a game by a table, in which rows and columns correspond respectively to the strategies of R (the Row player) and C (the Column player), and where two payoffs $\Pi_R(s_R, s_C)$ and $\Pi_C(s_R, s_C)$ are placed in the cell positioned on the intersection of the row s_R and the column s_C.

Example. *Head and Tail (or Matching Pennies) game.* Ruth and Charlie simultaneously put two coins on the table. If two coins are put in the same way (two heads or two tails), C pays to R one dollar. Otherwise, R pays to C one dollar (the procedure is similar to the game where R announces Tail or Head, and C throws the coin). This game can be represented by the following table

Table 1.1

C

	head	tail
head	1,-1	-1,1
tail	-1,1	1,-1

R

Let us stress here that the first number in a cell shows the winning of R (the Row player), and the second number shows the winning of C (the Column player).

Recall another well known children's game: C conceals a penny in one of his fists and R has to guess Left or Right. It is described by the same table with the strategies Left and Right instead of the strategies Head and Tail.

Yet another realization of this game is the Penalty-Shooting game: football player C shoots Left or Right, goalkeeper R dives Left or Right.

Example. *Rock-Paper-Scissors game.* This is a well known children's game when R and C have both three strategies. They simultaneously display their hands in one of three shapes denoting schematically a rock, a paper, or scissors. The rock wins over the scissors as it can shatter them, the scissors win over the paper as they can cut it, and the paper wins over the rock as it can be wrapped around the latter. A winner takes a penny from the opponent. If both players display the same, then the game is drawn. The game can be tabulated as

Table 1.2

C

		R	S	P
R	R	0,0	1,-1	-1,1
	S	-1,1	0,0	1,-1
	P	1,-1	-1,1	0,0

A quite different example of a game is supplied by chess (or Go or draughts). In the normal form of this game a strategy is a rule which prescribes the next move for a player for any possible position of the chessmen. The number of such strategies is so immense that it is impossible to represent this game in a table even by using the memory of the most perfect modern computers. So, using normal form is not an adequate tool for such games. Here the extensive form is more preferable that we will look at later.

Two classes of games are often considered separately, as they have some nice, but special, properties. These are strictly competitive and symmetric games.

Strictly competitive games (or *games with opposite interests*) are the games where the gain of one player always equals the lose of another one, i.e. $\Pi_R(s_R, s_C) = -\Pi_C(s_R, s_C)$ for all strategies of R and C. This equation can be rewritten in the form $\Pi_R(s_R, s_C) + \Pi_C(s_R, s_C) = 0$, which means that the joint payoff of two players always vanishes. Hence these games are also called *zero sum games*. Both examples on Tables 1.1 and 1.2 above are zero sum games. In case of strictly competitive games one can leave only the first number in each cell of the table of the game, as the second

number always differs by sign only. For instance, the game of Table 1.1 can
be specified by the reduced table

$$C$$

		head	tail
R	head	1	-1
	tail	-1	1

Symmetric games are the games where each player has the same set of
strategies S (i.e. the sets S_R and S_C coincide) and the payoff depends on
the pair of strategies only, and not on the name of the player that uses
them. In other words, $\Pi_R(s_1, s_2) = \Pi_C(s_2, s_1)$ for any pair of strategies
s_1, s_2 from S. For example, the Rock-Paper-Scissor game is symmetric,
because, say, Rock wins over Scissors independently on the players that
use these strategies. However, the game on Table 1.1 is not symmetric.
The table of a symmetric game is a square table, and the second number
in each cell coincides with the first number of the cell that is positioned
symmetrically with respect to the main diagonal (the diagonal going from
the upper left corner to the lower right corner). Hence, symmetric games
can also be described by reduced tables with only one (the first) number in
each cell.

Remark. The modern development of the theory was started at the beginning of the 20th century with the analysis of strictly competitive games.
The first texts for students were devoted mainly to this class of games. In
the future, however, the interest has moved to other classes of games. Now
these original games are called by some authors "the dinosaurs of the game
theory".

Of course, one is interested in solving the games, which broadly speaking
means finding reasonable strategies for players. To begin with, one should
realize that it is not even clear what should be meant by "reasonable strategies", or by "a solution to a game". If everything depended on your actions
only, then it would be clear (at least in principle) that you would choose the
action that gives you the best result (or the best possible approximation to
the objective you have), and this action (or, may be several such actions)
would represent "the solution to your problem". In case of a game, the
result depends on the (often unknown) action of your opponent. Say, if you
play the Head and Tail game described above, choosing Head does not give
you any prediction at all about the outcome of the game. So, what is good
and what is bad? There are various approaches to tackle this problem for
various kinds of games. We shall start the discussion with a (probably, the
most famous) example of a two-action game, called the prisoner's dilemma.

1.2 Prisoner's dilemma. Dominated strategies and Pareto optimality

Two crooks are arrested in connection with a serious crime. However, without a confession the police only have enough evidence to convict the two crooks on a lesser charge. The police offer to both prisoners the same deal (in separate rooms so that no communication between them is possible): if you confess that both of you together committed the serious crime and your partner does not confess, then your sentence will be suspended, and the other will spend 6 years in jail (4 for the crime and 2 for obstructing justice); if both of you confess, then you will both get the 4 year sentence; and if neither of you does so, you both will each spend 2 years in jail for the minor offense.

The table of this game is the following.

Table 1.3

		C	
		confess	not confess
R	confess	-4,-4	0,-6
	not confess	-6,0	-2,-2

What should each prisoner do? Imagine yourself sitting in a cell and thinking what is the best action for you. Of course, you do not know what your partner is going to do, but you can try to consider both possibilities and their results. If your partner does not confess, then you should confess because that gives you 0 years in jail rather than 2 years. On the other hand, if your partner confesses, then you should also confess, because that gets you 4 years in jail rather than 6. So whatever your partner is doing, you are better off if you confess your crime. But you both are in the same position. So the most reasonable strategy for both prisoners is to confess. Then the prisoners end up with the following "solution of the game": they both confess and get 4 years in jail.

Both experience and experimentation testify that this is in fact frequently the decision made by people in such circumstances. Where is the dilemma?

The point is that although each player tries to maximize his/her payoff, the end result for both of them is a payoff which is less than optimal, as the prisoners could end up much better off, if they both do not confess (they both get only two years in jail). This paradoxical result encapsulates the major difference between interactive and non-interactive decision models. One can notice that even if they had agreed before being arrested that they wouldn't confess, each prisoner has no way of ensuring that the other follows this agreement.

As our further examples will show, the prisoner's dilemma appears in disguise in various situations. But first let us make some general observations about the method of solution of a game exemplified above.

Let us say that a strategy s_R^2 of the player R is *strictly dominated* by a strategy s_R^1 of the same player (or s_R^1 *strictly dominates* s_R^2), if R is always better playing s_R^1 whatever the other player does, in other words, if

$$\Pi_R(s_R^1, s_C) > \Pi_R(s_R^2, s_C)$$

for all strategies s_C of C. Similarly one defines strictly dominated strategies for the second player C. Of course, it seems reasonable never to use (i.e. to eliminate) all strictly dominated strategies. If there is a (clearly unique) strategy of a player that strictly dominates all other strategies of this player, this strategy is called the player's *strictly dominant strategy*. If both players have a strictly dominant strategy, then this pair of strategies is called a *(strictly) dominant strategy equilibrium*. One also says in this case that the game has a solution obtained by *elimination of strictly dominated strategies*.

One sees that our solution to the prisoner's dilemma was based precisely on the elimination of strictly dominated strategies (not confess) for both players leading to the dominant strategy equilibrium (confess, confess).

Now, what was good about the better outcome (not confess, not confess)? One says that a pair of strategies (or a *profile*) (s_R, s_C) (and the corresponding outcome of a game) are *efficient* or *Pareto optimal*, if there is no other outcome of the game that makes the payoff of one of the players better than in (s_R, s_C) without making the payoff of another one worse. For example, one observes directly that the profile (not confess, not confess) is Pareto optimal, as any other outcome is necessarily worse for at least one of the players. It is clear that when players of a game are treated as a group, Pareto optimality is reasonable and desirable. Using the profile (not confess, not confess) represents some sort of cooperation between the players. What can be a mechanism leading to cooperation of players and to Pareto optimal outcomes? Under what circumstances can the profile (not confess, not confess) in the prisoner's dilemma be achieved? This remains one of the central questions in the modern game theory, which we shall touch upon later.

Of course, the concrete numbers in Table 1.3 are not important. In general, *a prisoner's dilemma* is a game with the table

Table 1.4

C

		defect	cooperate
R	defect	p, p	q, r
	cooperate	r, q	s, s

where $r < p < s < q$.

Exercise 1.1. Check that the order $r < p < s < q$ ensures that (defect, defect) is a unique dominant strategy equilibrium, which is however not Pareto optimal, the Pareto optimal solution being (cooperate, cooperate).

1.3 Prisoner's dilemma for crooks, warriors and opera lovers

The prisoner's dilemma can be used as a model of a variety of interactions. In this section we discuss some qualitative examples without specifying precise payoffs.

Honor among thieves. Suppose a smuggler is going to sell some goods to a criminal boss. Defecting here just means not fulfilling your part of the deal: e.g. the smuggler can bring a fake diamond instead of the real one, and the criminal boss can just shoot the smuggler on the place of exchange. This is precisely a conflict situation of the prisoner's dilemma type. Of course, defecting gives better payoffs to each of the players, but if both defect they both get a worse outcome than they would by cooperating. You can easily assign some reasonable payoffs for this game.

Dilemmas of the arms race and preventive war. This is possibly the most important example of a prisoner's dilemma type conflict, as in the atomic age the mere existence of our whole civilization depends on the outcomes of such conflicts. Suppose two nations are competing for influence in some region of the globe, and for this purpose are planning to develop some deadly new weapon. Each one is better with the new weapon: if my opponent has developed it, then I would better have it as well in order not to be humiliated; if my opponent has not developed it, then I am again better with it, because a mere threat to use it is a serious argument in support of my superiority. However, of course, both are better off without the weapon: the development requires lots of effort and (what is even more important) a possible use (or misuse) of the weapon by both sides can lead to disastrous outcomes. Clearly, this was the situation with the development of the atomic bomb and then H-bomb by the USA and USSR in 1940s and 1950s. Both sides developed the bomb sticking to the defecting strategy of this conflict, the "Mutually Assured Destruction" (or MAD) strategy!

If the weapon is developed, a conflict of the same type arises when you think about using it. In the 50s of the last century the discussion of preventive war was seriously under way in USA. This amoral but seductive doctrine was supported by the arguments of similar kind: if the Soviets bomb us, then we better bomb them as well, but if they don't, we are again better by bombing them first to destroy their power before it is too late. Among the resolute supporters of preventive war was John von Neumann, an outstanding mathematician of the 20th century, one of the founders of modern game theory, who also took an active part in the H-bomb project and in the creation of the first modern computers. His warmongering words from the time of the Cold War are worth citing: "With the Russians it is not a question of whether but when. If you say why not bomb them tomorrow, I say why not today? If you say today at 5 o'clock, I say why not one o'clock" (cited from [159]). We can only thank God that this time the defecting strategy (preventive war) was not used by either player of this conflict. Seemingly such conflicts are not the best testing grounds to check the validity of game theoretic prediction, not the least because even the assignment of concrete payoff values to such outcomes as war and peace is not obvious. It can be done only subjectively, and by such arbitrary (ad hoc) assignments you can basically justify any action you have in mind. Hopefully, our brief sketch of the problem provides at least some insights into its scope, at least enough for the reader not to be intimidated by the "scientific arguments" of an opponent. For a deeper discussion of the problem see in [31], [80], [170].

Tosca and Cavaradossi. This is a classical example for opera lovers. In the famous opera by Puccini Tosca's lover Cavaradossi is condemned to die before a firing squad. To save him, Tosca agrees to give herself to the henchman Scarpia, if he, in turn, agrees to order a fake execution of Cavaradossi using blank cartridges. However, in accordance with the solution to the prisoner's dilemma, each then breaks the agreement by choosing his or her dominant strategy, namely Scarpia secretly countermands the order to switch to blanks (thus killing Cavaradossi) and Tosca stabs Scarpia with a knife instead of giving herself to him.

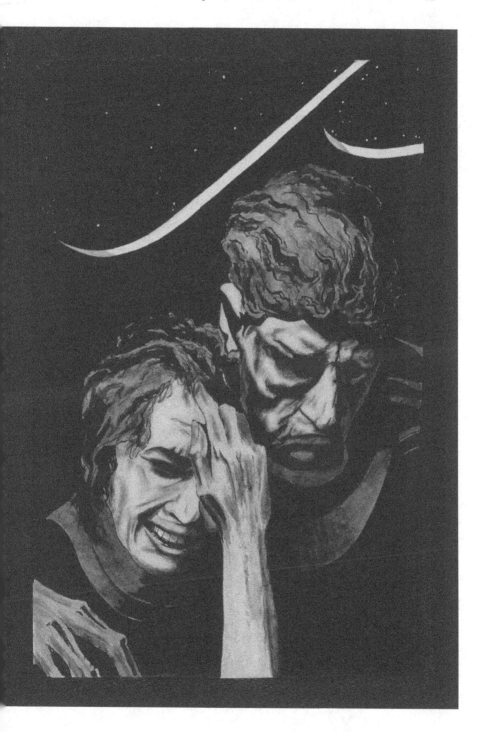

1.4 Discrete duopoly models and common pool resources; public goods

Two firms R and C produce and sell a product (identical for both firms, e.g. sugar of the same quality) on the same market. The price of the product is supposed to decrease proportionally to the supply, i.e. if Q_R and Q_C are the quantities of the product (in some units, say, thousands of barrels of oil, or hundreds of bottles of wine, or tonnes of sugar, etc) produced by R and C (say, in some fixed period of time), the market price for the unit of the product becomes $a - b(Q_R + Q_C)$, where a and b are some positive numbers. The cost of production of a unit of the product is c for both firms. Hence the payoffs are

$$\Pi_R(Q_R, Q_C) = Q_R[a - b(Q_R + Q_C)] - cQ_R,$$

$$\Pi_C(Q_R, Q_C) = Q_C[a - b(Q_R + Q_C)] - cQ_C$$

(assuming that everything is sold out). Since the number of strategies is infinite in this example, to honestly investigate the problem some calculus is required. But to catch the essence of the problem, it is enough to analyze the following simplified version of this game. Let us choose some concrete simple values of parameters, say $b = 1$, and a, c such that $a - c = 12$, and assume that both firms have a choice between only two levels of production: produce 3 or 4 units of the product. This reduces the discussion to the two-strategy symmetric game with the table

C

		$Q_C = 4$	$Q_C = 3$
R	$Q_R = 4$	16,16	20,15
	$Q_R = 3$	15,20	18,18

which is precisely of the form of the prisoner's dilemma game of Table 1.4. In this example "cooperate" means to produce less of the product and get better payoff. As we discussed above, this is efficient (Pareto optimal), but differs from the dominant strategy equilibrium and is not stable under unilateral deviations of either of the two firms.

Basically the same payoffs can describe a model of exploiting a common pool resource. Say, two persons fish in the same lake, or two persons hunt or look for mushrooms in the same wood. Denoting by Q_R and Q_C the level of their efforts in some units (say, days per week spent on the hunt), by c the cost of a unit of effort, and noting that as the pool is finite, the more effort they both apply, the worse becomes the reward on a unit of effort, we arrive to the same formulas for the payoffs $\Pi_R(Q_R, Q_C)$ and $\Pi_C(Q_R, Q_C)$ as above.

Another relevant example is given by the following "Public goods game". Ruth and Charlie are supposed to play the following game: both are given the same sum of money, say 20 pounds. Then each of them has a right to either keep this money or anonymously deposit it in a "public account". The money in the public account will be then increased by 50% and shared equally between the players. As one easily checks, this game can be described by the following table

		C	
		defect	cooperate
R	defect	20,20	35,15
	cooperate	15,35	30,30

where "cooperate" means here, of course, to deposit your money in a public account. This game (which is again a version of prisoner's dilemma) is instructive for a discussion of the Social Contract. For instance, one can interpret the strategy "deposit money in a public account" as paying taxes. Of course, you are better, if you do not pay taxes, but everyone else does. However, if no one pays taxes (everyone is as selfish as you are), then there is no financial support to community services (health care, schools, etc.), and everyone is worse off than if everyone had paid taxes. Here arises, of course, the usual dilemma for politicians: increase taxes and improve social services (cooperative strategy) or decrease taxes and allow everyone to do the best on his own (defective strategy). Further philosophical discussion can be found in [66].

1.5 Common knowledge, rationality and iterated elimination of strictly dominated strategies

The method of the elimination of dominated strategies is based on the assumption that the players are rational. One can go further by assuming that not only the players are rational, but they all know that the other players are rational, and then further that the players all know that the other players all know that they are rational, and so on ad infinitum. This chain of assumptions is called the *common knowledge of rationality*. It turns out to be rather practical, as it leads to the method of solving games by the *iterated elimination of strictly dominated strategies*, whose efficiency we shall show using the following game:

$$C$$

		1	2	3
R	1	5,0	5,4	0,3
	2	0,4	0,3	5,2

Here R has no dominated strategies, and consequently the game can not be solved by elimination of dominated strategies. However, for C the last strategy is strictly dominated by the second one and hence can be eliminated as a feasible strategy of C, which reduces the game to the table

$$C$$

		1	2
R	1	5,0	5,4
	2	0,4	0,3

Of course, R needs to use the Common Knowledge of Rationality (R knows that C is rational) to deduce that C will never play his strictly dominated strategy, and consequently to bring the game to the above reduced form. In this reduced form the second strategy of R is strictly dominated by the first one, so it can be eliminated (second step in the iterated elimination of strictly dominated strategies) and hence the game can be reduced to the table

$$C$$

		1	2
R	1	5,0	5,4

In this game the first strategy of C is strictly dominated by the second one and using "C knows that R is rational and that R knows that C is rational" the player C can now eliminate his first strategy, which leads to a conclusion that the game will be played with R playing the first strategy and C playing the second one. If applying the iterated elimination of strictly dominated strategies leads to a unique pair of strategies, this pair of strategies is again called a *dominant strategy equilibrium or a solution obtained by the iterated elimination of strictly dominated strategies*.

The use of common knowledge of rationality can be looked at as a logical version of an ethical principle: "do not do wrong to anybody, because if you do, he/she would feel the pain just like you would do".

1.6 Weak dominance; debtors and creditors

Another generalization of the method of the elimination of strongly dominated strategies is the method of the elimination of weakly dominated strategies. A strategy s_R^2 of the player R, say, is called *weakly dominated* by a strategy s_R^1 of the same player (or s_R^1 *weakly dominates* s_R^2), if playing s_R^1 is always not worse and sometimes better than playing s_R^2, in other words, if

$$\Pi_R(s_R^1, s_C) \geq \Pi_R(s_R^2, s_C)$$

for all strategies s_C of C and

$$\Pi_R(s_R^1, \tilde{s}) > \Pi_R(s_R^2, \tilde{s})$$

for at least one strategy $\tilde{s}$ of C. As in case of strictly dominated strategies, one can argue that it is reasonable never to use weakly dominated strategies. If a strategy of a player weakly dominates all other strategies, then this strategy (which is unique, as one can show) is called *the weakly dominant strategy*. If each player has the weakly dominant strategy, then this pair of strategies is called a *(weakly) dominant strategy equilibrium*. The following example is instructive.

C and R are creditors, and a debtor owes 3000 pounds to each of them, but he has only 5000. The cost of liquidation of his assets is 3000, so if he defaults on the debt, he loses all his money, but R and C each gets 1000 only. The debtor offers to both C and R 1001 pounds, if both of them agree to cancel his debt. If at least one of them does not accept the offer, he would declare a default. The table of payoffs clearly has the form

Table 1.5

		C	
		accept	not accept
R	accept	1001,1001	1000,1000
	not accept	1000,1000	1000,1000

so that "accept" is a (weakly) dominant strategy, which also yields (unlike the prisoner's dilemma) an efficient (Pareto optimal) solution. If R and C adhere to this strategy, then the debtor will retain 2998 for himself.

In the last section of this chapter we shall discuss the reasons why the solution (accept, accept) may not correspond to the real life resolution of this conflict situation, though formally this is of course the only reasonable solution of the game above.

Now we would like to stress that a serious problem with weakly dominated strategies can appear when applying iterated elimination of weakly dominated strategies. One can show that the results obtained by iterated elimination of strictly dominated strategies can never depend on the order of such elimination. But in case of iterated elimination of weakly dominated strategies, the situation is different and the final result can actually depend on the order of elimination (see e.g. [58], [188]). So the latter method should be applied with some caution.

1.7   Nash equilibrium

In discussing the prisoner's dilemma and related games we have developed several tools for solving games (even if a solution sometimes differs from the outcome one would expect or would like to achieve). As we illustrated, the games of prisoner's dilemma type are common in a variety of real life interactions. However, there are lots of other games that could not be solved by the elimination (even iterated) of dominated strategies. We still have no clue to the solutions of even the simplest games like scissors-paper-rock. The two corner stones of the modern game theory that together could yield a solution, at least in principle, to any finite game, are the notions of the Nash equilibrium and of mixed strategies. We shall discuss now the first of these notions, and the second one (which requires some understanding of probability) will be explored in an elementary way in Chapter 4.

To grasp the basic idea behind the Nash equilibrium let us look again at the profile (not confess, not confess) of the prisoner's dilemma of Table 1.3. Why is such an outcome not feasible, even if we allow for a preliminary agreement of the players to act this way, say, like in the example of Tosca and Scarpia in Section 1.3? Because an individual deviation from such an agreement would allow any player to do better. In other words, this situation is unstable, if players are allowed to independently explore their most profitable choices. Avoiding such instabilities leads to a general notion of an equilibrium (as a solution of the game) that we are going to define now. Namely, a pair of strategies $(\tilde{s}_R, \tilde{s}_C)$ of R and C is called a *Nash equilibrium* for a given game if

$$\Pi_R(\tilde{s}_R, \tilde{s}_C) \geq \Pi_R(s_R, \tilde{s}_C)$$

for all strategies s_R of R, and

$$\Pi_C(\tilde{s}_R, \tilde{s}_C) \geq \Pi_C(\tilde{s}_R, s_C)$$

for all strategies s_C of C, i.e., if each of the strategies $\tilde{s}_R$ and $\tilde{s}_C$ is the best reply to another one. In other words, neither player can do better by deviating unilaterally from the profile $(\tilde{s}_R, \tilde{s}_C)$. Saying it in yet another way, if such a pair of actions is agreed between players, then no one would have any reason to break this agreement. Nash himself stressed that the equilibria are the "no regret" outcomes, meaning that if the players played a game according to an equilibrium, then no one would have any regret about his/her action. After a game, ask each player, whether he would do differently given the action of the other players. The equilibrium is a profile, where everybody is happy with the way he/she played. It turns out that many games are in fact played according to a Nash equilibrium. What is more surprising, an equilibrium behavior manifests itself also in the biological context, i.e. in the interaction of animals, where one can not expect any rational calculations of the best responses: see Chapter 4.

Remark. The life of John Nash, an outstanding American mathematician who got the 1994 Nobel Prize in Economics, was popularized recently in a spectacular Hollywood film "The Beautiful Mind".

Nash equilibria can be easily found if a game is described by a table. Namely, a pair of payoffs in a cell of such table corresponds to a Nash equilibrium, if the first number is a maximum amongst all first numbers from the cells of the same column, and the second number is the maximum among all second numbers from the cells of the same row.

Exercise 1.2. Check that the profiles (confess, confess) and (accept, accept) are Nash equilibria in games given by Tables 1.3 and 1.5 respectively.

The main drawback of the notion of the Nash equilibrium is the possibility of having several equilibria, which leads to the problem of choosing which one among them is the most plausible. Moreover, though no strictly dominated strategy can obviously be a part of a Nash equilibrium (so, elimination of strictly dominated strategies can not lead to the loss of a Nash equilibrium), a weakly dominated strategy can be a part of such equilibrium.

Consider the following symmetric modification of the Head and Tail game. Namely, suppose that Tail always wins against Head and in case of identical outcomes both players lose, i.e. the table is

Table 1.6

C

		head	tail
R	head	-1,-1	-1,1
	tail	1,-1	-1,-1

One sees by inspection that the situations (head, tail), (tail, head), (tail, tail) are all Nash equilibria. At the same time, strategy "head" is clearly weakly dominated by "tail", and hence the elimination of weakly dominated strategies (both players eliminate "head") leads to the solution (tail, tail), which is the (weakly) dominant strategy equilibrium of this game. So in this example, elimination of dominated strategies allows one to choose seemingly the most reasonable Nash equilibrium. Equilibrium (tail, tail) is also more reasonable from the symmetry, because situations (head, tail) and (tail, head) can not be considered as fair, as they give a winning to one of the player at the expense of the other one under quite symmetric conditions.

However, it turns out that looking from another point of view shows that one can not claim that the profile (tail, tail) giving the payoffs $(-1, -1)$ is in fact absolutely the best for the players. First of all, this outcome is not Pareto optimal, as there are outcomes that are better for one of the players without making any additional harm to another one. At the same time, the "unfair" (non-symmetric) Nash equilibria (head, tail) and (tail, head) are Pareto optimal.

On the other hand, if one imagine that this game is played again and again several times, then the players can agree to play half of the times (tail, head) and another half of the times (head, tail), which gives to both player the average payoff zero, as each then wins and loses an equal number of times. This outcome is, of course, better than what each gets from the dominant strategy symmetric equilibrium (tail, tail). The possibility of alternating the choices leads to the second corner stone of the theory mentioned above, namely that of the mixed strategies. We postpone the discussions of this topic till Chapter 4.

The "innocent" game of Table 1.6 can actually serve as a model for a variety of conflict interactions in a variety of contexts (biological, psychological, etc.), and can be called a *sacrifice game*. Going to macabre extremes, suppose that a sadistic maniac caught you and your loved one and offers you to play this game, where the payoff −1 means you will be seriously injured (hand cut off or suchlike), and the payoff 1 means you will be set free. What would be your actions? In [159] a time dependent version of such a "game" is discussed, where two persons are held in separate cells and are given one hour of time during which each can pull a trigger (playing "head") that would bring a punishment on himself and set free another one. If no one would pull the trigger, then after an hour both triggers are automatically pulled bringing a punishment to both.

As another example, imagine two persons on a flooded area have a chance to jump onto a lifeboat ("tail" strategy) which can bear, however, only one person (i.e., two persons in the boat are doomed to die). Everyone can imagine lots of similar situations.

1.8 Battle of the sexes and Kant's categorical imperative

Consider the following game: R and C independently guess an integer between 1 and 3. If the numbers coincide, both win the amount they guess. Otherwise each loses the amount of his/her guess. The normal form of the game is

$$C$$

		1	2	3
R	1	1,1	-1,-2	-1,-3
	2	-2,-1	2,2	-1,-2
	3	-3,-1	-3,-2	3,3

Exercise 1.3. Convince yourself that the outcomes (1,1), (2,2) and (3,3) are the Nash equilibria in this game. Moreover, (3,3) is the only one which is also a Pareto optimal outcome.

Notice that in this game there are no dominated strategies, but the solution is easily found from the notion of Nash equilibrium combined with efficiency. In general, a game is called a *pure coordination game*, if it has a unique Nash equilibrium that is also Pareto optimal. These games are mostly simple, as more or less any reasonable player would play such a game according to the Pareto optimal Nash equilibrium.

Another famous example used to illustrate the notion of Nash equilibria is the "battle of the sexes" game that models a common problem for (especially young) families. Suppose husband and wife plan an evening's entertainment. They would prefer to stay together, but the husband wants to go to the football match, and the wife would prefer to watch a ballet in the theater. Assigning two units of pleasure to "being together" and one unit of pleasure to "spend time according to your choice" yields the two-player game with the table

Table 1.7

Wife

		football	ballet
Husband	football	3,2	1,1
	ballet	0,0	2,3

Exercise 1.4. Convince yourself that the outcomes (3,2) and (2,3) are both Nash equilibria, which are both efficient (Pareto optimal).

The game of Table 1.7 is not symmetric, but intuitively one feels some symmetry in the positions of wife and husband. In fact, one can make it symmetric just by changing the order of columns, which leads to the representation

Wife

		ballet	football
Husband	football	1,1	3,2
	ballet	2,3	0,0

of the battle of the sexes game. In this representation the first strategy means just to be selfish, and consequently a more general representation is

C

		selfish	non-selfish
R	selfish	1,1	3,2
	non-selfish	2,3	0,0

or even more generally it can be described by Table 1.4 with $s < p < r < q$ (to be selfish corresponds, of course, to defection in the prisoner's dilemma).

Even simple game theoretic models can be useful in testing and clarifying philosophical theories. For instance, recall famous Immanuel Kant's *categorical imperative*: "Act only on such a maxim through which you can at the same time will that it should become a universal law" (Metaphysics of Moral). For the prisoner's dilemma game this imperative clearly prescribe you "to cooperate" (as you do not want your opponent to defect, you shouldn't do it either) leading to a nice "moral" solution (cooperate, cooperate).

On the other hand, the battle of the sexes shows clearly the limitation of the application of the famous principle: not liking people to be selfish yields by this principle the outcome (not selfish, not selfish), which is the worst possible for both players. In the literature this situation is nicely reflected in a beautiful short story of O'Henry about a poor family, where (in order to make a Christmas present) the Wife sells her long hair to buy a chain to the old family watch of her Husband's, while the Husband sells this watch to buy appropriate combs for her hair.

Generally one can observe that the categorical imperative seems morally persuasive only for games with symmetric equilibria, see [40] or [66] for more extensive discussion.

1.9 Chicken game and the Cuban missile crisis

As it is connected with a crucial and a very dangerous moment in the history of mankind, this story is widely represented in the literature, see particularly books [170], [80], [31]. So we shall be very short. The classical *chicken game* is played by choosing a straight road and driving two fast cars towards each other by the middle of the road from opposite directions. Mutual destruction is inevitable, if neither of them changes the course. If one of the drivers swerves before the other, the other, as he passes, shouts "Chicken!" The driver who has swerved becomes an object of contempt. The table

Table 1.8

	not swerve	swerve
not swerve	-1,-1	2,0
swerve	0,2	1,1

yields a reasonable example that captures the main features of the conflict. The worst happens, if both will not swerve (both defect) and die in a crash: payoffs $(-1, -1)$. The best for a player is to show his superior courage by not swerving when another did so (and becomes a chicken with payoff 0). Notice the difference with the prisoner's dilemma where each player does always better by defecting. Here a player is better by defecting only if another one cooperates. As is easily seen, the Nash equilibria are asymmetric outcomes (2,0) and (0,2). How to choose between them? As we shall see later, using mixed strategies can give a reasonable answer for some types of similar interactions, if they are repeated several times. But it does not help in a single experiment. Of course, as in the case of the prisoner's dilemma, one can vary the entries of the table essentially without destroying the main features of this conflict. The general *game of a chicken* is a game with the table of the type

	defect	cooperate
defect	p, p	q, r
cooperate	r, q	s, s

where $p < r < s < q$.

A prime example of such a conflict was the nuclear stalemate between USA and Soviet Union during the famous Cuban Missile Crisis of 1962, where "not to swerve" strategy meant, of course, to continue escalating the conflict with nuclear war at the end of it (the same MAD strategy already mentioned).

The analogy between arms race and the chicken game is artistically presented by Adreano Celentano in the brilliant Italian comedy "Innamorato Pazzo".

1.10   Social dilemmas

Symmetric two-player, two-action games with some dilemma between cooperation and defection (or selfishness) are sometimes called the simplest social dilemmas. For an extended and exciting discussion of these dilemmas (with psychological aspects and various examples) we refer to [159] and [40]. We shall only describe them briefly. There are several types of these games. In some sense, they are all similar to the prisoner's dilemma, but each stresses some different aspects of conflict interactions. The prisoner's dilemma, the battle of the sexes and the game of chicken are the most famous among these dilemmas.

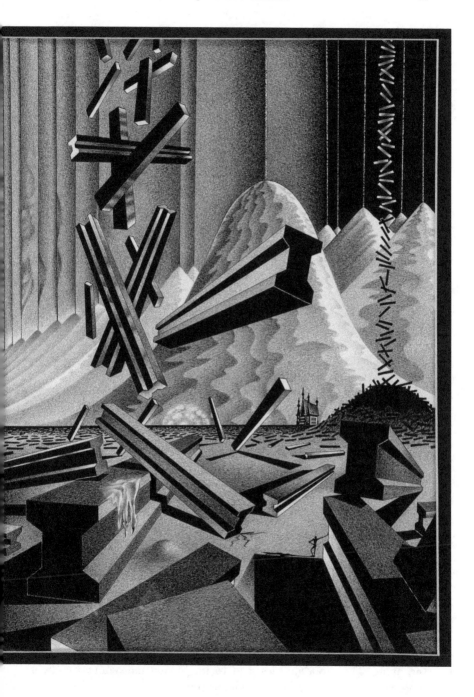

Let us mention two other dilemmas that are conceptually much simpler being examples of the pure coordination games. They can be exemplified by the tables

Table 1.9

	defect	cooperate
defect	2,2	3,0
cooperate	0,3	1,1

and

Table 1.10

	defect	cooperate
defect	1,1	2,0
cooperate	0,2	3,3

and are called the *deadlock* game and the *stag hunt* dilemma respectively. More generally, they are given by Table 1.4 with $q > p > s > r$ and $s > q > p > r$ respectively. The difference between these games consists basically in the fact that for the first one (defect, defect) is a Nash equilibrium, and for the second (cooperate, cooperate) is a Nash equilibrium. The poetic title "Stag hunt" for the second dilemma came from Jean-Jacques Rousseau's book [168], where hunting alliances in the first human societies were, in particular, described. To get a stag all hunters must act together (cooperate). If one of them were to defect by chasing a hare instead, then the stag would not be caught. Of course, everyone prefers a stag to hare, and a hare to nothing. However this "nothing" could be a payoff for a cooperative hunter in a company containing too many defectors, though (cooperate, cooperate) is, of course, a Nash equilibrium in this game, which is moreover Pareto optimal.

A distinguishing feature of the social dilemmas considered is the symmetry of the corresponding games. This means that both players share the same preference. In general, the preferences need not match. Say, one player can have preferences as in a chicken game and another those of a stag hunt. These possibilities lead to interesting nonsymmetric (hybrid) games that model a variety of human conflicts.

1.11 Guaranteed payoff, minimax strategy, hedge

It is worth mentioning yet another approach to a rational choice. A player may be interested in obtaining something for sure. A payoff P is called *attainable* or *guaranteed* for a player, if this player has a strategy that guarantees him a payoff not less than P whatever strategy would be used by the opponent.

It is not difficult to calculate the maximum guaranteed payoff in any game with finite number of strategies: it is the maximum over all available strategies (s_R for Ruth and s_C for Charlie) of the worst outcomes that can occur applying those strategies. Formally, the maximum guaranteed payoffs for Ruth and Charlie are

$$G_R = \max_{s_R} \min_{s_C} \Pi_R(s_R, s_C), \tag{1.1}$$

$$G_C = \max_{s_C} \min_{s_R} \Pi_R(s_R, s_C). \tag{1.2}$$

The corresponding strategies that yield the maximum in (1.1) or (1.2) are called the *minimax strategies* of Ruth and Charlie respectively.

Example. For the prisoner's dilemma of Table 1.4 the strategy "defect" is the minimax strategy leading to guaranteed payoff that coincides with what one gets in the Nash equilibrium. In the chicken game of Table 1.8 the minimax strategy for either player is swerve (i.e. cooperate), and the outcome (cooperate, cooperate) obtained when both use their minimax strategies does not correspond to any Nash equilibrium.

At this point you may start thinking like "Well, guys, you have so many notions of rationality, all of them yielding different outcomes, that you can justify everything (and hence effectively nothing) by choosing an appropriate approach". Well, of course, there are lots of different notions of rationality, they coexist together and usually compete in your mind (you may wish to have a guaranteed salary, but you may wish also to take a risk by playing lottery and hoping for a large win). One of the tasks of the game theory is to make you aware of various reasonable possibilities and to predict what you can get *given the objective* (or rationality criterion) *you have chosen.*

The notion of attainability or, equivalently, of a guaranteed payoff is central for various applications, in particular in financial mathematics (especially in the theory of *derivative securities*, also called *contingent claims*), where the guaranteed payoff (and also the corresponding strategy) is called a hedge. For game theory applied to hedging financial securities (which is a natural ground for game theoretic approach) we refer, for instance, to [23], [87], [136], [150], [124] and [35], see also Chapter 11.

1.12   Utility function

Talking about payoffs we tacitly assumed that we measured the outcomes of games by some monetary values. Of course, this is not always possible, an obvious example being the interactions of animals. To compare the outcomes in a general context one should assess the outcomes by some personalized measures of utility (degrees of happiness) that reflect the personal preferences of the players. A *utility function* of each player prescribes a numerical value of the utility of this player for all possible outcomes. Using the game theoretic jargon, we shall often keep referring to these utilities as payoffs, as we did above. The difference, however, should be kept in mind. For example, in biological context, the payoffs (or utilities) are usually expressed in terms of fitness of species, which in the simplest case represent just the reproduction rate. Even if an outcome is given by some amount of money, the face value and actual value of this amount can be quite different for different players. Suppose you have a debt of 10000 dollars that is due for payment tomorrow, and suppose that non-payment would result in serious consequences (arrest, dishonor, death, etc.). So for you the actual value of this sum is much greater than just 10000. Similarly, you can imagine that you need a quick medical treatment that costs 10000. To try to make some comparison between actual and face value of this amount of money, imagine now that you won this amount in lottery just in time to fix your problem. You are happy, of course. Suddenly you win the same amount again. Will this second win double your joy?

The difference between actual rewards and individual utilities has important consequences for game theory, especially for its applications and interpretations. An implausible outcome predicted by game theory is often due to an incorrect identification of individual utilities with monetary outcomes. Personal utility functions can reflect human perceptions of fairness, national or personal preferences (or prejudices), etc. For example, the outcome of the Debtor-Creditors game described above is not plausible, because the creditors would probably consider such an outcome as unfair and an additional pound would not suffice to make them happy. Actual experiments with various games could in fact be useful for identifying personal utility functions and for catching and understand some psychological aspects of human behavior. This leads, in particular, to psychological applications of game theory, see e.g. [161], and [58] for a recent research.

1.13 General objectives of game theory; Pascal's wager

Now is just the right time to talk about the general aims and objectives of the game theory. In this introductory chapter we discussed the simplest examples of games and their elementary properties. In the future we shall make a deeper analysis of these games, and some more involved models will be analyzed. But the reader should see already the main point: we shall discuss the interaction of individual preferences, where these preferences as well as the results of any concrete actions are described by certain numerical characteristics.

Roughly speaking, the general objectives of game theory are the following:

(i) to work out the concepts of reasonable (rational, stable, equilibrium) strategies of behavior and to find when these strategies exist and how to calculate them; to analyze these strategies in situations when a cooperation and a coalition formation is possible;

(ii) to analyze the optimal (equilibrium) strategies dynamically, when a game is developing in time (e.g. chess, pursuit) and one has to correct the behaviour subject to permanently varying circumstances;

(iii) to describe and to assess the methods of the dynamical transition from an arbitrary state to an equilibrium (say, the ways of moving an economy from a crisis to a stable development);

(iv) to clarify the laws of the creation of cooperation from purely individual preferences; these laws can be views as certain social analogs of the laws of the creation of order from chaos.

One can distinguish two approaches to this analysis: theoretical (mathematical) and practical (applied). In the first one, the stress is on the development of general methods (analytical or numerical) that are meant to be suitable for application to wide classes of problems. In the second approach one starts an analysis from concrete real life situations (so-called case studies). Surely these two approaches are closely connected (in biological language they are in a symbiotic relationship), as for practical purposes general techniques are required, and a theoretician needs concrete models to test his/her general methods.

From a historical perspective it is worth noting that the development of game theory was often closely linked with the military applications. This history can be traced back to the ancient Chinese book "The Art of War" written by the philosopher and military leader Sun Tsu about 6 hundred years B.C. This book is still studied in detail in many higher education institutions of contemporary China. Though it does not contain any formulae, its arguments on the strategy (see chosen quotation in [47]) anticipate many formal constructions of the modern game theory. The first translation of this book in a European language appeared in France in 1782, which allows one to suggest that it was known to Napoleon.

The second (historically) main impetus to the development of game theory was given by the practical analysis of the games of chance, which also stimulated the development of the probability theory. As one of the first serious treatises in this direction one should mention the book "Liber de ludo aleae" (1560) of Girolamo Cardano, whose name is well known in mathematics also in connection with the formulas of the solutions of the third order algebraic equations and in engineering for the "Cardan shaft", which allows the transmission of rotary motion at various angles (still used in modern car industry).

We conclude this introductory chapter with Pascal's wager, where the famous French philosopher, theologist and natural scientist B. Pascal argued the advantages of religious faith with an imaginary rationalist, approaching with game theoretic reasoning arguably the most important and intriguing question of the human civilization: the question of the existence of a benevolent God. Pascal compares human life with a game, where a coin is tossed that will turn up at the end of the human life: "Let us weigh gain and loss in calling heads that God is. Reckon these two chances: if you win, you win all; if you lose, you lose naught. Then do not hesitate, wager that He is" (cited from [47]). Pascal could not use the modern language of game theory (as it did not exist in the 18th century), but it is easy to cast his arguments as a formal game of two players, a human being and Nature, that has previously determined whether there exists a God who cares about human beings (see [47] or [37] for details): as an error with God's existence has infinite payoffs (eternal life or eternal damnation), any finite "payments" resulting from belief would be insignificant cover the expenses implying that the minimax strategy of a self-interested rationalist is to believe in Pascal's God.

Chapter 2

Auctions and networks

2.1 Several players; the volunteers' dilemma

The concepts of the theory of two-person games described above can be extended straightforwardly to the case of an arbitrary number of players, which leads to the notion of general noncooperative games. A *noncooperative game* with an arbitrary, finite number of players $A, B, C, ...$ in *normal form* can be described by the sets S_A, S_B, $S_C, ...$ of possible strategies of these players and by their payoff functions $\Pi_A(s_A, s_B, s_C, ...)$, $\Pi_B(s_A, s_B, s_C, ...)$, $\Pi_C(s_A, s_B, s_C, ...), ...$ These functions specify the payoffs of the players $A, B, C, ...$ for an arbitrary profile $(s_A, s_B, s_C, ...)$, where a *profile (or a situation)* is any collection of strategies s_A from S_A, s_B from S_B, s_C from S_C, etc.

Strictly and weakly dominated strategies, dominant strategies and Nash equilibria are defined in the same way as for two players. In particular, a situation $(s_A^\star, s_B^\star, s_C^\star, ...)$ is called a *Nash equilibrium*, if none of the players can win by deviating from this situation, or, in other words, if the strategy $s_A^\star$ is the best reply to the collection of the strategies $s_B^\star, s_C^\star, ...$, the strategy $s_B^\star$ is the best reply to the collection of the strategies $s_A^\star, s_C^\star, ...$, etc. In formal language this means that

$$\Pi_A(s_A^\star, s_B^\star, s_C^\star, ...) \geq \Pi_A(s_A, s_B^\star, s_C^\star, ...)$$

for all s_A from S_A,

$$\Pi_B(s_A^\star, s_B^\star, s_C^\star, ...) \geq \Pi_B(s_A^\star, s_B, s_C^\star, ...)$$

for all s_B from S_A, etc. Zero sum games and symmetric games are defined as for two players.

Most of the games considered in the previous chapter have natural extensions to a game with an arbitrary number of players.

Exercise 2.1. Try to describe the analogs of the common pool resources game and the public goods game of the previous chapter for three
or more players. One can find out that similar prisoner's dilemma type
contradictions between cooperative and noncooperative behavior arise for
these games that are also called N-person prisoner's dilemma.

As a philosophical application let us note that an N-person prisoner's
dilemma model makes clear the distinction between *general will* and *the
will of all* made rather vaguely (and thus causing much trouble for further
commentators) in one of the most influential books in political and social
philosophy, "The Social Contract" of Jean-Jacques Rousseau: "there is
often a great difference between the will of all and the general will; the latter
regards only the common interest; the former regards private interests, and
is merely the sum of particular desires" (cited from [40]). In the prisoner's
dilemma the difference between "common interest" (everyone cooperates)
and the "sum of particular desires" (to defect) is quite apparent, see [40]
for a fuller discussion.

A multi-person version of the chicken and sacrifice games is the "volunteer's dilemma" that can be illustrated by the following story. Suppose
that after a storm the electricity and the telephone connection are cut off
in a small village you live in. Then someone has to drive to the electricity
company to ask them to fix the problem. Lots of people would prefer to
save effort by letting someone else to do the job. As another version of the
same conflict, you can imagine lots of military situations, in which someone
has "to pay the price" in order to save other people's lives.

This situation is also reflected in a well known anecdote about four
brothers called Anybody, Somebody, Everybody and Nobody who agreed
to carry out some job with the following result. Anybody could have done
the job. Everybody thought Somebody would do it, but Nobody did it.

It is possible to play a volunteer's dilemma as a party game. Ask all
participants to submit a bid (written secretly on a piece of paper) asking
for either 2 or 5 pounds. If, say, not more than 20% of participants ask
for 5, then everyone gets what he/she asked for. Otherwise no one gets
anything (version: in the last case, everyone pays a 1 pound fine).

2.2 An example on iterated elimination of dominated strategies

Consider the following simple game with an arbitrary number of participants. All n participants announce simultaneously an integer between 1 and 100, say, $k_1, k_2,...,k_n$. A winner is a player, whose number turns out to be the nearest (in magnitude) to the half of the average of these numbers, i.e., to the number

$$\frac{k_1 + \cdots + k_n}{2n}.$$

Notice that there can be several winners, but let us exclude from the analysis (for simplicity) those rare events when there are no winners. All losers (those who are not winners) pay a dollar, and the obtained sum is equally divided between the winners. The following exercise shows that this game can be solved by iterated elimination of dominated strategies.

Exercise 2.2. (i) First show that the strategy "announce 100" is strictly dominated by the strategy "announce 99", and hence it can be excluded as a reasonable strategy of the game. Secondly show that when the strategy "announce 100" is eliminated, the strategy "announce 99" is strictly dominated by the strategy "announce 98", and hence "announce 99" also can be excluded as a reasonable strategy of the game using the second iteration of the method of elimination of strictly dominated strategies. Going on with this procedure (i.e. eliminating successively 100, 99, 98, 97, etc.) show that the only strategy left is the strategy "announce 1", which is therefore the only dominant strategy in this game.

(ii) One may be tempted to think that for arbitrary $k < l$ the strategy "announce l" (i.e. the bigger number) is strictly dominated by the strategy "announce k", and hence choosing the smallest number is always optimal, and consequently to choose the dominant strategy "announce 1" one does not need any iterations of the method of elimination of dominated strategies. Show that this is not the case. (Hint: consider three players announcing numbers 2, 30, 100; half of the average is 22 and the winner is the player that announces 30, and not the smallest number 2.) Show, however, that the case of only two players is quite special. In that case, for arbitrary $k < l$ the strategy "announce l" is in fact strictly dominated by the strategy "announce k", and hence the dominant strategy "announce 1" can be obtained by just one step of the method of eliminations.

(iii) Play the game with your friends. Will they stick to the dominant strategy?

2.3 Second price and increasing bid auctions

There are n bidders that submit a single bid, in secret, to the seller of an object, i.e. each player i (with $i = 1, ..., n$) sends a letter to the seller announcing a bid for the object (a positive number), say, v_i. The winner of the object is the player with the highest number among all v_i, but he/she pays not the price he/she announced, but the next highest bid. This auction is sometimes called a Dutch auction, or a Vickrey auction.

Remark. W. Vickrey got the 1996 Nobel Prize in Economics, in particular, for his contribution to the theory of the Dutch auction. Unfortunately, W. Vickrey died three days after receiving the news that he had been awarded the Nobel Prize.

Assuming for simplicity that draws are excluded (i.e., excluding the situations when the two highest bids are equal), let us show that truth telling, which means to each player i to announce as his bid his true value t_i (i.e., the price which he is willing to pay for the object) is the (weakly) dominant strategy and, in particular, truth telling for all players is the unique Nash equilibrium in this game. To this end we shall compare the strategy t_i of player i with other strategies v_i. We have to consider two cases. (i) Suppose i wins the object with the bid t_i, and suppose the second highest bid that he actually has to pay is $u \leq t_i$. Then if his bid were $v_i \geq u$, he would get the same win for the same price, but if his bid were $v_i < u$, then he would lose the auction thus getting worse payoff than with t_i. (ii) Suppose i loses the auction (payoff is zero) with the bid t_i, and suppose the highest bid was $h > t_i$. Then if player i bid $v_i \leq h$, he loses anyway, thus changing nothing. If he bid $v_i > h$, then he wins the auction but has to pay $h > t_i$, i.e. more than he would like to, thus having negative payoff, which is again worse than bidding t_i.

Exercise 2.3. Compare the Dutch auction with the usual auction, where the winner is again the player with the highest bid, but he pays the price he announced. Show that in this case truth telling is not a dominant strategy.

One can analyze similarly the *flower auction* in the Netherlands, where one starts with a very high price, which is decreased. The first hand up gets the flowers.

Another similar example is the so called *increasing bid auction*. There are n bidders in an open auction. The auctioneer begins by bidding at some initial price (stated in dollars, or thousands of dollars, etc.), and raises the price at the rate of one per second. All bidders willing to buy at the price stated put their hands up simultaneously. When only one hand is put up, this highest bidder is the winner and has to pay the price stated.

Exercise 2.4. Assume for simplicity that a strategy of any player i is to choose a price v_i and keep a hand up until the auctioneer's price goes higher than v_i. (In fact, it is easy to show that any other strategy is dominated by a strategy of this type.) Show that the optimal strategies are the same as for the second bid auction, i.e., the dominant strategy for each player i is to choose as v_i the actual amount he is willing to pay for the object.

2.4 Escalating conflicts

Escalating conflicts can be illustrated by the *dollar auction game* devised in 1971 by Martin Shubik. The rules are as follows. An auctioneer auctions a prize of value v (say, US dollars) and the players (an arbitrary number of them) take turns in some fixed order. Each player either increases the bid by one or drops out of the game. The games ends when only one player remains. The difference with the usual auction is that both the highest bidder (who gets the prize v) and the second highest bidder (who gets nothing) must pay the auctioneer the amounts of their last bids. A strategy p_i of player i is, of course, to keep bidding until the level of spent money goes beyond p_i (say, p_i is the amount of money in your pocket) or until everyone else has dropped out.

Exercise 2.5. Show that all Nash equilibria have the following form: one bidder i chooses $p_i \geq v - 1$ and all others $j \neq i$ choose $s_i = 0$. Bidder i then gets v.

The structure of Nash equilibria imply that if two or more players started to bid positive amounts, they would continue to escalate their bids without reaching an equilibrium thus sustaining considerable losses of money. The history of this game was analyzed in [39], where it was called the Macbeth Effect by quoting Shakespeare: "I am in blood, Stepped in so far that, should I wade no more, Returning were as tedious as to go o'er". One can observe that the dollar auction game can serve as a model of a variety of real conflicts, not the least important being the arms race that we already mentioned above in connection with prisoner's dilemma. We refer to [40] for an appealing description of experimental tests of this game and for further examples from politics and economics, like the long participation of the US in the Vietnam war, where the usual argument against disengagement was that US had "too much invested to quit".

Example. *Moscow dollar auction.* This game and its variations were once very popular on the streets of Moscow. (When organized by a crook, it can bring lots of free money.) An auctioneer declares to several "players" that together they won a prize v (say, in dollars), but they have to compete for this prize by the following procedure. The players take turns in some fixed order. Each player can either add 10 dollars to the prize, or drop out of the game. The game ends when only one player remains. This player receives the whole prize.

Exercise 2.6. The game above has no Nash equilibria. Hint: in any outcome, those who did not win, could act better than they did (in principle, of course).

The analysis of auctions is a popular theme in the modern game theory that we have only touched upon. See, e.g., [61] and references therein for more recent developments.

2.5   Braess paradox

This well known paradox discovered in [28] demonstrates a surprising phenomenon in traffic behavior, when increasing the size of the network can lead to a decrease in its performance. We shall give a simple example, following [28] and [14]. In the network of five one-way roads in Figure 2.1 six drivers have to go from B to E. The time needed to traverse each segment depends on the number of cars, n, using this segment and equals (in some units) $10n$ for BC and DE, $50 + n$ for BD and CE, and $10 + n$ for CD.

The drivers wish to minimize their driving time by choosing (simultaneously and independently) one of the three possible routes: BCE, BDE, $BCDE$. This leads to a six-player game, whose outcome can be compactly described by three integers x_1, x_2, x_3: the number of drivers that have chosen the first, the second and the third route, respectively. For any such outcome the time of travel clearly becomes $10(x_1 + x_3) + 50 + x_1$ for the first route, $50 + x_2 + 10(x_2 + x_3)$ for the second one, and $10(x_1 + x_3) + 10 + x_3 + 10(x_2 + x_3)$ for the last one. It is not difficult to see that in a Nash equilibrium all these times should be equal (otherwise some of the drivers could do better by switching to another route) leading to the system of equations

$$10(x_1+x_3)+50+x_1=50+x_2+10(x_2+x_3)=10(x_1+x_3)+10+x_3+10(x_2+x_3),$$

$$x_1 + x_2 + x_3 = 6.$$

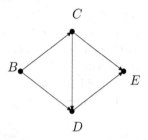

Figure 2.1

Exercise 2.7. Show that this system has a unique solution $x_1 = x_2 = x_3 = 2$ yielding the unique Nash equilibrium with total driving time 92.

Suppose now that segment CD is closed and hence the third route is not available. Similar (and simpler) considerations show that the unique Nash equilibrium becomes $x_1 = x_2 = 3$ leading to a total driving time of 83. Thus reducing the network leads to a better result or, the other way round, increasing capacity yields a worse equilibrium. This effect is called the *Braess paradox*.

The situation is similar to the N-person prisoner's dilemma. If, in the first network, the drivers could agree to minimize the maximum of the three total driving times (leading to a Pareto optimum), the result would be $x_1 = x_2 = 3$ like in the Nash equilibrium of the second network (with route CD). But such an outcome (corresponding to the cooperative outcome in prisoner's dilemma) is not stable against cheating, as using individual preferences leads to a different (and overall worse) outcome. Individual choice of shorter (or cheaper) routes, often at the expense of overall performance of the network, is called *selfish routing* in the literature.

The importance of the consequences of the Braess paradox are obvious for the planning and construction of networks (new roads in congested cities, internet communications, etc.). See also [38] for analogs in quite different, mechanical and electrical contexts. This has led to extensive work on the performance and detection of the Braess paradox (see e.g. [15] for a non-technical account and [167] for a general review). Unlike our example above, in real network modelling it is usually assumed that an agent controls only a negligible fraction of the overall traffic network (there are too many cars in a big city to look at them individually) thus leading to (game theoretic) models with infinitely many players. We shall touch upon this subject in the next section.

2.6 Wardrop equilibria and selfish routing

Consider again the network in Figure 2.1, where now one unit of traffic that consists of infinitely many individuals (like points on the interval $[0,1]$) has to pass from B to E. Suppose the cost of transportation for an individual is zero along CD, is one for BD and CE and equals the proportion x of the traffic along the segment for BC and DE. Again there are three paths and if x_1, x_2, x_3 indicates the proportions of traffic driving again these routes, then the costs of driving along these routes become $x_1 + x_3 + 1$, $x_2 + x_3 + 1$, $x_1 + x_2 + 2x_3$ respectively. Unlike the example of the previous section, there are infinitely many players and the x_j stand for proportions thus satisfying the normalizing condition $x_1 + x_2 + x_3 = 1$. The analog of the Nash equilibrium in such networks is occasionally called the *Wardrop equilibrium*. It is not difficult to understand (as for the Nash equilibrium above) that, in a Wardrop equilibrium, the proportions of traffic going through each possible route are such that the costs of following each route are the same (for a rigorous discussion see, e.g., [67]). As easily seen the system of equations

$$x_1 + x_3 + 1 = x_2 + x_3 + 1 = x_1 + x_2 + 2x_3, \quad x_1 + x_2 + x_3 = 1$$

has the unique solution $x_1 = x_2 = 0$ and $x_3 = 1$, leading to a cost of transportation of 2 along each route.

Now, if segment CD is closed and the third route is not available, similar considerations lead to the solution $x_1 = x_2 = 1/2$ with the cost of transportation being $3/2$ (again better than with the road CD). The ratio of costs under the Nash (or Wardrop) equilibrium (2) and the optimal flow $(3/2)$ is $4/3$. The remarkable and surprising result of the theory, see [167], states that the ratio $4/3$ is rather universal and holds under very general assumptions about the networks (latency of edges is a linear function of their congestions). See also [16] for the existence and essential uniqueness of Nash equilibria in such networks.

Chapter 3

Wise men and businessmen

3.1 Wise men and their wives, imp in the bottle, unexpected punishment

To define complex dynamic games through their normal form is not an effective procedure, because the whole dynamics becomes hidden inside the strategies. In this chapter we are going to systematically exploit the dynamics of games, the main instrument for this analysis being the method of backward induction (and, closely connected with it, the graphical method of the representation of games by means of game trees). This method is based on the application of the Common Knowledge of Rationality (discussed in Chapter 1) to dynamics or, roughly speaking, choosing actions on foreseeing the best replies of the opponent: "if I do this, then my opponent would do this". We shall try to get used to the logic of backward induction through a series of instructive examples, some of them leading to quite paradoxical conclusions.

We shall start with the following story.

In a city there live several wise men with their wives. Some of these wives are unfaithful. There is a law in the city that demands that whenever a husband learns that his wife is unfaithful, he must stab her that same night and bury her publicly on the next day. However, though each wise man knows the truth about all other wives (the wives discuss their affairs between themselves and each wife can't help telling these stories to her husband hiding, of course, her own story, if any), none of the wise men knows whether his wife is faithful or not. As the wise men never gossip, the cruel law is never applied in practice. But one day an inveterate gossip (scandal-monger) appears in the city, who declares openly on the central square that he knows for sure that there exists at least one unfaithful wife in the city. Suppose he tells the truth and everyone believes that he tells the truth. Does it change the situation? Will the unfaithful wives be punished according to the law?

Exercise 3.1. Give an affirmative answer to this question by arguing in the following way.

1. Suppose there is only one unfaithful wife in the city. Convince yourself that she will be stabbed by her husband on the first night after the arrival of the gossip. (Hint: her husband knows everything about other wives, so he knows that they all are faithful, consequently only his wife can be unfaithful.)

2. Suppose that there are only two husbands, say A and B, with unfaithful wives. Try to understand that both of them will stab their wives on the second night after the arrival of the gossip. (Hint: A is not going to kill his wife on the first night, as he can assume that the only unfaithful wife is that of B. The wife of B is not killed on the first night either by the same reason. But after the first night A concludes by the common knowledge of rationality that his wife is unfaithful, because otherwise B would have to kill his wife on the first night.)

3. Similarly if there are three unfaithful wives, they all would die on the third night. By extrapolating (extending these arguments from two or three to an arbitrary number of unfaithful wives) convince yourself that if there are 50 (or 1000) unfaithful wives they all would die on the 50th (respectively 1000th) night (of course, if they are not clever enough to run away or to poison their husbands).

Remark. This story migrates from book to book in various disguises, in particular, there are versions with unfaithful husbands (punished by their wives by various methods, usually rather macabre) or with moles (traitors) in a secret service team.

Exercise 3.2. *Liars and truth tellers.* In an isolated island there are two cities. One of them is called the city of truth tellers, as its citizens always tell truth, and the other is called the city of liars, as its citizens always lie. The citizens of both cities are friends and often visit each other, so that an occasional individual you meet on the streets of each of these cities can equally probably belong to either of them. A traveler happened to land on this island and entered one of these cities. However, he does not know which of the two cities he is in. Then he met a pedestrian. Can he pose him a single question (with possible answer just "yes" or "no") so that after receiving an answer (of course, the answer could be a lie or the truth depending on whether this pedestrian is a citizen of the city of liars or truth tellers) he would learn in which city he has found himself? (Hint: the question should be: "Do you live in this city?" Convince yourself that the answer "Yes" would imply that this is the city of truth tellers, independently of whether the pedestrian lives in this city or is a guest. Another, less elegant way of asking, but revealing the same logic more explicitly, will be: "If I would ask you whether this were the city of truth tellers, would you answer Yes?")

Let us recall a curious example from literature. In a story of Robert Louis Stevenson "The Bottle Imp" a man was offered to purchase a bottle in which dwelt the devil himself who would fulfil any wish of the owner of the bottle. However, the owner can sell the bottle only for a price that is strictly less than what he paid for it and for a price that can be expressed by an integer number of coins. If the owner dies before selling the bottle to another person (to whom he has to explain clearly all the rules), he would be thrown directly to hell.

Suppose you are offered a bottle for a price of, say, 100 (smallest possible) coins. Well, you can think that you have a good chance to get rid of this bottle afterwards as there is a long way to the minimum price of 1 coin.

Exercise 3.3. Show that backward induction reasoning forbids buying a bottle at any price. (Hint: you can not buy a bottle at the price of one coin as it is then forbidden to sell it and you inevitably ends in hell; but then you can not buy it for two coins either, because you would be allowed to sell it only by the price of one coin, and no one would agree to buy it for this price, etc.).

Remark 3.1. In the real Stevenson's story the main character buys the bottle for the lowest price of one coin prepared eventually to go to hell. However, later on finding himself in a different country he manages to change his coin for several coins of another currency thus allowing him to get rid of the devil.

Let us complete the section with another story (told to authors by D. Sobol) revealing some kind of a gap in the conclusions of backward induction. A criminal is sentenced to a punishment (e.g. death penalty or any other kind of milder execution) on a Sunday and the judge declares that he will now be placed in prison and will be executed in the morning of one of the days of the coming week, but the one he will not be able to expect. The criminal is acquainted with logic and contemplates his fate by the backward induction. He thinks that they cannot execute him on Sunday, because this will be the last day of the week, hence their last chance to execute him, and hence he would be able to predict it (and thus this will be expected). Since Sunday is absolutely impossible, the last real chance is Saturday, but then they cannot do it on Saturday either by the same reasoning, as this again can be predicted. Continuing the argument by the logic of backward induction implies that they cannot execute him in any day of the week. Thus the criminal relax and feels safe in his prison. However, on Friday they come and execute him to his full unexpected surprise!

3.2 King Solomon's wisdom

In the famous Bible story two women came to king Solomon with a child. Each woman claimed that it was her child. Solomon had to decide which of them is the true mother. "Cut the child into two pieces, let each take one half", proposed Solomon. One of the women then cried: "O my lord, give her the living child, and in no wise slay it", and another said: "Let it be neither mine nor thine, but divide it." "So the true mother has revealed herself," said Solomon and told to the first woman: "Take your child, mother".

Of course the imposter in the story was both extremely cruel and stupid, otherwise Solomon would not have had such an easy success. Glazer and Ma [59] offered more effective (and at the same time more peaceful) procedure to find out the truth under these circumstances. To make it work one needs just some rough numeric estimates for the values V_t and V_f of the child to the true mother and the false mother respectively. Of course, $V_t > V_f$. Then the women play "a game" with all rules and payoffs explained beforehand.

Step 1. The first woman is asked "Is it your child?" If the answer is "No", the child is awarded to the second woman, and the game is over.

Step 2. If the answer is "Yes", then the same question is posed to the second woman: "Is it your child?" If the answer is "No", the child is awarded to the first woman. If the answer is "Yes", the child is awarded to the second woman, but they both get a punishment (money fine, or a prison sentence, or whatever), the punishment of the second woman being of value P and that of the first woman being p, where the values P and p are chosen in such a way that $0 < p < V_f < P < V_t$.

If the first woman is the true mother, the course of the game can be described graphically by the following *game tree*.

imagine (of course, assuming that the monopolist is rational) that he would not fight, and hence he can enter the market to obtain the payoff 1 (instead of zero, if not entering). Hence (enter, no fight) is the solution of this game. Even if the monopolist were to threaten to fight, this threat would not be credible, as to fight is not a rational behavior (at least, according to the rules of the game).

Selten [172] proposed the following extension of this game that seemingly was not a small part of the argument for his nomination to the 1994 Nobel Prize in economics which he got, together with Nash and Harsanyi.

Suppose a monopolist has stores in 20 cities, and in each of them he must deal with a different competitor. The above market-entry game begins in the first city and is repeated city by city (say, one game per month) against the next potential opponent. It seems reasonable to argue that by fighting with the first intruder the monopolist, though sustaining a loss in utility, would get a reputation of a fighter which would prevent the other intruders to enter the market in other places. The chain store paradox is revealed through the understanding that such a fighting strategy would contradict the ultimate logic of backward induction. In fact, let us consider the last city, or in other word, the last of these 20 identical games. This last game is just the simple market-entry game considered above and represented by Figure 3.1. So the solution is (entry, no fight), as there are no future possible games that could change this situation. So, in the last city the intruder can assume that the monopolist should accept his entry without a fight, and hence enter the market. We can now consider the previous game, in the 19th city. But as the solution (entry, no fight) to the last game is known anyway, it can not influence the previous game. So, to the 19th game the same considerations are applied, which again lead to the solution (entry, no fight). Arguing in the same manner backward we shall conclude that in all cities the solution is identical (entry, no fight). And the answer remain the same if you consider 20, or 100, or 1000 cities. Of course, this rational solution is not plausible. So, can we rely on the assumption that the players are rational?

A similar paradoxical conclusion can be obtained by the method of backward induction in the so called *Centipede game*. In this game R and C alternate rounds in the game described by the following game tree:

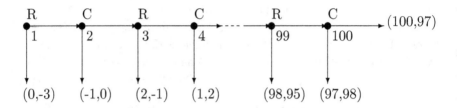

At each node the corresponding player can either finish the game by choosing the downwards arrow, or move to the right offering the next choice to the other player. For example, on the first node, R can either stop the game obtaining zero payoff and leaving C with even more unpleasant -3, or move to the right. The game finishes either if someone chooses a downwards arrow, or on the hundredth move.

Exercise 3.5. Show that the backward induction arguments necessarily imply that for each player the optimal strategy is to go downwards as soon as possible. In particular, R should finish the game on the first step. Hint: Imagine you are player C and you have arrived at the last (hundredth) move. Of course, you should then choose the downwards arrow. Now imagine your opponent on the 99th step of the game. He knows that you should choose the downwards arrow on the next step, so he is much better off choosing the downwards arrow at this position, than if he allows you to choose your downwards strategy on the next position. So the first player on the 99th step should also choose the downwards strategy, etc.

Of course, such a solution contradicts common sense. Experiments also show that the game is not played like that. Players choose the right moves (which represent a cooperative strategy) long enough, and choose a downwards move (defecting strategy) somewhere close to the end of the game. The discrepancies between theoretical predictions and experimental results are common in such situations and are called the *paradoxes of backward induction*. Are the players irrational? Later on we shall see some ways out of this paradoxical situation.

3.4   Normal and extensive forms of a game; battle of the sexes revisited

As above examples show, dynamic games (that evolve in time somehow) can be conveniently described by game trees, also called *extensive forms of games*. In these trees one just denotes by a point each possible position of a game indicating also which player has the right to move in each particular position, and then connects these points by lines if there is a legitimate move of a player which allows a transition between these positions. A description of a dynamic game in a normal form is also possible, but as we mentioned above, the normal form does not catch the dynamics explicitly and hence is not very illustrative. As an instructive example consider a dynamic version of the battle of the sexes game given by Table 1.7. Namely, let us assume that the players do not make their movements simultaneously, but, say, the husband makes the first choice, and the wife makes her decision already taking into account the choice of her husband. The game tree of this new game is the following.

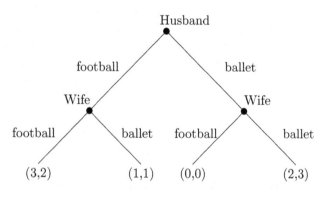

Figure 3.2

The backward induction arguments lead to the following solution. The husband can conclude that if he chooses football, then the wife would also chose football (as this choice gives her the better possible payoff 2) leading to the outcome (3,2). On the other hand, if he chooses ballet, then the wife would chose ballet leading to the outcome (2,3). Clearly the husband prefers (3,2) to (2,3), and hence he must chose football leading to the outcome (3,2).

The model looks like a joke, but seriously speaking it is not. It suggests that the ability to make the first move can be a serious advantage in certain situations. This happens in various contexts (see the chain store paradox above) and in particular in the biological context of the "battle of the sexes". This explains, for instance, why in the world of mammals and birds it is much more often that a male leaves a female with the child to rear it alone, than vice versa. Both males and females are interested in their children being reared (to pass their genes to next generations), and if left alone with a child each of them would prefer to rear it than to let it die. However a male has the advantage of the first move (at least, when the egg is inside a female, she can not leave it). Knowing (by the Common Knowledge of Rationality!) that the female would rear his child anyway, the optimal behavior for a male (having as an objective to pass the maximum number of his genes to further generations) becomes to go away and try to copulate with other females. On the other hand, in the world of fishes an opposite tendency is observed, which is due to certain physical traits that give the first move advantage to the females, see [46].

To assess the correspondence between normal and extensive forms of a game, it is instructive to represent this version of the "Battle of the Sexes" also in the normal form. For this, we have first to clarify what are the available strategies for both players. The husband still has only two strategies: football (f) and ballet (b). But the wife has now four strategies: ff, fb, bf, bb, where the first letter indicates her reaction to her husband's choice f, and the second letter indicates her reaction to her husband's choice b. The normal form of the game defined by Figure 3.2 is therefore

		Wife			
		ff	fb	bf	bb
Husband	f	**3,2**	**3,2**	1,3	1,1
	b	0,0	2,3	0,0	**2,3**

There are three Nash equilibria, given in bold. However, only one of them, (f,fb), corresponds to the solution found by backward induction. Hence we conclude that the usual notion of the Nash equilibrium is not sufficient for a successful analysis of dynamic games. In the next section we shall discuss a modified notion of Nash equilibrium that is suitable for the dynamic context.

3.5   Dynamic games and subgame perfection; pursuit games

Having discussed above some examples of dynamic games, it seems to be a good time to introduce the general notion. An arbitrary *dynamic game G* is specified by (i) a number of players; (ii) a set of its possible positions that are divided into two classes: terminal and intermediate positions; (iii) with each intermediate position P there is associated a *local game* G_P usually specified by its normal form with its set of (*local*) strategies and payoffs and a rule that specifies the next position of the game for each profile of local strategies applied at P. The game is then played in the following way: it starts at some initial position P_0, then the players choose some admissible local strategies, which define both their local payoffs and the next position, in the next position the procedure repeats and goes on either ad infinitum or till it reaches a terminal position, where the game stops. In our examples above, only one player has a non-empty set of strategies in each local game, but it needs not be necessarily so (see repeated prisoner's dilemma below). Anyway, the method of backward induction can be applied for any finite game (that always terminates after a finite number of local games) starting the argument from all terminal positions.

But there is another method to distinguish the solution obtained by backward induction, that is more general and can be applied also to infinite games (in some sense it is "backward induction in disguise"). This method is based on the notion of *subgame perfection*. A *subgame* of a dynamic game is a game that is conducted by the same rules as the initial game but is started not necessarily from the initial position of the original game, but at any position that can be in principle reached by playing the original game. A Nash equilibrium in a dynamic game is called *subgame perfect* if it remains a Nash equilibrium when reduced to any subgame.

For example, for a game shown by the tree on Figure 3.2, there are three subgames: initial one and the two "small" subgames starting at two positions that can be reached by two possible moves of the husband. Clearly, Nash equilibria (f,ff) and (b,bb) are not subgame perfect, as they include the actions "answer by f to b" and "answer by b to f" respectively, which are not optimal for the corresponding small subgames. The notion of subgame perfection is in fact rather powerful and can be applied in many situations where backward induction is not applicable (see Chapter 5).

The apparatus of backward induction and subgame perfection can be applied to the strategic analysis of a variety of conflict situations from real life, literature and history. We discussed only few of them. Many other beautiful examples can be found in [138] including the Madness of Odysseus going to Trojan War, the Trojan Horse, the last battle of Richard the Third from the War of Roses, the complicated treaty between Goethe's Faust and Mephisto, general duels, as well as the discussion of the "literary birth" of game theory in the famous novel "The Glass Bead Game" by Hermann Hesse. An exciting discussion of a variety of natural and social science oriented problems (including quite recent research) can be found in [58]. Other good sources for various models and problems are the textbooks [141] and [177].

An important class of dynamic games constitute the games of pursuit and evasion. The foundation of the theory of these games was laid out in [76], with a motivation coming mostly from military origins, where a typical question was as follows: what strategy of pursuit should be chosen in order to catch an aircraft in the shortest time. Games of pursuit belong to a more general class of differential games, where the notion of equilibrium is basically the same as for usual games, but the positions evolve in continuous time, the dynamics being described by differential equations. Hence we postpone the discussion of these games to Chapter 11, as it requires at least an elementary knowledge of differential calculus. However, some elementary problems can be analyzed by hands. As an illuminating brain teaser we suggest the following problem.

Exercise 3.6. A policeman is placed in the middle of a square and a gangster is in its corner. The policeman can move freely inside the square, and the gangster can move only along its boundary, but the speed of the policeman is two times less than that of the gangster. The objective of the policeman is to achieve a position, where he finds himself on the same side of the square as the gangster. Show that the policeman has a strategy that allows him to achieve his goal. Moreover, the maximal time needed equals the time the policeman moves from a corner of the square to a neighboring one. Hint: assume the gangster is place in the corner A of the square $ABCD$; the strategy of the policeman should be as follows: for any step of the gangster along AB to the direction of B, the policeman makes the (half of this) step in the orthogonal direction towards the side AB, and for any step of the gangster along AB back to A, the policeman makes the (half of this) step in the same (parallel) direction (and similarly if the gangster chooses to move along AD).

A famous game of pursuit and evasion is the following *Lion and Man game*. It plays on a unit disc on a plain. The Lion starts in the middle O and the Man at any other point A in the interior of the disc. They have equal speeds and cannot touch the boundary of the disc. The Lion wants to catch the Man, that is to turn out in the same point of the disc. The problem of whether the catch was possible has been discussed for some period of time, and it was a surprise for many that the Man has a strategy allowing him to avoid the catch in any finite time. The full proof is a bit technical (see [272]), but the main idea is quite clear intuitively, assuming that pursuit is organised by the following simplified discrete setting. The Man moves a distance l_1 from A to a point A_1 and then the Lion moves the same distance l_1 from O to the point O_1, which is the nearest possible to A_1. Then the Man moves a distance l_2 from A_1 to some point A_2 and the Lion moves the same distance l_2 from O_1 to the point O_2, which is the nearest possible to A_2, and so on. The winning strategy of the Man is as follows. For any j, at a point A_j he chooses A_{j+1} at a distance l_{j+1} in such a way that the segment $A_j A_{j+1}$ is orthogonal to the segment $O_j A_j$. Moreover, the sequence of distances $\{l_j\}$ is chosen so that $\sum_{j=1}^{\infty} l_j^2 < 1$ (this condition ensures that the Man always remains in the interior of the disc), but $\sum_j l_j = \infty$, which ensures that the pursuit will last for ever.

The situation is quite different if the Lion and the Man are not considered as points, but as the bodies of some size. Here one says that a catch occurs if the distance between the Lion and the Man becomes smaller than some fixed number ϵ (say, the size of the Man). It is then possible to show that for any given ϵ the catch is possible for two objects of equal speed moving in any convex bounded closed subset of a plane. Even convexity is not a necessary condition for this to hold. But some condition is required. For instance, if the pursuit is conducted in a band between two concentric rings, the catch on equal speeds is of course impossible.

It is curious to note that one of the first calls to apply differential calculus to the analysis of large scale conflicts in military, historical and social processes belongs to the Russian writer Leo Tolstoi: "Only accepting an infinitely small unit of observation — the differential of history, i.e., the homogeneous attractions and inclinations of people, and then mastering the art of integration (taking the sums of these infinitely small values), one can hope to grasp the laws of history" (War and Peace, v. 3, part 3:1).

At the dawn of the 20th century one of the first English car engineers, Frederick Lanchester, became a pioneer in the actual application of differential equations in modeling the processes of military confrontations. In modern times, his models and their extensions are applied mostly in the economics context, say, when modeling market wars; see an extensive discussion in [79].

For a modern exposition of differential games from various point of view we can refer the readers, for instance, to [14], [180], [139], [105], [122], [151], [157]; see also [153], [154], [152], [119], [123]. Let us mention finally that in order to model deception or evasion in a game of pursuit (where an evading object can create false objectives or hide in order to deceive the pursuit), the notion of nonlinear Markov games was suggested in [98].

3.6　Fair division and the ultimatum game

The fair division is an outstanding problem in economics and politics. To start a discussion one can pose the following question: how can you help two children to divide a cake into two pieces so that both would consider it as a fair division (due to the possibility of subjective evaluation of pieces by children, this is not the same as to divide 1 pound, where the fair division yields doubtlessly 50 pence to each).

A well known procedure for such a division (which really works well with children) is the following: let one child divide the cake into two pieces, and then let the other choose one of them. For a full discussion of similar problems (say, how to divide into seven parts), we refer the reader to [33].

The *ultimatum game* that we are going to discuss is designed to solve a similar problem under slightly different rules. Suppose two persons have to play the following game. The auctioneer gives 10 dollars to the first player that he/she has to divide with the second player by offering him/her any integer amount of dollars (e.g. $1, 2, 3, \ldots, 10$). If the second player agrees to take the offer, the first player keeps the rest for himself. However, if the second player rejects the offer, then the whole amount is returned to the auctioneer. Clearly first player has 10 strategies (to offer $1, 2, \ldots, 10$). The number of the strategies of the second layer is 2^{10} (a strategy of the second player is characterized by a subset of the set of first 10 numbers that he/she is willing to accept).

Exercise 3.7. Convince yourself that there are lots of Nash equilibria in this game (in fact 2^{10}), for instance (offer 6, accept only 2, 5, or 6), but there is only one subgame perfect equilibrium that can be also obtained by the arguments of backward induction, namely (offer 1, accept any offer).

This ultimatum game is almost as popular in game theory as the prisoner's dilemma, and a lot of experimental work has been carried out on this game. Even the comparison between the behaviors of people living in different parts of the globe was investigated, see [166]. These investigations show that almost nobody plays this game according to the theoretically predicted subgame perfect equilibrium. There is a wide discussion of this phenomenon in the literature. One of the natural ways out (also supported by questioning people involved in the experiments) is based on the observation (that we made already in the last section of the first chapter) that the face value of money does not need to coincide with its actual value. In particular, the idea of fairness, which is somehow present in the human mind, should be reflected in payoff designations. We refer, for example, to [58] for a serious discussion of this problem, noting only that here we are touching already on social psychology, suggesting the use of game theory to test hypotheses quantitatively.

3.7 Cooperation by threat and punishment; infinitely repeated games

A natural idea to search for cooperative outcomes of the prisoner's dilemma type game is by repeating it several times. One can imagine that by several repetitions the players could be able to learn that they both do better by cooperation. In particular, a defecting strategy can be punished by a similar strategy of an opponent in the next round of the game. In fact, lots of work was done on experimental testing of the outcomes (see e.g. [159], [161]). These experiments show that (at least under some assumptions on payoffs) the players turn out to be cooperative most of the time. At the same time, any use of the cooperative strategy contradicts the backward induction arguments that can be applied to this situation, much like the centipede game or the chain store paradox. Thus we have another manifestation of the paradox of backward induction. In fact, suppose two persons are to play the game with Table 1.4, where $r < p < s < q$, 100 times. At the last stage, the dominant strategy is, of course, defect. So, if the players are going to play according to their dominant strategies, they would surely defect at the last stage. In other words, using the method of elimination of weakly dominated strategies, both player should eliminate any strategy that includes cooperation at the last stage. Consider then the previous step. As their behavior at the next stage is fixed anyway, no threats of punishment could be credible or reasonable at this step either. In other words, the logic of backward induction inevitably demands that both players should now delete the cooperative strategies on the second step from the end. Natural extension of this argument leads to the conclusion that the only dominant strategy is "defect" all the way through.

However, one can observe that the mechanism of backward induction which prevents cooperation (makes it unstable, does not allow players to achieve it as a subgame perfect Nash equilibrium) depends crucially on the existence of the fixed last stage of the game (where one can start the backward induction arguments). So one can hope to avoid the "curse of backward induction" by considering a situation where the end point is not precisely fixed. And in fact, in real life, though one usually understands that the interaction may not continue forever, one does not know precisely when the last stage will occur: firms competing each year in the same market, families living together day after day, etc. To model this indefiniteness one can introduce some probability for a game to stop after each stage of interaction.

A slightly different (but similar) approach is to consider the game lasting infinitely many times but *discounting* each stage by a fixed discounting positive factor, say $\delta < 1$. More precisely, at each stage of the game all payoffs are multiplied by δ compared with the previous stage. Thus, instead of the one stage prisoner's dilemma described by Table 1.4, one is lead to consider a game where the players play infinitely many such games, but when the game is played for the k-th time, its payoff matrix is given by the table

Table 3.1

		C	
		defect	cooperate
R	defect	$\delta^k p, \delta^k p$	$\delta^k q, \delta^k r$
	cooperate	$\delta^k r, \delta^k q$	$\delta^k s, \delta^k s$

Since δ^k becomes smaller and smaller, as k becomes large, the influence of each further game on the overall payoff becomes smaller and smaller (tends to zero). This rule describes a situation, which is very similar to the game with a very large but unknown number of repetitions. The introduction of a discount over time is very natural and commonly used in practical calculations of an income. It reflects the general philosophy of "Only God knows what happens after ten years anyway". (Recall the story about Khadgy Nasredin, who took a job from a king promising to teach his donkey to talk during the next 20 years, while thinking that during this period either donkey would perish or the king would die. Possibly similar reasoning was behind N.S. Khruschev's promise to build communism in the USSR in 20 years.) It is also a practical fact that any fixed amount of money obtained today is quite different from the same amount obtained next year. In fact, putting it in a bank account, say, you can increase this amount for the next year.

We shall denote here by c and d respectively the cooperative and defective strategies of the prisoner's dilemma (notice that in the previous chapter these strategies were denoted d and h for dove and hawk). As in the previous chapter our game will be symmetric and we shall again simplify the general notations for the payoffs writing simply $\Pi(d,d)$, $\Pi(c,d)$, $\Pi(d,c)$, $\Pi(c,c)$ for the payoffs of the player (irrelevant if the first or the second) playing respectively d against d, c against d, d against c, c against c. Thus, for Table 1.4, that we shall deal with in this chapter,

$$\Pi(d,d) = p, \quad \Pi(d,c) = q, \quad \Pi(c,d) = r, \quad \Pi(c,c) = s.$$

In general, the function $\Pi(s_1, s_2)$ for symmetric games indicates the payoff to the player that plays s_1 against s_2.

To begin our discussion of such an infinitely repeated game, observe first that if both players are confined to use always one and the same strategy, i.e. either playing "cooperate" all the time, which strategy we denote by C, or playing d all the time, which strategy we denote by D, then the payoff table of this game becomes the same as for a one stage game, but with all payoffs divided by $(1 - \delta)$. In fact, say,

$$\Pi(C,C) = \Pi(c,c)(1 + \delta + \delta^2 + \delta^3 + ...) = \Pi(c,c)\frac{1}{1 - \delta}.$$

Similarly $\Pi(D,D) = \Pi(d,d)/(1 - \delta)$ and $\Pi_R(C,D) = \Pi_R(c,d)/(1 - \delta)$.

Remark. We used above a well known formula $(1 + \delta + \delta^2 + \delta^3 + ...) = 1/(1 - \delta)$ for the infinite sum of a geometric sequence. A simple way to discover this formula is to denote the unknown sum by x, i.e. $x = 1 + \delta + \delta^2 + \delta^3 + ...$, and then observe that $x = 1 + \delta(1 + \delta + \delta^2 + \delta^3 + ...)$, and hence $x = 1 + x\delta$. This simple equation clearly implies the required formula for x.

Of course, the main interest of an infinite game is the possibility to vary the action from stage to stage. It turns out that unlike a finitely repeated game, threats to punish can now become credible, which can lead to stable cooperation. To see this, let us define the so called *trigger strategy* C_T: start by cooperating (playing c) and continue to cooperate until your opponent defects (plays d), then defect forever after. More precisely, it specifies that if you or your opponent defect at least once, then you defect forever after. One sees directly that if each player plays either C_T or C, the game develops as if both have adopted the cooperating strategy.

Theorem 3.1. (C_T, C_T) *is a symmetric Nash equilibrium in the above infinitely repeated game (based on the one-shot prisoner's dilemma specified by Table 1.4) whenever*

$$\delta \geq \frac{q - s}{q - p}.$$

In particular, for Table 1.3 we have $\delta \geq 1/2$.

Proof. We have to show that C_T is the best response to C_T, i.e. that

$$\Pi(C_T, C_T) \geq \Pi(S, C_T)$$

for any strategy S. We know already that playing C against C_T yields the same payoff as playing C_T against C_T. So we only need to consider a situation when S includes d on some stage of the game. But whenever the first player plays d, the second player (with the trigger strategy) will play d for ever afterwards, so in order to give a best reply, the first player has also to play d forever afterwards. In other words, the best responses strategies S can be only of the form S_k (with some integer k): play c the first k steps, and then switch to d forever after. Consequently, in order to convince ourselves that C_T is the best response to C_T we only need to check that

$$\Pi(C_T, C_T) \geq \Pi(S_k, C_T),$$

which is equivalent to

$$\Pi(c, c)(\delta^k + \delta^{k+1} + ...) \geq \Pi(d, c)\delta^k + \Pi(d, d)(\delta^{k+1} + \delta^{k+2} + ...),$$

because

$$\Pi(C_T, C_T) = \Pi(c, c)(1 + \delta + \delta^{k-1}) + \Pi(c, c)(\delta^k + \delta^{k+1} + ...),$$

$$\Pi(S_k, C_T) = \Pi(c, c)(1 + \delta + \delta^{k-1}) + \Pi(d, c))\delta^k + \Pi(d, d)(\delta^{k+1} + \delta^{k+2} + ...).$$

Thus we have to check that

$$\Pi(c, c)\delta^k(1 + \delta + \delta^2 + ...) \geq \Pi(d, c)\delta^k + \Pi(d, d)\delta^k\delta(1 + \delta + \delta^2 + ...).$$

Canceling δ^k and using the formula for the sum of a geometric series yields

$$\Pi(c, c)\frac{1}{1 - \delta} \geq \Pi(d, c) + \Pi(d, d)\frac{\delta}{1 - \delta},$$

or equivalently

$$\frac{s}{1 - \delta} \geq q + \frac{p\delta}{1 - \delta}.$$

This can be rewritten as

$$s \geq q(1 - \delta) + p\delta,$$

which leads to $\delta > (q - s)/(q - p)$, as required.

So, a nice solution to the problem of the natural appearance of cooperation is found.

Exercise 3.8. Show that under conditions of the above theorem C_T is also subgame perfect. (Hint: subgames that differ from the original one, are specified by the previous round of the game that can be (c, c), (d, c), (c, d), (d, d). From the first of these positions you would play as from the initial one, and from the other three position C_T turns to D, which is again a Nash equilibrium.)

A drawback of the proposed approach lies in the observation that there exists a tremendous number (in fact, infinite) of Nash equilibria in the repeated game considered.

Exercise 3.9. Suppose each player alternates $c, d, c, \ldots$ as long as other player does the same, and each would change to d forever whenever the opponents deviates at least once from the adopted pattern $c, d, c, \ldots$ Show that these pair of strategies also define a subgame perfect symmetric Nash equilibrium for δ being close enough to one.

The most famous strategy for repeated games (see the next section for its merits) is the so called *Tit-for-Tat strategy* T. Tit-for-Tat means "I give to you, what you give to me" or "tooth for tooth and eye for eye" and is defined as a strategy that starts with playing c and then in each stage plays the strategy that the opponent played on the previous stage.

Exercise 3.10. Show that (T, T) is a Nash equilibrium, which is however not subgame perfect. (Hint: to show the latter, consider the subgame that follows a round specified by situation (c, d). Then (T, T) specify permanent alternation (d, c), (c, d), (d, c), etc. But alternating $d, c, d, \ldots$ is not the best reply to the opponent strategy that alternates $c, d, c, \ldots$.)

Exercise 3.11. Show that neither (T, T) nor (C_T, C_T) is an ESS (evolutionary stable strategy described in Chapter 4). (Hint: As "always cooperate" strategy in a society of, say, Tit-for-Tat players does precisely the same as the Tit-for-Tat itself, Tit-for-Tat players do not do better than unconditional cooperators, and can be invaded by mutants with "always cooperate" strategy without even noticing it. Hence Tit-for-Tat is not ESS.) It is easy to imagine that in the society of the Tit-for-Tat players the members can get used to permanent cooperation, steadily forget their original strategy and at the end turn into unconditional cooperators. This would leave them unprepared for an invasion of more aggressive players.

Remark. Despite the results of the above exercise, Tit-for-Tat turns out to be very close to the evolutionary stability, see the next section.

Remark. For those acquainted with probability theory, let us note that the same results can be achieved without any discounting, but if one assumes that after each stage of the game it will stop (never occurs again) with some probability $p > 0$ (which implies, in particular, that almost surely the game will last only a finite number of times). You can show that all payoffs (of course, better to say their expectations) are the same as in case of discounting with $\delta = 1 - p$.

In justifying cooperation among humans playing a prisoner's dilemma one can now argue that the human mind treats a large number of repetitions practically as infinity and refer to the above theory. Another argument claims that it is too much to ask from an individual to carry out large number of rounds of the iterated eliminations of weakly dominated strategies. There are other approaches: see e.g. [148], [108] for the dilemmas in a spatial environment, [186] for models with learning, [77] for self-organized criticality, [81] for use of private information, [83] for games under pressure from a third party, [58] for altruism and assortative interactions.

3.8 Computer tournaments; the triumph of the strategy Tit-for-Tat

An interesting computer experiment was conducted by R. Axelrod, see [10]. He invited well known scientists (game-theorists, sociologists, psychologists, economists) to submit iterated prisoner's dilemma strategies for a computer tournament. Each strategy had to be a computer program that specifies precisely what to do on each step of a game repeated 200 times given the complete history of the interaction. The tournament was organized as a sequence of pairwise contests where each strategy submitted was played 200 times in a repeated game specified by the table

		C	
		defect	cooperate
R	defect	1,1	5,0
	cooperate	0,5	3,3

against all other submitted strategies, against itself, and against a program that chooses cooperation or defection randomly. To get an overall score, Axelrod averaged each strategy's scores. As the maximum win is 5 in each round, the scores could be any positive number not exceeding 1000.

Fourteen strategies were submitted to the first tournament, some of them quite complicated. The highest score of 505 points was achieved by the Tit-for-Tat strategy submitted by A. Rapoport.

Afterwards Axelrod organized a second computer tournament in which the entrants were informed about the results of the first one. The implicit task was to beat Tit-for-Tat. He got 62 entries from six countries.

Amazingly enough, Tit-for-Tat was again a winner leading to the conclusion that it is, in fact, optimal or nearly so. Axelrod observed further that
Tit-for-Tat usually does not beat anybody, but it still wins the tournament,
which is a bizarre way of winning (as if you win a chess tournament by never
beating anybody).

At last, in the third tournament, Axelrod introduced an artificial "natural selection" for his computerized strategies. He ran a series of tournaments
("iterated" iterated prisoner's dilemma), where after each round of iterated
pairwise contests, all participating strategies reproduce the number of identical offspring proportional to the scores they have got. The next round of
the tournament would then the start with a new generation that include
more copies of successful strategies. The evolution of the number of various
strategies was quite curious and displayed features that one could expect to
observe in the biological context. However eventually, Tit-for-Tat became
the most common strategy thus showing itself again as a superior strategy,
fitting well to the biological idea of evolutionary stability (see Chapter 4).
The experiments showed that Tit-for-Tat also worked well in other social
dilemmas. For further discussion we refer to [149].

3.9   Logical games; limits of the sequences

This section is mostly meant for readers having interest in logic and mathematics per se.

The language and the methods of game theory can be used as a general tool for logical constructions of complicated abstract models. In other
words, one can apply game theory to mathematics! An introduction to this
subject in given in [71] (which is not a simple reading, however). Another
point of view with applications to probability theory is developed in [174].
Not going into any serious discussion, we shall just illustrate on an example
how the language of game theory can be used to give insightful representations for familiar mathematical objects. We shall discuss a notion of the
limit of a sequence, which is central in mathematical analysis. First we
explain this notion for those not acquainted with it.

Let us start with an example. It is more or less an obvious observation
that the sequence of numbers $a_1 = 1$, $a_2 = 1/2$, $a_3 = 1/3,..., a_n = 1/n,...$
becomes smaller and smaller as the integer n becomes larger and larger. In
other words, it tends to zero as n tends to infinity.

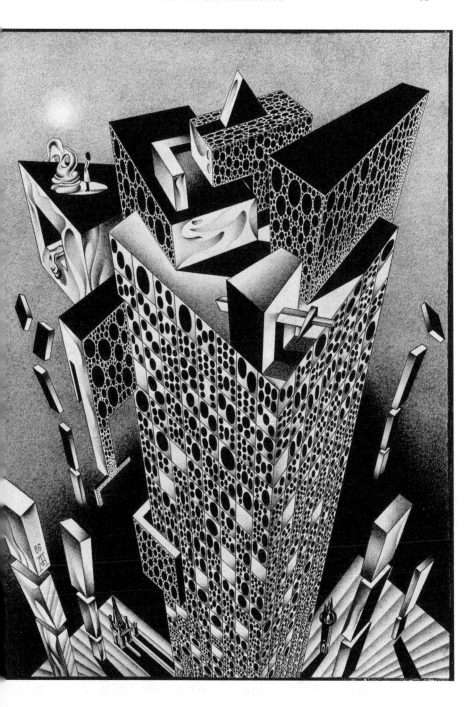

This example suggests the following definition. A sequence of numbers a_1, a_2,..., a_n,... *tends to zero as n tends to infinity* (notation $a_n \to 0$ as $n \to \infty$), if for an arbitrary $\epsilon > 0$ there exists an integer number N such that all terms of the sequence with numbers greater than N are smaller than ϵ in magnitude, i.e. $|a_n| < \epsilon$ for all $n > N$.

For instance, in the above example of the sequence $a_1 = 1$, $a_2 = 1/2$, $a_3 = 1/3$,..., $a_n = 1/n$,... choosing N so large that $1/N < \epsilon$ implies that $|a_n| = 1/n < 1/N < \epsilon$ for all $n > N$. Hence this sequence tends to zero as n goes to infinity in the sense of the definition given.

Analogously one can define convergence to an arbitrary number, say A, which is not necessary zero. One says that a sequence a_1, a_2,..., a_n,... *tends to A*, if $a_n - A \to 0$ as $n \to \infty$. In this case, one says also that the *limit of a_n is A* (notation $\lim_{n\to\infty} a_n = A$).

Of course, not all sequences have a limit (convince yourself that the sequence $1, 0, 1, 0, ...$ has no limit), and it is not always obvious whether a sequence has a limit (tends to some number). For instance, it is already rather difficult to show that the sequence of numbers $a_1 = (1 + 1)$, $a_2 = (1 + 1/2)^2$, $a_3 = (1 + 1/3)^3$,..., $a_n = (1 + 1/n)^n$,... has a limit. Note that this sequence (not at all remarkable from a superficial view) plays an important role in the human society describing the growth of a capital according to the law of compound percents. It turns out that the limit of this sequence is a remarkable number $e = 2.718281828...$ that plays the central role in calculus. It is denoted e in honor of a famous Swiss mathematician L. Euler, who spent most of his life working in St. Petersburg Academy. Like the number π (describing the length of the circumference) the number e is surd (it can not be expressed as a finite or periodic decimal fraction).

Connoisseurs of world literature can easily remember the first 10 numbers, as 2.7 is followed by two times the date of birth of the famous Russian writer Leo Tolstoi. (This is, however, a long way to a place in the Guinness book of world records, where one can find a person knowing thousands of numbers in the decimal representation of the numbers π and e.) You can get e and π with lots of precision by looking at the picture on the previous page (guess how!).

Now, what this all has to do with games? To see the connection, the notion of guaranteed payoff from Chapter 1 is required. Moreover, for games with an infinite number of strategies, a slight modification is useful, namely: a payoff P is called *attainable (or guaranteed) with arbitrary precision* for a player, if this player can ensure a payoff which is arbitrary close to P: i.e., for an arbitrary $\epsilon > 0$, this player has a strategy that guarantees him a payoff not less than $A - \epsilon$, whatever strategy is used by his opponent.

Now consider the two-player zero-sum game with an infinite number of strategies, given by the tree

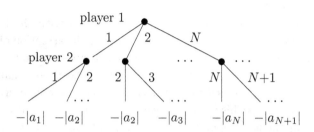

Figure 3.3

where only the payoffs of the first player are given (as the game is zero-sum, the payoffs of the second player are just negations of the first player payoffs). In other words, the first player can choose any positive integer, say N, and then the second player can choose any integer, say n, that is not less than N. The payoff to the first player becomes $-|a_n|$, which means that he has to pay $|a_n|$ to the second player. The following observation allows for a game theoretic interpretation of the notion of the limit.

Exercise 3.12. Convince yourself that the sequence a_n tends to zero as $n \to \infty$ according to a definition given above if and only if the payoff zero is attainable with arbitrary precision for the first player in the game given by Figure 3.3.

3.10 Russian Roulette; games with incomplete information

Till now we have studied the dynamic games with complete information, i.e. games where each player knows with certainty his/her position when choosing the next move. In many situations such complete information is not available, the corresponding games being called *games with incomplete information*. In particular, this may happen in the presence of chance in the system, when the position of a player can possibly be only predicted with some probability. These circumstances lead to the so called *stochastic games* (and hence to a special branch of game theory that is closely connected with the probability theory). In fact these situations are not only easy to imagine, but the presence of chance should be taken into account in almost all real life interactions (no human can predict the behavior of the environment with full certainty). We shall not go into any serious analysis in this direction, but confine ourselves to describing a classical "game" of "Russian Roulette" that captures in an elementary way these new features of conflict interactions.

The historical context of this game is connected with the fate of two officers of the Russian Tsarist army competing for the attention of a certain pretty girl. Instead of a duel or a simultaneous attempt to win her affection, they decide to settle the matter by the following "peaceful" game. They load a bullet at random into one of the chambers of a six-shooter, and then start to take turns. At his turn, a player can either chicken out (leaving the girl for his opponent) or point the gun at his head and pull the trigger. If he kills himself (the chamber turned out to contain the bullet), his opponent is left alone to win the maiden's heart. Otherwise (the chamber turned out to be empty), he rotates the chambers of the shooter on one position and passes it to his opponent for the next turn. This game clearly finishes after at most 6 moves, as during this time all 6 chambers would be used and one of the officers would be dead, if no one chickened out earlier. Hence each officer can have at most three moves, and consequently he has four possible strategies SSS, SSC, SC, C (where the letter S stands for shooting and the letter C for chickening out).

Exercise 3.13. Denoting by W (win), D (dead) and Sh (shame for being a chicken) the possible outcomes of the game, draw a 4×4 table (normal form) and a tree (extensive form) of the game assuming that the bullet turns out to be in the third chamber.

Of course the main point of this game is that the players do not know in which chamber the bullet is placed (incomplete information). So to analyze the problem properly one has to assign some probabilities to the six possible random events (say, 1/6 for each), calculate the probabilities of various outcomes under various profiles (choices of strategies of both players) and then look for optimal strategies assuming certain numerical values for the possible outcomes W, D and S (see e.g. [25] for a more comprehensive analysis).

Chapter 4

Hawks and doves, lions and lambs

4.1 Fitness and stability in population biology (general ideas)

In the first book on game theory [145] this theory was mainly meant to become a universal tool for the analysis of economics. Later the possible applications to war strategies and psychology were exploited. In the 80's the doors were open to a wide application of game theory to biology. This development could probably be surprising for the founders of the theory, because the assumption of the rationality of the players was at the heart of all earlier constructions, and one could not expect that animals would be involved in any calculations of optimal strategies to choose their behavior.

The first idea behind this development is the perception that the types of behavior can be passed by parents to their offspring through the genetic code like physical traits. The behavior of the fittest, i.e. those that produce the largest number of offspring is therefore likely to persist. According to the second key idea, the survival of the fittest means roughly speaking the survival of the stable phenomena (unstable species are to decay and hence cannot survive). This leads to the central notion of the *evolutionary stable strategies* (abbreviated ESS) proposed in [135] and later developed in more detail in monograph [134]. Loosely speaking, ESS are the strategies that are stable under the invasion of a small group of mutants. In other words, if a small number of mutants (with a different strategy) occasionally appears in a population dominated by an ESS, then they would do worse than the average member of this population and consequently are doomed to eventually die out. This implies, in particular, that since ESS does better against itself than any single mutant, it is the best reply to itself, and consequently a symmetric Nash equilibrium.

In the next sections we shall give a more precise exposition of these ideas. But let us point out now that the development of these ideas led to an appealing explanation of the various types of cooperation and competition among species. We refer to [46] for a serious treatment of the problem from the biological point of view, and simply note here that similar dilemmas between cooperation and defection exist in the wild nature as in the human society. Examples of cooperation (or altruism) are numerous among living beings. For instance, many small birds when seeing a dangerous predator like a hawk give a certain alarm call, turning the attention of the predator on themselves but allowing the whole flock to evade the attack. Even more amazingly, there are birds called ziczacs that enter the mouths of crocodiles to eat parasites, and the crocodiles do not eat them. (Similar "cooperation" exists between some fish species, so the fable about a wolf asking a crane to pull out a bone that is stuck in his throat has a firm ground in the real world.) Could the crocodiles' brains perceive the intellectual argument that if all these ziczacs would be eaten, then no one would be able to clean their mouths? Not very probable. So why does'nt an individual crocodile "defect" by eating an easy meal? Game theory offers a plausible explanation to such a phenomena.

We shall further work mostly with only two-move games for simplicity, but notice here that more general games are relevant for biology as well. Even the simplest three actions game Rock-Paper-Scissors has analogues in real life. This game is "played", for instance, by the side-blotched lizards *Uta stansburiana* (see [58] for more detail). The males of these lizards has three types: orange-throats, blue-throats, and yellow-striped. The orange-throats are aggressive and control large territories and large harems. The blue-throats are more peaceful and keep control of small territories and small harems. The yellow-striped are even weaker than blue-throats, but they look like females, and can secretly infiltrate the big territories of orange-throats and copulate there with females. Thus, orange-throats beat blue-throats, blue-throats beat yellow-striped, but yellow-striped beat orange-throats, i.e. this is precisely the situation described by Table 1.2. A serious analysis of this "game" leads to a conclusion (supported by field observations) that the fractions of orange-throats, blue-throats and yellow-striped oscillate according to a certain law. However, this analysis requires some calculus and is beyond the scope of the present exposition.

4.2   Hawk and Dove games as social dilemmas

We shall return now to the analysis of the symmetric two-player and two-move games (e.g. prisoner's dilemma and more general social dilemmas considered in Chapter 1). In this chapter, however, we are going to stress the biological context. In this context, it is natural to consider defect-cooperate dilemma as a dilemma between an aggressive and a peaceful behavior, where an aggressor is supposed to win over a peacemaker in a one-shot game. A further convention is to associate these behaviors with some concrete animals. Appropriate pairs of animals are Hawk and Dove, Lion and Lamb, Wolf and Hare. We shall further use the first pair, which is the one most often used in literature. Thus, in a general symmetric two-actions game we shall choose the titles Hawk and Dove for the strategies in such a way that playing Hawk against Dove yields a better payoff than vice versa. Thus any two-player two-strategy symmetric game can be described by the table

	hawk	dove
hawk	p, p	q, r
dove	r, q	s, s

with $r \leq q$. As shifting (adding a number to all entries of the table) does not change the content of the conflict, we can further subtract r from all entries leading to the table

Table 4.1

	hawk	dove
hawk	H, H	$V, 0$
dove	$0, V$	D, D

with $V \geq 0$. Written in this form, a symmetric game is often referred to as the general *Hawk-Dove game*, which will be the subject of analysis in this chapter. A usual interpretation of this game in the biological context is the following. Recall first that in this context all payoffs express the increase (or decrease) of the fitness of an individual, i.e. the expected number of offspring produced by this individual. Suppose now that two individuals in a population compete for some resource (food, female, etc.) of value V. A dove surrenders the full resource to a hawk without a struggle. Two hawks always fight for the resource, the cost of fighting (say, injuries) leading to a decrease in the average value of the resource. Thus the title "Hawk-Dove"

game is often restricted to the subclass of general symmetric games of
Table 4.1 specified by additional assumptions $V > 0$ (e.g. V is strictly
positive as compared with general $V \geq 0$) and $H < V$. These assumptions
catch the most natural situations for biology. The original Hawk-Dove game
(see [134]) had even further restrictions, namely that the cost of fighting is
greater than that of the resource leading to H being negative. At last, any
two doves share the resource between themselves without a fight, which can
usually lead to a decrease of its value (say, if they just share the resource
equally), but not necessarily, as there can be some increase in fitness due
to being a peacemaker.

Using the reduced form of the table for symmetric games (i.e. omitting
the symmetric payoffs for the second player), the game of Table 4.1 can be
described by a reduced table

Table 4.2

	hawk	dove
hawk	H	V
dove	0	D

We shall write further h for hawk and d for dove. Due to the symmetry
one can further simplify the general notations for the payoffs (from Chapter 1) writing simply $\Pi(h, h)$, $\Pi(h, d)$, $\Pi(d, h)$, $\Pi(d, d)$ for the payoffs of the
player (irrelevant if the first or the second) playing respectively h against h,
h against d, d against h, d against d. Thus in the table above, $\Pi(h, h) = H$,
$\Pi(h, d) = V$, $\Pi(d, h) = 0$, $\Pi(d, d) = D$.

All social dilemmas of Chapter 1 are particular cases of this general
Hawk-Dove game with the strategies "defect", "cooperate" being replaced
by "hawk" and "dove". To move our analysis further than in Chapter 1,
in particular, to make a discussion of ESS more precise, the introduction
of mixed strategies now becomes indispensable. This is a bit more of an
involved notion compared to all we dealt with earlier, as it requires some
understanding of probability theory. We shall try to give an elementary
(hence necessarily incomplete) account of this topic in the next section.

To conclude the section, let us note that similar to aggression-peace
dilemma, one can analyze other pairs of opposite behavior of animals, say
be coy or fast for a female, or be faithful or a philanderer for a male, see [46].

4.3 Mixed strategies, probability and chance

To begin with, let us return to our first examples of games: Matching Pennies and Rock-Paper-Scissors. Our methods still suggest no clue to the solutions of these simple games. At the same time, every child knows how to play them. Namely, if you are to play them repeatedly several times, you should choose all strategies randomly and with equal probability. This random switching of (pure) strategies defines a *mixed strategy* and this is exactly what we are going to discuss now.

However, the precise meaning of the key words "randomly" and "probability" is not obvious. It is in fact rather non-trivial. Of course, "randomly" means "avoiding all possible patterns". If an opponent recognizes any pattern in your choices he would be able to predict them and hence win. However, behavior without a pattern is a notion, which is as vague as randomness. In fact, one can imagine infinite number of rather complicated patterns, so how to avoid them all? Putting the question another way round, how do we judge whether there is no pattern or you didn't find one?

Turning to probabilities: in the every day language, probability is of course associated with frequency. Roughly speaking, playing three strategies with equal probability means to play them with equal frequency, i.e. for any given number n of rounds you play each strategy approximately $n/3$ times. The catch here is in the word "approximately". How approximate should it be? You can not have the frequency at precisely $1/3$ without breaking the randomness, because, say, if you play each strategy once in any three rounds, this is already a pattern that you have to avoid by all means (to be random).

Precise connection between frequency and probability is quite nontrivial and is given by the strong *law of large numbers*, which is a famous result of probability theory proved by A. Kolmogorov. Loosely speaking, it states that for a typical series of independent identical experiments (notice that we do not specify here what do we mean by "typical series"), the frequency tends to the probability (i.e. becomes closer and closer to the probability, rigorous definition of the notion of limit being given in Section 3.10), as the number of these experiments goes to infinity.

Having discussed the difficulties in precise understanding of the notions of probability and chance, let us indicate that one can get a simple idea of what is going on (adequate enough for practical purposes of the game theory, at least to begin with) by associating randomness with some well known models of random experiments like throwing a die or a coin.

Namely, suppose you like to model a series of independent random experiments with two possible outcomes, say A and B. If they are supposed to be equally probable, i.e. each has a probability of $1/2$, you can just throw a die and choose A when you get Head and B otherwise. If A is supposed to occur with a probability of $1/6$ (respectively $1/3$), you can throw a die and choose A when you get 1 (respectively if you get 1 or 2). Generally, if you like A to occur with probability being an arbitrary rational number m/n (with $m < n$, of course), you should draw a roulette with n equal sectors and choose A, if a pointer would turn out among the first m sectors. Similarly one can model experiments with three or more outcomes. There are also lots of computer programmes that can supply you with random sequences with prescribed probabilities without any hassle. For instance, in case of an experiment with two outcomes A and B one can get a series of sample events with the outcome A occurring with probability p, where p is an arbitrary number from the interval $(0, 1)$ (not necessarily rational).

For further discission of the notions of chance and probability, we refer readers to the textbooks in probability, a rather comprehensive exposition for beginners being given e.g. in [4] and [165].

Let us now return to games. Playing a *mixed strategy* $\sigma = (p, 1 - p)$ (where p is any number between 0 and 1) in a two-action two-player game means to choose the first of the two strategies randomly with probability p (say, by using one of the standard random devices described above). The original strategies clearly correspond to the values $p = 0$ or $p = 1$ and are called pure strategies.

Let us consider the symmetric game given by Table 4.1 or 4.2. What should be the payoffs $\Pi(h, \sigma)$, $\Pi(d, \sigma)$ for a player that uses pure strategies against a mixed strategy σ? Well, it is clear that if pure strategies are switched, then the payoffs are different in different rounds of the repeated game. Thus it is natural to define a payoff as an average of payoffs over several rounds of the game. Suppose you are playing the game n times in such a way that you use only a fixed pure strategy s (which is, of course, either h or d), and your opponent used n_h times the strategy h and $n_d = n - n_h$ times the strategy d. Then your total payoff in n games is

$$n_h \Pi(s, h) + (n - n_h) \Pi(s, d)$$

and consequently the average payoff per game is

$$\frac{n_h}{n} \Pi(s, h) + (1 - \frac{n_h}{n}) \Pi(s, d).$$

This depends on the number of rounds n. However, if your opponent plays the mixed strategy $\sigma = (p, 1 - p)$, we know (by the law of large numbers, see above) that the frequency n_h/n of using h tends to the probability p, as n tends to infinity. Consequently, the average payoff tends to

$$p\Pi(s, h) + (1 - p)\Pi(s, d).$$

This leads to the following key definition: the *payoff to a pure strategy s against a mixed strategy* $\sigma = (p, 1 - p)$ is defined as

$$\Pi(s, \sigma) = p\Pi(s, h) + (1 - p)\Pi(s, d). \tag{4.1}$$

Similar arguments imply that *playing a mixed strategy* $\eta = (q, 1-q)$ *against a mixed strategy* $\sigma = (p, 1 - p)$ should *yield the payoff*

$$\Pi(\eta, \sigma) = q\Pi(h, \sigma) + (1 - q)\Pi(d, \sigma), \tag{4.2}$$

and consequently

$$\Pi(\eta, \sigma) = qp\Pi(h, h) + q(1-p)\Pi(h, d) + p(1-q)\Pi(d, h) + (1-q)(1-p)\Pi(d, d). \tag{4.3}$$

The conclusions of this long discussion are the following. To any two-player and two-action symmetric game there corresponds another two-player game with each player having the infinite number of (mixed) strategies $\sigma = (p, 1 - p)$ parametrized by numbers p between 0 and 1. The payoffs for playing one of such strategies $\eta = (q, 1 - q)$ against another one $\sigma = (p, 1 - p)$ is given by formula (4.3). The strategies with $p = 0$ or $p = 1$ coincide with the original strategies h and d respectively and become to be called pure strategies. The strategies with p being different from 0 or 1 are called sometimes pure mixed (or strictly mixed).

There are two (naturally connected) interpretations to this lifting of an original two-action game to a new one with possibly mixed strategies. One interpretation comes from allowing repeated games with random switching of pure strategies (as we discussed above), where probability p of strategy h equals approximately to the frequency of using h (and coincides with the limit of this frequency as the number of experiments tends to infinity). The second interpretation is more "population biology" oriented. In this approach, in order to imagine playing a pure strategy against a mixed strategy $\sigma = (p, 1 - p)$ you have to imagine yourself playing a game with a randomly chosen representative of a large population with a fraction p of hawks and the fraction $1 - p$ of doves.

Observe there is nothing pathological in mixed strategies, where an individual sometimes plays a hawk and sometimes a dove. In fact, even humans are sometimes aggressive, sometimes more peaceful. Thus the probability p of being a hawk reflects in some sense the level of patience of an individual.

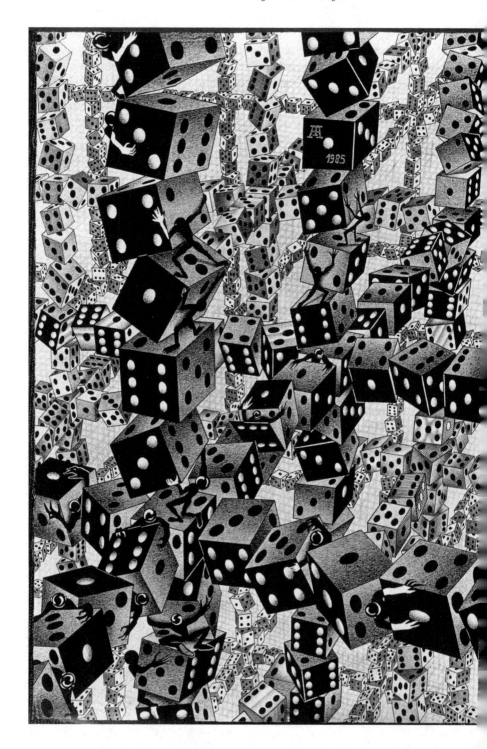

In the next two sections we shall discuss some relevant questions, postponing to the Section 4.7 the application of mixed strategies to the Hawk-Dove game.

4.4 The theorems of Nash and von Neumann

Though we shall not not deal further with non-symmetric games, it is worth noting that the notion of mixed strategies work similarly for general games. Say, for a game between Ruth and Charlie with two strategies 1_R, 2_R of Ruth and two strategies 1_C, 2_C of Charlie as defined in the first section of Chapter 1, mixed strategies for both of them are defined again as pairs of positive numbers $(p, 1 - p)$, so that if Ruth plays $\eta = (q, 1 - q)$ and Charlie plays $\sigma = (p, 1 - p)$, their payoffs are

$$\Pi_R(\eta, \sigma) = qp\Pi_R(1_R, 1_C) + q(1 - p)\Pi_R(1_R, 2_C)$$

$$+ p(1 - q)\Pi_R(2_R, 1_C) + (1 - q)(1 - p)\Pi_R(2_R, 2_C),$$

$$\Pi_C(\eta, \sigma) = qp\Pi_C(1_R, 1_C) + q(1 - p)\Pi_C(1_R, 2_C)$$

$$+ p(1 - q)\Pi_C(2_R, 1_C) + (1 - q)(1 - p)\Pi_C(2_R, 2_C).$$

Similarly, the notion of a mixed strategy can be extended to games with an arbitrary finite number of players and (pure) strategies.

In Chapter 1 we defined dominated strategies and Nash equilibria for games with arbitrary sets of strategies. Consequently, when an original two-action game is enlarged to a new game with possibly mixed strategies, in this new game one can define *dominated strategies and Nash equilibria* in a usual way.

One of the main achievements of John Nash was the *Nash theorem* stating that for any such game there exists at least one Nash equilibrium (in mixed strategies). Thus, as we already mentioned earlier, the ideas of Nash equilibria and mixed strategies provide (if taken together), at least some solution, to arbitrary finite games. There are generalization of this fundamental result to various classes of game with infinite numbers of players or actions.

A special case of the Nash theorem that concerns two-player zero-sum (or strictly competitive) finite games was obtained much earlier by J. von Neumann. It turns out moreover that in this particular case, the Nash equilibrium is always realized by mixed minimax strategies (the definition of mixed minimax strategies is the same as for pure minimax strategies that were discussed in Chapter 1), and hence prescribes a unique optimal (and guaranteed) payoff to each of the players.

The proof of the Nash theorem can be found in many books on game theory (its generalization is presented in part II, Section 11.2). It is a consequence of a rather deep mathematical fact that does not have much to do with the game theoretic intuition that we are trying to develop here.

4.5   Expectation and risk; St. Petersburg game

This section takes us aside of the mainstream of this chapter. We just want to stress here that using only average payoffs (as in Nash' s theorem above, and as we are going to do further) is not a "once and for all approach" in games and optimization. It becomes non-adequate when risk evaluation becomes critical. This can be demonstrated, for instance, just by mere existence of insurance companies: they make profit from your money, therefore your average payoff is negative whenever you take an insurance policy, and hence you never maximize your average payoff by doing this. However, no one calls you irrational.

An illustration to the same idea is supplied by the game proposed in the 18th century by D. Bernoulli to St. Petersburg Academy. This game is conducted as follows. You throw a coin one time after another until you get Head. Then the game stops and you receive the payoff 2^k, where k is the number of times you have thrown the coin. The question is: what should be a reasonable price for a privilege to play this game? A remarkable feature is the possibility to win an arbitrary large amount of money, and even more crucial observation is that the expectation of your payoff (your average payoff) is infinite.

Remark. To get this infinity is a simple exercise in probability. Let us indicate where it comes from for those not acquainted with probability theory. Generalizing our formulas for expected payoffs of Section 4.4 one shows that the average payoff is the sum of all possible payoffs multiplied by their respective probabilities. It is not difficult to show that the probability to get 2^k is $2^{-(k+1)}$ and hence the average is

$$\frac{1}{2} + \frac{2}{4} + \frac{4}{8} + \ldots = \frac{1}{2} + \frac{1}{2} + \frac{1}{2} + \ldots,$$

that equals infinity as was claimed.

Consequently, if one is interested only in average payoffs, the cost of the right to play such a game should be a fortune (infinity), which, of course, contradicts any common sense.

The general conclusion is as follows. The risk (its probability and value) is often not a less important indicator in the assessment of a game, than the average payoff. In the theory of finances, a well balanced analysis of risk and expectation simultaneously gave rise to the Nobel prize winning theory of H.M. Marcowitz on the portfolio optimization.

4.6 Symmetric mixed strategies Nash equilibria

In symmetric games, of special interest are symmetric Nash equilibria, i.e. the equilibria (σ, σ) where the strategies of both players in the equilibrium coincide. It seems to be a more difficult task to find Nash equilibria among mixed strategies. In fact, in case of finite number of strategies, you could just explicitly check for each pair of strategies whether it represents a Nash equilibrium. But in case of infinite numbers of strategies you have to have some tool to calculate these Nash equilibria. We shall explain how these calculations are carried out for symmetric Nash equilibria of symmetric games (though for non-symmetric games and equilibria everything works more or less in the same way).

Theorem 4.1. *(The equality of payoffs lemma). A profile $(\sigma^\star, \sigma^\star)$ with $\sigma^\star = (p^\star, 1 - p^\star)$ such that $p^\star \neq 0$ and $p^\star \neq 1$ is a symmetric mixed strategy Nash equilibrium in a symmetric game of Table 4.2 if and only if*

$$\Pi(h, \sigma^\star) = \Pi(d, \sigma^\star). \tag{4.4}$$

Proof. By the definition of Nash equilibrium, the profile $(\sigma^\star, \sigma^\star)$ with $\sigma^\star = (p^\star, 1 - p^\star)$ is a symmetric Nash equilibrium if and only if $\sigma^\star$ is the best reply to itself, i.e. if

$$\pi(\sigma^\star, \sigma^\star) \geq \pi(\sigma, \sigma^\star) \tag{4.5}$$

for all strategies $\sigma = (p, 1 - p)$. From (4.2) it follows that (4.4) implies that

$$\Pi(\sigma, \sigma^\star) = \Pi(h, \sigma^\star) = \Pi(d, \sigma^\star) = \Pi(\sigma^\star, \sigma^\star) \tag{4.6}$$

holds for all σ, and consequently (4.5). Vice versa, suppose (4.4) does not hold, for instance,

$$\Pi(h, \sigma^\star) > \Pi(d, \sigma^\star).$$

Then

$$\Pi(h, \sigma^\star) = (p^\star + 1 - p^\star)\Pi(h, \sigma^\star) > p^\star\Pi(h, \sigma^\star) + (1 - p^\star)\Pi(d, \sigma^\star) = \Pi(\sigma^\star, \sigma^\star),$$

which contradicts (4.4). This contradiction proves our assertion and also the following remarkable fact.

Corollary. *If* $\sigma^* = (p^*, 1 - p^*)$ *is a symmetric strictly mixed (i.e. with* $p^* \neq 0$ *and* $p^* \neq 1$*) Nash equilibrium, then equality* (4.6) *holds for all* σ. *In other words, the inequality* (4.5) *for all* σ *implies the corresponding equality* (4.6) *for all* σ.

Using the equality of payoffs lemma, one can now easily calculate all symmetric Nash equilibria in game of Table 4.2. In fact, due to (4.1), (4.4) reads as

$$p^* H + (1 - p^*)V = (1 - p^*)D,$$

or

$$p^*(H + D - V) = D - V,$$

and hence finally

$$p^* = \frac{D - V}{H + D - V}, \tag{4.7}$$

if $H + D - V \neq 0$. Consequently, subject to the latter condition, there can be at most one symmetric strictly mixed Nash equilibrium. For this equilibrium to exist the number p^* given by (4.7) should be between 0 and 1. A simple analysis of this formula leads to the following conclusion.

Theorem 4.2. *In a symmetric two-action two-player game given by Table 4.2 with either* $D \neq V$ *or* $H \neq 0$ *there can be at most one symmetric strictly mixed Nash equilibrium. This equilibrium exists and is given by formula* (4.7) *only if either*
(i) $D > V$, $H > 0$, *or*
(ii) $D < V$, $H < 0$.

Proof. In case (i) or (ii) p^* is well defined and $0 < p^* < 1$. Otherwise, p^* is outside the open interval $(0, 1)$, as one sees by inspection.

Observe that the two cases of the theorem corresponds to the stag hunt dilemma and to the game of chicken respectively. The possibility of these symmetric outcomes sheds a new light on how these games can be played and are played in reality.

As there can be at most two symmetric pure strategy Nash equilibria in two-action two-player games, it follows that there can be at most three (pure or mixed) symmetric Nash equilibria in a game given by Table 4.2 with either $D \neq V$ or $H \neq 0$.

4.7 Invasion of mutants and evolutionary stable strategies

Suppose an infinite (or very large) population consists of individuals that in pairwise contests described by Table 4.2 play either hawk, or dove, or some mixed strategy $\sigma = (p, 1 - p)$. A *population profile* is a pair of non-negative numbers $\nu = (q, 1 - q)$ such that q (respectively $(1 - q)$) is the probability with which the strategy h (respectively d) is played in this population. Such a profile can be realized by a variety of ways, for example, by (i) *monomorphic population*: all members of population use the same mixed strategy $\sigma = (q, 1 - q)$, or by (ii) *polymorphic population*: the population consists only of pure hawks and pure doves with a fraction q of hawks. For simplicity we shall talk only about monomorphic populations.

We are now approaching the main goal of this chapter. We shall define the notion of ESS (vaguely introduced at the beginning) and calculate it for the basic Hawk-Dove game. Suppose all individuals are playing a strategy $\sigma^\star = (p^\star, 1 - p^\star)$, and hence the population is monomorphic with the profile $\nu = \sigma^\star$. Suppose a mutation has occurred that caused a small proportion ϵ of individuals to change its strategy to some mutant strategy σ thus changing the profile ν to the new (so called *post-entry*) profile ν_ϵ. A mixed strategy $\sigma^\star$ is called *evolutionary stable* (ESS) if there exists $\tilde{\epsilon} > 0$ such that

$$\pi(\sigma^\star, \nu_\epsilon) > \pi(\sigma, \nu_\epsilon) \tag{4.8}$$

for all $\sigma \neq \sigma^\star$ and all ϵ such that $0 < \epsilon < \tilde{\epsilon}$. In other words, if the proportion of mutants is sufficiently small, ESS does strictly better (e.g. produce more offspring) against the post-entry population than the mutant strategy. This implies that any mutants (if appeared in a small proportion) should eventually die out.

Theorem 4.3. *The main criterion for ESS. The strategy $\sigma^\star$ is ESS if and only if for all strategies $\sigma \neq \sigma^\star$ either*
  (i) $\Pi(\sigma^\star, \sigma^\star) > \Pi(\sigma, \sigma^\star)$
  or
  (ii) $\Pi(\sigma^\star, \sigma^\star) = \Pi(\sigma, \sigma^\star)$ *and* $\Pi(\sigma^\star, \sigma) > \Pi(\sigma, \sigma)$.

Proving this result is a more or less elementary exercise for a mathematician, but it requires some knowledge of linear algebra and probability and will not be given here. We shall just apply it to the analysis of ESS for main examples of Hawk-Dove games (i.e. social dilemmas).

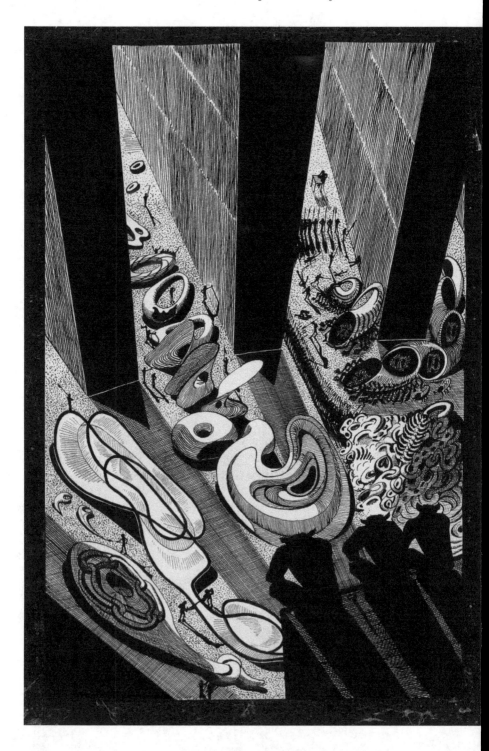

Observe first that the theorem implies that if a strategy $\sigma^\star$ is ESS, then necessarily (4.5) holds for all σ and hence the profile $(\sigma^\star, \sigma^\star)$ is a symmetric Nash equilibrium. This leads to the basic connection between classical and evolutionary game theories. Practically it implies that in order to find all ESS in, say, Hawk-Dove games, one just have to investigate whether each of (at most) three symmetric Nash represent an ESS.

Example. *The game of chicken given by Table 1.8.* Here the strategies "swerve" and "not swerve" correspond to the strategies d and h respectively of the Hawk-Dove game of Table 4.2. Pure strategies Nash equilibria are (h, d) and (d, h) and are not symmetric. Hence (as this game fits to condition (ii) of Theorem 4.1) we conclude that in this game there is a unique symmetric Nash equilibrium $(\sigma^\star, \sigma^\star)$ with $\sigma^\star = (1/2, 1/2)$. To check whether this is ESS, we have to check the conditions of Theorem 4.2. But as $\sigma^\star$ is a strictly mixed Nash equilibrium we see by Corollary to Theorem 3.1 that (4.6) holds for all σ. Hence to check that $\sigma^\star$ is ESS, we have to check that

$$\Pi(\sigma^\star, \sigma) - \Pi(\sigma, \sigma) > 0$$

for all $\sigma = (p, 1 - p) \neq \sigma^\star$. Using (4.3) we find that

$$\Pi(\sigma^\star, \sigma) = -\frac{p}{2} + (1 - p) + \frac{1 - p}{2} = \frac{3}{2} - 2p,$$
$$\Pi(\sigma, \sigma) = -p^2 + 2p(1 - p) + (1 - p)^2 = 1 - 2p^2,$$

and consequently

$$\Pi(\sigma^\star, \sigma) - \Pi(\sigma, \sigma) = 2(p - \frac{1}{2})^2,$$

which is obviously positive for all $p \neq 1/2$. Hence $\sigma^\star$ is ESS.

Example. *Deadlock game of Table 1.9.* Of course, defect $= h$ and cooperate $= d$ here. By Theorem 4.1 there are no strictly mixed strategy Nash equilibria, and the pure strategies profile (h, h) (or (defect, defect)) is the unique symmetric Nash equilibrium in this game. To check whether this is ESS by Theorem 4.2 we find that $\Pi(h, h) = 2$ and $\Pi(\sigma, h) = 2p$ for $\sigma = (p, 1 - p)$. Hence $\Pi(h, h) - \Pi(h, \sigma) = 2(1 - p)$, which is positive for all $p \neq 0$, i.e. for all $\sigma \neq h$. Hence (h,h) is ESS by condition (i) of Theorem 4.2.

Exercise 4.1. Check that for a prisoner's dilemma type game

Table 4.3

	hawk	dove
hawk	1,1	3,0
dove	0,3	2,2

the only symmetric Nash equilibrium is the pure profile (h, h) and that the hawk strategy is ESS.

Exercise 4.2. Check that for the stag hunt dilemma of Table 1.10 there are three symmetric Nash equilibria: (defect, defect), (cooperate, cooperate) and the mixed equilibrium $(\sigma^\star, \sigma^\star)$ with $\sigma^\star = (1/2, 1/2)$. Check that the mixed equilibrium does not yield ESS, but the other two do so.

4.8 The sex ratio game

So far we studied the simplest population games, where the interaction of each individual with the population was realized only through pairwise encounters with other individuals. In nature one can observe other types of interactions of an individual with the whole population, which can be also analyzed by the game theoretical methods (through some complication of the formalism). A remarkable example of these (not pairwise) interactions is modeled by the so called "Sex ratio" game, that is to explain, why animals usually produce an equal number of males and females approximately, or in other words the probability of giving birth to a male is about $1/2$. Note first that this is by no means obvious at the first sight, because one male could in principle copulate with several females and one could expect that females had to be born much more often.

We are not going to plunge into formal details of the a model, but we shall explain qualitatively, where this $1/2$ comes from.

The main point is again the assumption that each female tends to maximize the appearance of her genes in the future generations and the choice of male or female offspring is the means to achieve this objective. Of course, "the choice" should be understood not in the sense of a rational strategy, but in the sense of a stable behavior (ESS) that is passed by parents to their offspring through the genetic code.

Suppose the population consists of marriages of one male and ten female (pure polygamy). Then for a female to produce a male leads to ten daughters-in-law and hence to many grandchildren. For a female to produce a female leads to only one son-in-law and only a few grandchildren (produced by this female daughter). Hence in the next generation the ratio 1:10 will be changed leading ultimately to the ratio 1:1. A dynamic point of view on this game is given in Section 12.4.

Chapter 5

Coalitions and distribution

5.1 Distribution of costs and gains; the core of the game

Till now we considered interactions, in which each player acts independently, whilst at the same time trying to explain the appearance of cooperation as a result of some sort of a rational egoism. Here we shall discuss another elementary brick of interaction, where the possibility of cooperation follows directly from the essence of a problem.

As a typical example consider a construction by neighboring cities (or houses, or firms), of a supply system (of some product of common use, like water, electricity, gas, etc.). For simplicity, we shall discuss the case of three cities only. Suppose the costs for the construction of the system for the three cities A, B, C are given by $c_1 = 20$, $c_2 = 20$, $c_3 = 50$ (thousands of dollars, or rubles etc.) respectively. However, if the three cities join their efforts they can manage the same construction with the better cost $c_{123} = 60$, than $c_1 + c_2 + c_3$ independently. And at last, any pair of cities can create a coalition and construct a system on their own with the costs $c_{12} = 15$, $c_{13} = 54$, $c_{23} = 54$ for the first and the second, first and third, second and the third cities respectively. Such a distribution of costs describes, for example, a situation, when the first and the second cities are closer to the source of supply than the third city. The question is now the following: whether it is reasonable for the three cities to join their efforts (which seems intuitively clear), and what should be a fair distribution of the cost c_{123} between them.

The first idea is to share the cost equally so that each city pays $c_{123}/3 = 20$, which of course is unreasonable, because the first two players can do much better forming a coalition without the third player. In fact $c_{12} < 40$. The next more sophisticated idea could be to calculate the global cost saving $c_1 + c_2 + c_3 - c_{123} = 30$ and share it equally between the cities giving the allocation of costs $20 - 30/3 = 10$ for the first and the second city, and $50 - 30/3 = 40$ for the third one. But this distribution should again be rejected by the same reason as above, namely the first two cities can do better forming a coalition with the cost $c_{12} = 15$, than with what they should pay $20 = 10 + 10$ according to the distribution proposed above. This discussion leads us to a conclusion that any reasonable distribution of costs should at least satisfy the *stand-alone principle*: any coalition should not pay more than what it needs to pay to provide service by itself. This motivate the following definition.

A *cost-sharing game* for three players is specified by a set of costs (non-negative numbers) c_1, c_2, c_3 that each player should pay on its own, the set of costs c_{12}, c_{13}, c_{23} that all possible coalitions of two players should pay, and the cost c_{123} that is required if all players join their efforts. A cost allocation is any collection of non-negative numbers x_1, x_2, x_3 such that $x_1 + x_2 + x_3 = c_{123}$. One says that such a cost allocation belongs to the *core of the game* if the *stand-alone principle* is satisfied, namely if

$$x_1 + x_2 \leq c_{12}, \quad x_2 + x_3 \leq c_{23}, \quad x_1 + x_3 \leq c_{13},$$

$$x_1 \leq c_1, \quad x_2 \leq c_2, \quad x_3 \leq c_3.$$

Let us calculate the core of the game in our example above. The above conditions become

$$x_1 + x_2 \leq 15, \quad x_2 + x_3 \leq 54, \quad x_1 + x_3 \leq 54,$$

$$x_1 \leq 20, \quad x_2 \leq 20, \quad x_3 \leq 50.$$

For calculations it is convenient to rewrite the problem in terms of the gain-sharing, namely in terms of $y_1 = c_1 - x_1 = 20 - x_1$, $y_2 = c_2 - x_2 = 20 - x_2$, $y_3 = c_3 - x_3 = 50 - x_3$ so that

$$y_1 + y_2 + y_3 = c_1 + c_2 + c_3 - c_{123} = 30,$$

$$y_1 + y_2 \geq c_1 + c_2 - c_{12} = 25,$$

$$y_2 + y_3 \geq c_2 + c_3 - c_{23} = 16,$$

$$y_1 + y_3 \geq c_1 + c_3 - c_{13} = 16,$$

$$0 \leq y_1 \leq 20,\ 0 \leq y_2 \leq 20,\ 0 \leq y_3 \leq 50.$$

In order to see this set graphically one can express y_3 in terms of y_1 and y_2 by the formula $y_3 = 30 - y_1 - y_2$, which leads to the following conditions:

$$0 \leq y_1 \leq 14,\quad 0 \leq y_2 \leq 14,\quad 25 \leq y_1 + y_2 \leq 30,$$

and the set of y_1, y_2 satisfying this conditions (which describes the core of our game) is the triangle (with vertices designated by black circles) drawn on the following picture.

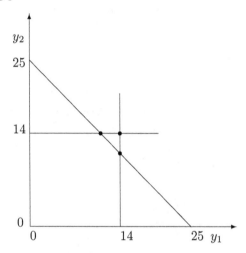

Taking the numbers somewhere near the center of this triangle seems to be a fair distribution of costs among the three cities.

However, the size of the core is sensitive to the parameters of the game. By changing the costs, one easily finds the situations when the core is large, or when it is empty.

Exercise 5.1. Check that if in the above example we change the value c_{12} from 15 to 12 (leaving other values the same), the core of the game would consist of one point only: $y_1 = y_2 = 14$, $y_3 = 2$. If we further make c_{12} smaller than 12, the core would become empty.

Exercise 5.2. Show that the core in a cost-sharing game for three players is non-empty if and only if

$$c_{123} \leq c_1 + c_2 + c_3,$$

$$c_{123} \leq c_{12} + c_3,\quad c_{123} \leq c_{23} + c_1,\quad c_{123} \leq c_{13} + c_2,$$

$$c_{123} \leq \frac{1}{2}(c_{12} + c_{13} + c_{23}).$$

The transformation from a cost-sharing problem to a gain-sharing problem that we carried out above, leads to the following general concept.

A *gain-sharing game* (also called a *game with transferable utility*) for three players is specified by a set of payoffs (non-negative numbers) p_1, p_2, p_3 that can be obtained by each player separately, the set of payoffs p_{12}, p_{13}, p_{23} that can be obtained by all possible coalitions of two players, and the payoff p_{123} that can be gained if the three players join their efforts. An allocation is any collection of non-negative numbers y_1, y_2, y_3 such that $y_1 + y_2 + y_3 = p_{123}$. One says that such an allocation belongs to the *core of the game* if the *stand-alone principle* is satisfied, namely if

$$y_1 + y_2 \geq p_{12}, \quad y_2 + y_3 \geq p_{23}, \quad y_1 + y_3 \geq p_{13},$$

$$y_1 \geq p_1, \quad y_2 \geq p_2, \quad y_3 \geq p_3.$$

As we did in our example above, it is easy to rewrite a cost-sharing game as a gain-sharing game and vice versa.

Exercise 5.3. Show that the core in a gain-sharing game for three players is non empty if and only if

$$p_{123} \geq p_1 + p_2 + p_3,$$

$$p_{123} \geq p_{12} + p_3, \quad p_{123} \geq p_{23} + p_1, \quad p_{123} \geq p_{13} + p_2,$$

$$p_{123} \geq \frac{1}{2}(p_{12} + p_{13} + p_{23}).$$

The situation is similar in the case of games with an arbitrary number of players, but the calculations of the core and conditions of its existence become more involved. A much more nontrivial problem is to find out what should be the fair distributions in the case of a large core, or an empty core? Notice that the core can be empty even if the cooperation is still effective, i.e. if all but the last inequality in the condition of Exercise 5.2 hold. The theory of games offers several reasonable solutions to the problem of fair distribution in these cases. But this deeper analysis is beyond the scope of our present exposition. However, we shall present at the end of Chapter 9 the main definitions of the cooperative games solutions (core and Shapley value) for an arbitrary number of players, see also [34], [127], [126] for some specific related models.

In the next section, in order to have a short break from mathematics, we are going to shortly discuss the general ethical and philosophical principles which lie in the heart of the problem of fair distribution.

5.2   General principles of fair distribution

There are two related approaches, one concerns a situation when the players have to decide a fair solution between themselves. These kind of problems are usually referred to as bargaining problems (see a good introduction in [58]). We shall discuss a situation where a solution can be supplied by an external advisor, sometimes called a *benevolent dictator* in the literature. For a developed exposition one can consult a recent monograph [143].

Philosophical basis for the theory of distribution justice is supplied by Aristotle's thesis: "Equal should be treated equally, and unequals unequally, in proportion to the relevant similarities and differences". (Note that this was written two thousands years before the famous motto from the satirical Orwell's novel "Animal farm": "All animals are equal, but some of them are more equal than others".) Four simple ideas govern the practical applications of this principle revealing more clearly what should be meant by "relevant similarities and differences". These ideas are *exogenous rights, compensation, reward and fitness*. These ideas could be instructively demonstrated on a classical flute assignment problem that goes back to Plato. Namely, a flute is to be given to one of four children in a class. What should be a fair decision of the teacher, who is taking into account the following arguments. The first child is the poorest and has significantly fewer toys than the others. By the *compensation principle*, he should get the flute. But the second child did much on cleaning and fixing the flute, hence he should get it as a *reward*. The third child is the son of a man who brought the flute to the school, so he has the *exogenous right* to claim it. At last, the fourth child is a flutist, and by *fitness* argument he should get the flute as all enjoy good music.

Roughly speaking, the fitness argument demands that resources go to whomever makes the best use of them for the benefit of all. Ideologically it is close to the program of classical *utilitarianism* that favors a distribution that maximizes the overall utility of a communion (say, the sum of all individual utilities) ignoring the needs of particular members of a community. Exogenous rights are irrelevant to the consumption of resources. The basic example is the fairness principle of equality in allocation of certain rights, say political, which follows the program of classical *egalitarianism*. Unequal exogenous rights are also common, say shareholders in a publicly traded company, or parties of different size in a parliament should have unequal shares of decision power.

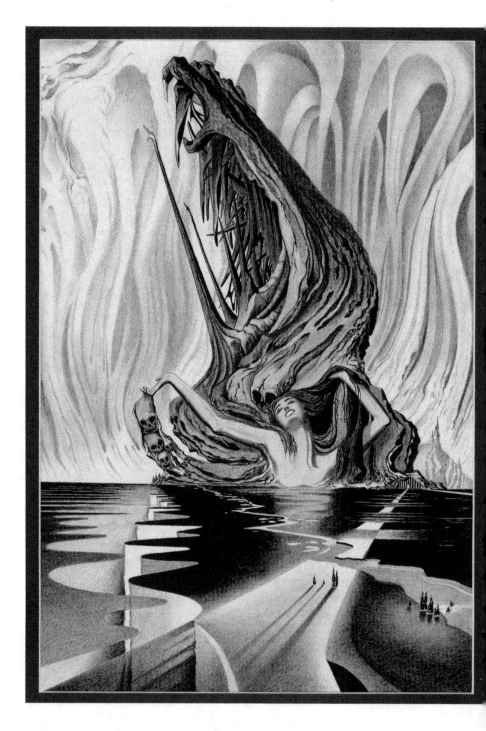

Another classical example that illustrates the four basic principles of distribution justice is the problem of distribution of water or food in a besieged town. Here exogenous right supports either strict equality, or is based on a fixed ranking depending on a social status. Compensation principle gives priority to the sick and the children. The principle of reward supports those who made the major contribution to the organization of the defence of the town, and fitness favors those who are currently the most active in fighting with the enemy.

As seen from these examples, the four principle act mostly as swan, lobster and pike in the Krylov's fable, where they decided to drive a cart-load but pulled in opposite directions ("swan is longing for the sky, lobster draws back and pike pulls into the water ..., and things haven't budged an inch"). The problem of mathematical modeling (based on microeconomic thinking and game theory) is to analyze quantitatively the results of systematic applications of the basic principles, as well as to find a reasonable compromise between them.

Exercise 5.4. Suppose there is a shortage of a medical supply (or care) in a military hospital. Identify the principles, on which the following policies of distribution are based:

favor the most severely wounded,

favor the bravest soldiers,

give priority according to a military rank,

maximize the number of recoveries that allow to return to fighting (in particular, favor several lightly wounded soldiers and sacrifice a badly wounded one).

Exercise 5.5. *The lifeboat story.* There is a limited number of seats in a boat that can save passengers after a shipwreck (recall Titanic). Describe various policies that can be used to select people to be taken on the boat, and for each policy specify one of four basic principles, which this policy is based on. What would be your policy, if you would have to decide?

Exercise 5.6. In his "Republic" Plato proposed to place philosophers at the reigns of government. Which of the four basic principles can be used to support such a proposal?

Finally let us note that the problems discussed here get new importance in the modern era of the development of artificial intelligence, when it becomes necessary to program robots to make decisions, in particular in critical situations. The emerging conflict of interests is reflected in a recent film 'I Robot', inspired by a book of I. Azimov, where a casuality is described (car is falling from a bridge) with a policemen and a child being involved. A robot arrived to help, but was able to save only one person, and it chose to save the policeman (to the dismay for many!). What principle of fair distribution was programmed in the robot?

5.3   Utilitarianism and egalitarianism; compromise set

An acknowledged classic of the philosophy of *utilitarianism* was Jeremy Bentham, see e.g. [20], also credited with the further development of liberalism. Karl Marx in his 'Capital' called Bentham "the genius of bourgeois stupidity" not withstanding the fact that Robert Owen, a student of Bentham, became later a founder of the socialism.

Roughly speaking the *utilitarianism* calls for the assessment of collective actions of a society only on the basis of the utility levels of its members (free wills of individuals). Not discussing here the justification or utility of utilitarianism for the Society in general (that we touched upon in the previous section), we note only that it leads to an easily apprehended and mathematically formulated problem (to which we permanently return in this book) to work out a systematic reasonable way of building a *collective utility function (CUF)* on the basis of individual ones (see Chapter 1 for the concept of utility function) in order to choose afterwards an alternative (outcome, way of action, distribution, etc.) that maximizes this CUF. In other words, given n utility functions (e.g. payoffs) $U_1, ..., U_n$ on a set of outcomes X (so that an agent i prefers an outcome x to an outcome y whenever $U_i(x) > U_i(y)$) the problem is to work out a CUF U on X. The maximum of this U would be then considered as the best outcome for the group of n agents. Of course, there are many ways of building a CUF (in some sense, this whole book is devoted to the analysis of possible alternatives). Assuming that all U_i are expressed in the same units the *classical utilitarianism* suggests to choose the CUF

$$U(x) = U_1(x) + ... + U_n(x), \tag{5.1}$$

i.e. to think only of the total utility. Another reasonable choice is the *egalitarian CUF*

$$U(x) = \min(U_1(x), ..., U_n(x)), \tag{5.2}$$

that concentrates on the best outcome for the weakest agent (see e.g. [162] for a philosophical discussion). Of course, one should be careful when choosing between these alternatives, as applying classical utilitarian CUF could lead to the situation, when the most capable members gets the lowest

rewards (thus urging them to hide their talents) and using egalitarian CUF
to its extremes could lead to blocking the work of the whole system due
to a weakness of a single member. Detailed mathematically worked out
examples could be found e.g. in [142].

Egalitarian CUF seems to lead to the most 'peaceful' resolution, as
those in the worst position get the best they can and the losses of those
in the better position are (at least partially) compensated by the general
consensus implying certain stability.

Applying egalitarian CUF leads to a certain useful and easy to calculate
concept of the solution to n-person games that was not yet mentioned above.
Notice first that when talking about losses rather then gains, egalitarian
program would lead to the distribution that minimizes the highest losses
(which is equivalent to maximizing the smallest gains as in (5.2)).

Assume now that the payoffs to the agents on a set of outcomes X
are given by certain functions $H_1, ..., H_n$. For instance, in the setting of
n-person games in normal form, X would stand for the set of all profiles
(all possible collections of the strategies of the players). Considering the
losses of the players to be the differences between their actual gains and
their maximal possible gains, i.e.

$$W_i(x) = \max_y \{H_i(y)\} - H_i(x), \qquad (5.3)$$

leads to the following definition based on the egalitarian program. The
compromise set (CS), or the *compromise solution* (for the problem specified
by n payoff functions H_i on a set X, e.g. for an n-person game with the set
of profiles X) is the set of all outcomes x from X, where

$$\min_x \max_i W_i(x) \qquad (5.4)$$

is attained. This set can be rather large, can be even the whole set X
(Exercise: give an example!), so that further improvements are in order.
For instance once finding the CS one can then reduce it by optimizing the
next worst payoff, etc. This leads to the so-called *ultimate compromise set*
or a solution based on maximizing with respect to the *leximin order*, see
e.g. [122] or [142].

Some elementary properties of CS are collected in the following statement which for simplicity deals only with the case of two players with equal
maximal payoffs (as for symmetric games in a normal form).

Proposition 5.1. *Suppose* $n = 2$ *and* $\max_y\{H_1(y)\} = \max_y\{H_2(y)\}$.
Then (i) CS consists of those points, where

$$\max_x \min_i H_i(x) \qquad (5.5)$$

is attained. (ii) Each point x from CS is weakly Pareto optimal in the sense that there does not exist any other point y such that $H_1(y) > H_1(x)$, $H_2(y) > H_2(x)$. (iii) Each point x from the ultimate CS, which is defined as a subset of CS where $\max_i H_i(x)$ attains its maximal value, is strongly Pareto optimal in the sense that there does not exists any y such that $H_1(y) \geq H_1(x)$, $H_2(y) \geq H_2(x)$ and at least one of this inequality is strong.

Proof. Since $\max_y\{H_1(y)\} = \max_y\{H_2(y)\}$, one can take $-H_i$ instead of W_i in (5.4) and (i) follows because

$$\min_x \max_i(-H_i(x)) = -\max_x \min_i H_i(x).$$

(ii) and (iii) are more or less obvious and are left as an Exercise.

Example. In prisoner's dilemma (see Tables 1.3, 1.4) the CS consists of the single profile (cooperate, cooperate).

As we mentioned, looking for the compromise set is appropriate when the players are looking for a compromise, when their interests are not opposite, say when they are agents of the same firm. As examples, one can consider the solutions of classical optimization problems subject to multi-criteria assessment.

Example. Postman problem (multicriteria variant). The birthday of the postman problem could be also taken as the birthday of the graph theory. The first work on this theme was published by L. Euler (which we mentioned already in this book) in 1736 in 'Trudy' of St. Petersburg Academy of Sciences. Euler's investigation was conducted in connection with the so called problem of the *Königsberg bridges*. Königsberg (now Kaliningrad) was built on the sides of two rivers and on two islands. There were 7 bridges that join the islands with other parts of the city. Was it possible for a citizen of Königsberg to make a round walk starting from his house that would cross each of the bridges precisely one time? The answer was negative. The general postman problem is the following. There are several given cities (mathematically — points in Euclidean space), joined by roads (arcs). The problem is to find a path that starts in a given city (where the post office is situated) goes through each of the roads precisely once and returns back (a postman has to deliver the post to the villages situated along all the roads). If such a path does not exist (as is the case in the Königsberg problem), one has to find the shortest path covering each road at least once. Assume now that the journey of a postman (or a selling representative of a firm) is organised by several agents (or departments of

a firm). One has to choose a path taking into account the interests of all
agents. For instance, different paths require different consumption of the
fuel (or other costs) that has to be minimise. Selling agents would like
to organise it in the shortest time. Insurance agents may think about the
safety of a path, etc. As these agents belong to the same firm, it is natural
to look for a compromise that could be chosen on the basis of the concept
of the compromise solution introduced above.

5.4 Equilibrium prices

Here we touch upon the classical model of pure exchange, which was introduced in the 19th century in an attempt to justify the basic principles
of market economy (all problems are solved by the reasonable allocation of
prices). The mathematical theory was finalized in the first half of the 20th
century and since then was used as a basis for a wide range of developments
and improvements.

Suppose there are l types of commodities on a market, whose values
(per unit) are specified by the collection of prices $p = (p_1, ..., p_l)$, $p_i > 0$
(expressed in terms of pounds, or dollars, etc.), and there are m players
(agents), each player $i \in \{1, ..., m\}$ having some initial collection of commodities

$$\omega^i = (\omega_1^i, ..., \omega_l^i),$$

so that his/her capital is

$$r^i = (p, \omega^i) = p_1 \omega_1^i + ... + p_l \omega_l^i.$$

Budget set $B(r^i)$ of a player i is the set of all collections $y = (y_1, ..., y_l)$ of
goods the player can buy with the capital r^i, i.e. such that

$$p_1 y_1 + ... + p_l y_l \leq r^i = p_1 \omega_1^i + ... + p_l \omega_l^i.$$

Budget sets describe the possibilities of players. Their desires (or measures
of preferences) are usually specified by the utility functions (already introduced in Chapter 1). Namely, a utility function (or index) of the player i is a
mapping $U_i(y)$ from all collections of l nonnegative numbers $y = (y_1, ..., y_l)$
to positive numbers such that $U_i(y) > U_i(y')$ if and only if the player i
prefers the allocation y to the allocation y'. Usual (simplifying, but quite
natural) assumptions on each U_i are the following:

(U1) U_i is continuous (if all y_j are closed to y_j', then $U_i(y)$ is closed to
$U_i(y')$),

(U2) U_i is an increasing mapping (i.e. all products are desirable)

(U3) the levels of U_i are strictly convex, i.e. if $U(y) = U(y')$ for some $y = (y_1, ..., y_l) \neq y' = (y'_1, ..., y'_l)$, then

$$U(\frac{y_1 + y'_1}{2}, ..., \frac{y_l + y'_l}{2}) > U(y) = U(y'),$$

so that a proportional mixing of equally desirable collection of commodities increase their desirability.

It is not difficult to show (though requires some mathematics that we omit) that under this assumption for each $r > 0$ there exists a unique collection of commodities $d_i(p)$ that maximizes the utility function U_i on the budget set $B(r)$. The mapping $p \mapsto d_i(p)$ is called the *function of individual demand* and the mapping $p \mapsto Z(p) = (z_1, ..., z_l)(p)$ with

$$z_k = (d_k^1(p) - \omega_k^1) + (d_k^2(p) - \omega_k^2) + ... + (d_k^l(p) - \omega_k^l)$$

is called the *excess demand*. As one easily deduces from the definition this function enjoys the equation

$$p_1 z_1(p) + p_2 z_2(p) + ... + p_l z_l(p) = 0,$$

which is called the *Walras law* discovered by L. Walrus in 1874.

Recalling the famous toast ("let us drink wishing to each other that our desires would always coincide with our possibilities!") one is interested in the situations when all z_k vanish, $k = 1, ..., l$, i.e. when the supply and the demand coincide. In these cases (i.e. when $Z(p) = 0$) the collection $p = (p_1, ..., p_l)$ is called the *equilibrium prices*. The basic nontrivial mathematical result (obtained by J. von Neumann and A. Wald, see e.g. [82] for a proof) states that the equilibrium prices always exist. In [91] a method of solving the market game under non-equilibrium prices is suggested via the introduction of a second currency (*coupons*).

Though this is not obvious from the first sight, the model discussed above is closely connected with cooperative games touched upon in Section 5.1. One can show that the allocations (or imputations) $d_i(p)$ specified by equilibrium prices belong to the core of certain cooperative game (whose definition is a slight generalization of the models from Section 5.1), i.e. it can not be blocked by (appropriately defined) coalitions. Hence two quite different approaches to the analysis of human behavior meet happily in this model: cooperation and coalitions on the one hand and egoistic optimization of personal wealth on the other hand. This is quite appealing: as long as the "right" prices are chosen, no conflicts arise. Of course, the reader can easily notice the shortcomings of the idealized assumptions of this model. One of the simple critical observation is that the real world business players (firms, supermarkets, etc.) never aim at any equilibrium distributions, but just at an infinite expansion by all possible means, see e.g. [58] for a constructive modern discussion of possible generalizations and improvements of this classical Walras model, based on game theoretic arguments. The development of the theory of competitive equilibrium was marked by Nobel prizes in economics to Arrow (1972) and Debreu (1983).

5.5 Linear models and linear programming

In this section we consider the examples of the optimal distribution problems belonging to the class of the so called linear programs.

Example. Diet problem. Assume there are three different types $i = 1, 2, 3$ of chicken food that supply 3 basic nutrients: 1) protein, 2) minerals, 3) vitamins. The minimal requirement for the nutrient j, $j = 1, 2, 3$, is c_j units, the positive numbers a_{ij} specify the content of the nutrient j in the unit of food i, and b_i is the price of the unit of food i. The question is to minimize the cost of the food that would supply the required amount of each nutrient. In other words one looks for the collections y_1, y_2, y_3 of three kind of foods that minimizes the total price $b_1 y_1 + b_2 y_2 + b_3 y_3$ subject to the conditions

$$y_1 a_{11} + y_2 a_{21} + y_3 a_{31} \geq c_1$$

$$y_1 a_{12} + y_2 a_{22} + y_3 a_{32} \geq c_2$$

$$y_1 a_{13} + y_2 a_{23} + y_3 a_{33} \geq c_3.$$

Example. Transportation problem. Assume there are three production centers of certain commodity containing s_1, s_2, s_3 units respectively, and two destinations to ship this commodity that require at least r_1 and r_2 units respectively. Let c_{ij} denote the cost of shipping of one unit from center i ($i = 1, 2, 3$), to destination j ($j = 1, 2$). The problem is to identify the quantities x_{ij} of the commodity to be shipped from the production center i to destination j that satisfies all requirements and minimize the total cost. In mathematical notations one seeks to find non-negative numbers x_{ij} minimizing the expression

$$x_{11} c_{11} + x_{21} c_{21} + x_{31} c_{31} + x_{12} c_{12} + x_{22} c_{22} + x_{32} c_{32}$$

subject to the conditions

$$x_{11} + x_{12} \leq s_1, \quad x_{21} + x_{22} \leq s_2, \quad x_{31} + x_{32} \leq s_3$$

and

$$x_{11} + x_{21} + x_{31} \geq r_1, \quad x_{12} + x_{22} + x_{32} \geq r_2.$$

In real life problems the number of centers (or nutrients, etc.) in above problems could be rather large making everything much more complicated and leading to the general linear programming problem that consists in finding a collection of positive numbers $x_1, ..., x_k$ maximizing or minimizing a linear form $c_1 x_1 + ... + c_k x_k$ subject to linear constraints. This problem appears in economics in a variety of contexts. The major breakthrough in its effective solution (duality, simplex algorithm) was signified by the award of the Nobel prize in economics 1975 to L. Kantorovich.

Though the above inequality surely would remind the reader the similar inequality that appeared in Section 1 when calculating the core of a cooperative game, in the formulation of the above linear programming problems there were no explicit game theoretic flavor (no competitive preferences). This demonstrates a unifying character of mathematical language, allowing to use the same tool for seemingly unrelated problem. But even more remarkable in this situation is the following fact: the linear programs are ultimately linked with simple zero-sum games of two players: one of the major tool in linear programming, the famous duality, is nothing else but a clever game theoretic reformulation of linear programs. We shall address this issue in Chapter 9.

Chapter 6

Presidents and dictators

6.1 Collective choice; problems of voting

Collective choice means the choice of an action of a group of people on the basis of their personal opinions. The problem of collective choice was outstanding for the mankind during the whole history of its existence. In the modern world this problem seems to be as actual as never before. Historically the problem has often been solved by the appearance of a certain leader (king, dictator, president), whose choice (opinion) has basically been identified with the collective choice. Systematically (and scientifically) the problem of collective choice started to be discussed in the 18th century on the eve of the great French revolution, which was meant to substitute a hereditary leader by laws that had to put into effect the decisions based on the collective choice. How these laws should look like, or how to work out a decision on the basis of individual preference? This was the question put forward in the classical works of Jean-Charles de Borda and Jean-Antoine the marquise de Condorcet. Both these scientists realized tremendous conceptual difficulties in creating such laws. As is well known, the French revolution did not manage to solve this problem in practice, and after several bloody years of "collective choice" performed under the ominous shade of the guillotine, the country returned to monarchy. Condorcet himself, an inspirer and active participant of the revolution, took poison hidden in his ring and died in prison preferring this death to the guillotine that waited for him next morning.

So, what are the difficulties of the collective choice? Following Condorcet [41] and de Borda [27] let us consider here the problem of *voting*, when several agents (or electors) must make a choice between a given collection of candidates (to choose one of them). Candidates are understood here in a broad sense as people (say, candidates for presidency), variants of a law, etc. If there are two candidates, then the ordinary majority voting is clearly the fairest and the most reasonable rule. The difficulties begin when the number of candidates becomes more than two. For simplicity, we shall reduce our discussion wherever possible to the case of three candidates, say a, b, c. The simplest and the most popular extension of the majority voting to the case of several candidates is the *plurality rule*, when each agents casts a vote for only one candidate and the winner is the candidates with the maximum number of votes. Both Condorcet and de Borda realized the drawbacks of this rule. To illustrate these drawbacks consider the following table (or profile) of preferences

$$
\begin{array}{ll}
4 & a > b > c \\
3 & b > c > a \\
2 & c > b > a
\end{array}
\qquad (6.1)
$$

Sign $>$ here and further on denotes the preference, and the whole table means that there are four agents preferring a to b and b to c, three agents preferring b to c and c to a, and two agents preferring c to b and b to a. By the plurality rule the winner is a. The arguments of de Borda and Condorcet are based on the observation that for the majority of agents (5 out of 9) the candidate a is the worst choice, and consequently the choice of a does not actually reflect correctly the opinion of voters. Hence the plurality rule can not be considered fair and reasonable. According to the idea of Condorcet, the winner should be b, because b beats both c (7 out of 9) and a (5 out of 9) in a pairwise contest carried on by the majority voting. This approach of Condorcet leads to the general notion of the *Condorcet winner*, which is a candidate (obviously necessarily unique) that beats all other candidates in a pairwise contest carried on by the majority voting. This notion sounds reasonable if leaving aside the fact that a voting that can specify the Condorcet winner must be much more complicated than the plurality rule. However, after some thought it becomes clear that there are too many profiles where the Condorcet winner simply does not exists. The simplest example of this situation called sometimes the *voting paradox* (or the *Condorcet paradox*) is the profile

$$
\begin{array}{ll}
1 & a > b > c \\
1 & b > c > a \\
1 & c > a > b
\end{array}
$$

Of course, this situation is completely symmetric with respect to all candidates, and even the plurality rule can not identify a winner. However, when the number of agents is large (as compared to the number of candidates), equal distributions are not very probable, and, say, on the profile

$$
\begin{array}{ll}
10 & a > b > c \\
11 & b > c > a \\
12 & c > a > b
\end{array}
$$

the plurality rule will identify a winner (of course, c), but the Condorcet winner still does not exist. So, choosing the Condorcet winner can not be called a proper voting rule. However it seems reasonable to look for *Condorcet consistent rules*, i.e. such rules that would choose the Condorcet winner in case it exists. One of such approaches was proposed by Condorcet himself, who suggested to break the cycle at its weakest link, in other words, to ignore the majority preference supported by the smallest majority. In the example above, the preference $b > c$ is the weakest link, because it is supported by 21 agents, while $c > a$ is supported by 23 agents and $a > b$ is supported by 22 agents. Thus Condorcet proposes to ignore the preference "b is better than c", and hence to elect the candidate c, which coincides in this example with the choice made by the plurality rule.

It must be clear from this discussion that the choice of the voting rule is absolutely crucial for the results. We are going to address now the following general questions: how to compare the different rules by their quality and what properties these rules have to possess in order to be considered fair or reasonable, and from what point of view. But first we introduce several common voting rules applied in practice, and then we use them as examples on testing our criteria of fairness. The first two rules below are simpler and are used more often than the last two, which are heavier for implementing.

6.2 Four examples of voting rules

Plurality with runoff. The first round is organized like usual plurality voting. But the results are interpreted differently. If a candidate wins a strict majority (i.e. got more than half of all votes), he is considered as a winner. However, if there is no such a candidate, then the first two are chosen and the second round is organized to choose one of them by the ordinary majority for two candidates.

This method is used, for example in presidential elections in France and

Russia. It is almost as simple as the plurality rule, but at the same time it would not allow the candidate, who is the worst for the majority, to win the voting. For example, under the profile (6.1) the winner will be b, and not a as by the plurality rule.

Sequential majority comparisons. The candidates are first ordered somehow a, b, c, d,..., which can be done, say, by a chairman of a session of the parliament. Then a and b are compared by the rule of majority and the loser is dropped out from the contest. The winner is then compared with the next candidate c and the loser is again dropped out. In case of a draw, the loser is considered to be c (the candidate which was the last to enter the contest). The procedure goes on until only one candidate remains.

This procedure is widely used in the congress of USA when voting for a new law and its amendments.

The Borda rule (a scoring voting rule). Each agent declares its full preference on the list of all candidates putting them in order from the best to the worst, say, $a > b > c$ in case of three candidates. Then each candidate gets the corresponding points: the worst gets 0, the next worst gets 1, etc. Then the *Borda score* is calculated for each candidate, which is the total score (sum of points obtained from all agents). The winner is the candidate with the highest total score.

The Copeland rule. For each pair of candidates the pairwise contest by majority is organized. For each candidate, say a, the *Copeland score* is defined as the difference between the number of candidates x that a beats in the pairwise contest and the number of candidates y that beat a in a pairwise contest. The winner is the candidate with the highest Copeland score.

Before departing on a systematic discussion of the quality of the voting rules, let us discuss the connection of the methods above with the approach of Condorcet. Namely, it is easy to see that the sequential majority comparisons and the Copeland rule are Condorcet consistent. At the same time, the plurality with runoff is not, as the following simple profile shows

$$2 \quad a > c > b$$
$$2 \quad b > c > a$$
$$1 \quad c > a > b$$

Under this profile c is the Condorcet winner as it beats both a and b in a pairwise contest with score 3 out of 5. However, a and b enter the second round and a wins. Neither the Borda rule is Condorcet consistent, as the profile

$$3 \quad b > c > a$$
$$2 \quad c > a > b$$
$$1 \quad c > b > a$$
$$1 \quad a > b > c$$

shows, where the Condorcet winner is b, but the Borda score for candidates a, b, c are 3, 8, 9 respectively, and hence c wins.

It is worth observing that in most of the rules of voting the draw can occur, i.e. two or more candidates can have equal rights for a winning, say, the Copeland or the Borda scores of several candidates coincide, or two candidates get the same number of votes in the majority voting. Thus these rules do not always define a single winner, but several winners. However, in many situations this is a rare event, and we shall not draw much attention to this problem for simplicity.

Exercise 6.1. (from [178]) For the profile

$$1 \quad a > b > c > d > e$$
$$4 \quad c > d > b > e > a$$
$$1 \quad e > a > d > b > c$$
$$3 \quad e > a > b > d > c$$

show that the Borda scores and the Copeland scores arrange the candidates in opposite orders.

6.3 Criteria of quality of voting rules

Now we shall formulate precisely what properties a voting rule has to possess in order to be considered a fair and reasonable rule. The next two properties or axioms seem to reflect our basic notion of fairness:

Anonymity. The names of agents are irrelevant to the result of voting, i.e. if any two agents exchange their votes (opinions), the result of voting will not be changed.

Neutrality. The name of candidates are irrelevant, i.e. if any two candidates a and b exchange their places in an arbitrary profile of the preferences of agents, the results of voting will change accordingly: if a was a winner, then b becomes a winner, if b was a winner, then a becomes a winner, if a winner was different from both a and b, then he remains a winner.

Clearly these properties must hold, if we want all agents and candidates to have equal rights. And, of course, any reasonable rule must satisfy the following unanimity property:

Effectiveness or Pareto optimality. If a is better than b for all agents, then b can not win the elections.

Let us check the validity of these basic properties (axioms) on our model examples given above.

It is easy to see that the plurality rule, plurality with runoff, the Copeland and the Borda rules all satisfy the axioms of anonymity, neutrality and efficiency.

On the other hand, the method of sequential majority comparisons (though being anonymous) is neither neutral nor efficient. To see the latter consider the profile

$$1 \quad b > a > d > c$$
$$1 \quad a > d > c > b$$
$$1 \quad c > b > a > d$$

with four candidates and three agents. If the pairwise comparison is organised in the order $abcd$, then d wins, but a is better than d for all agents. To see why the neutrality breaks down, consider the simplest profile (6.1) of the Condorcet paradox. It is clear then that the result depends on the order in which candidates are considered. Say, a chairman that declares the agenda (the order of candidates), can achieve any result he personally prefers. Hence the method of sequential majority comparisons can not be considered as a fair method.

We shall now consider an important property of a voting rule that ensures that the chances of a candidate to win can only increase by increasing a support to this candidate:

Monotonicity. Suppose a candidate a wins under a certain profile of preferences, and then this profile is changed in such away that the position of a is improved and the relative comparison of any other pair of candidates remains the same for all agents. Then a should win under the new profile as well. Shortly, improving the position of a candidate can not lead to the defeat of this candidate.

This property sounds as obvious. In fact, it is easy to check that the plurality rule, the rules of Copland and Borda are monotone.

Exercise 6.2. Convince yourself that the latter statement is true.

However, it turns out (surprisingly enough) that the plurality rule with runoff is not monotone, which the following two profiles show

6	$a > b > c$	6	$a > b > c$
5	$c > a > b$	5	$c > a > b$
4	$b > c > a$	4	$b > c > a$
2	$b > a > c$	2	$a > b > c$

In fact, under the first profile, a and b go to the second round and a wins. The second profile is the same but for two agents who change $b > a > c$ to $a > b > c$ so that a becomes better than before. At the same time, under this profile a and c go to the second round and c wins.

At last, let us consider the following two axioms.

Reinforcement. If two disjoint groups of agents consider the same set of candidates, and both these groups choose a candidate a (voting independently of each other), then a should win if these groups vote together.

Axiom of participation. Suppose a group of agents chooses a candidate a from a given set of candidates. Suppose another agent joins this group of agents. Then the new group will choose either a or a candidate which is better than a in the opinion of the new agent.

The reinforcement axiom seems to be quite reasonable when a choosing organ is subdivided on several sections (chambers of a parliament, regional committees of a party, etc.). And the axiom of participation ensures that for any agent it can only be useful to take part in voting and to express frankly his/her opinion. The following result constitutes a serious trouble to the supporters of the Condorcet approach to voting: there is no Condorcet consistent voting rule (i.e. the one that always chooses the Condorcet winner whenever he exists) that satisfies either the reinforcement axiom or the axiom of participation. We refer to [142] for the proof of this fact and for the references to the original papers.

The non-fulfilment of the axioms of monotonicity, reinforcement or participation is a key to the game theory involvement in the analysis of voting, because this indicates that telling truth is not always the best strategy to achieve one's goal. Finding an optimal strategy for a person or a coalition becomes a real question from the theory of games. This leads to *strategic voting*, see the pioneering book [51] on this subject and further discussion e.g. in [40] and [29]. It is also worth citing a pamphlet [48] of L. Dodgson (better known as Lewis Carroll), considered as classics in the field.

6.4 The minority principle; dictators

So far, we discussed the quality of voting rules from the general point of view that a fair rule should reflect the preferences of the majority of voters in the best way. This point of view can not serve as a sole guide for fairness. The majority of, say, 51% of agents ignoring the opinion of the minority can become a real dictator. "Cependent une minorité ne peut pas etre à la merci d'une majorité: la justice qui est la negation de la force, veut que la minorité ait ses garanties" [160]. ("However the minority can not depend completely on the majority: justice which is the negation of force, demands that minority has its guarantees.") In the same way, it is surely not fair, when a superpower invades a small or weak country and then rules it (in the full agreement with the principle of majority). This arguments lead, in particular, to the search of voting rules satisfying *the minority principle*, which states that any coalition, whatever small, can exercise at least some influence on the decision making. We refer to [142] for a full discussion, and only give an example of a voting rule, which allows each agent to influence the final decision.

Voting by successive veto. The agents that are ordered in some way, declare their preferences. Then the least preferred candidate of the first agent is eliminated, then from the remaining set the worst candidate of the second agent is eliminated, etc. This continues until only one candidate remains. He is declared the winner.

This rule allows any coalition of k agents to exclude any set of k candidates, whenever k is less than the number of candidates.

Our process of the selection of the best rule of voting so far was a bit similar to the selection of a bridegroom by a girl in a Gogol's play "Marriage", where she thought like "if the lips of Nikanor Ivanich could be put to the nose of Ivan Kusmich, and if to add some off-handedness from..., I would make my choice." In fact, each of the rules turns out to be unsatisfactory from one point of view or another. An obstruction to a discovery of perfect rule lies in a series of deep negative results in the theory of collective choice. We shall mention a couple of the most important ones.

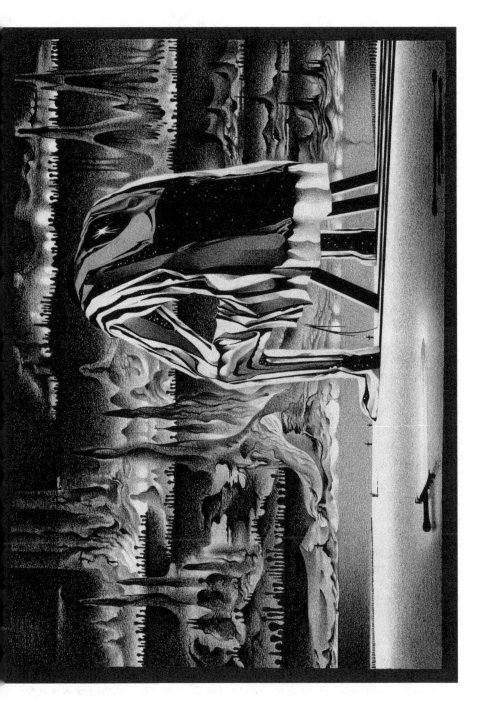

Recall that the main problem with the axioms of monotonicity or participation is the fact that a break down of these properties opens the road to the possibility (or, may be better to say necessity) for voters to manipulate. I.e., in order to achieve their goals in voting, they should not always declare their preferences frankly. A natural general question arises, whether there exist voting rules which are protected from manipulations. I.e. the ones, where truth telling is always the best strategy. The following result (the theorem of Gibbard and Satterthwaite) gives a rather strange answer to this question: in case of more than two candidates, the only rule that is protected from manipulation is the *dictator rule*, i.e. the rule which elects the candidate that is preferable to some fixed agent (*dictator*). We refer to [142] for a precise formulation and a proof of this theorem, which is similar to the result we discuss below with more detail. Namely, instead of voting for one candidate from a collection of the profiles of preferences of all agents, one can discuss a more general question of choosing a "collective profile of preferences" from a collection of individual profiles of preferences of all agents.

This problem is an important element of the theory of mechanisms design. Let the set of agents $I = \{1, \ldots, N\}$, and the set of alternatives $A = \{x, y, \ldots, z\}$ are given. Each agent has its preference on the set $A = \{x, y, \ldots, z\}$, which can be described by permutation of the set $A = \{x, y, \ldots, z\}$. The set of such permutations is denoted by $L(A)$.

The point $R(N) = (R_1, \ldots, R_N)$ in the product $(L(A))^N = L(A) \times \cdots \times L(A)$ (N times) is called the preference profile. Building the map $f \colon (L(A))^N \longrightarrow L(A)$ is the *preference aggregation problem*. The natural conditions for such *preference aggregation function* f are:

Condition 1 (Pareto, or a unanimity condition). If the preferences of all agents are the same and equal to R, that is $R_i = R$ for all i, then $f(R(N)) = R$.

Condition 2 (Independence of Irrelevant alternatives). Group preference $f(R(N))$ on any pair of alternatives (x, y) depends only on the preferences of agents on the same pair of alternatives (and thus the rule f is protected from manipulations).

The following Arrow's theorem on the aggregation of preferences is valid.

Theorem 6.1. *Let the number of alternatives be not less than three. Under the conditions 1 and 2, $f(R(N)) = R_i$, where i is some fixed agent, that is, the only way to aggregate the preferences of society is to choose a leader representing its interests (which is a well established practice for* *many centuries, both in the animal world and in any society).*

Proof. We shall write $x >_i y$, if the agent i prefers x to y. For a coalition K (a subset of the set of agents) we write $x >_K y$ if $x >_i y$ for all $i \in K$.

Let us denote by $W(x, y)$ the set of coalitions K such that $x >_K y$ implies $x > y$. For example, $N \in W(x, y)$ because of the conditions of unanimity. Let $W = \cup_{x,y} W(x, y)$ and C be a minimal coalition of W (that is, no proper subset of C belongs to W). Then C belongs to some $W(x, y)$. We claim that C consists of a single agent. Indeed, let $i \in C$ and z be an alternative other than x and y. Let us consider the following profile $R(N)$:

$$x >_i y >_i z, \quad z >_{C \backslash \{i\}} x >_{C \backslash \{i\}} y, \quad y >_{\bar{C}} z >_{\bar{C}} x,$$

where $\bar{C}$ is the complement of C in $\{1, \dots, N\}$ and where the preferences on other agents are not indicated, since they are irrelevant by Condition 2. Let us write $x > y$ if x is preferred to y by the collective preference $f(R(N))$. As $x >_C y$ and $C \in W(x, y)$, then $x > y$. Since $z >_{C \backslash \{i\}} y$ and otherwise for other agents, and $C \backslash \{i\} \notin W(z, y)$, the preference $z > y$ is not satisfied, so that $y > z$. By transitivity, $x > z$. But $x >_i z$ and otherwise for other agents. Hence $\{i\} \in W(x, z)$ for any $z \neq x$. Thus C is not minimal leading to contradiction.

It remains to check that the agent i is decisive for any pair of alternatives. Take two alternatives y, z other than x, and consider the profile $Q(N)$:

$$y >_i x >_i z, \quad z >_{N \backslash \{i\}} y >_{N \backslash \{i\}} > x.$$

Let us now write $x > y$ if x is preferred to y by the collective preference $f(Q(N))$. We have $x > z$ (since $\{i\}$ belongs to $W(x, z)$) and $y > x$ by unanimity and independence from exterior alternatives). Hence $y > z$ and thus $\{i\} \in W(y, z)$, so that i is the leader.

Let us complete this section by the quotation from [301], which aimed to address the social aspects of centralism applied by Arrow's result: "The strong side of centralism is harmony, consistency, proportionality in its parts of the whole social whole. Without centralism, there can be no planned character of the whole, since this planned character can only be brought into society by the subordination of separate components to its single central plan. Only under this condition does the social system become entirely the product of the rational will of man and most fully adapts to the needs and requirements of the person. In general, the advantageous sides of centralism far outweigh the disadvantageous ones."

Chapter 7

At the doors of quantum games

7.1 Quantum bits and Schrödinger's cat

In our time it became popular to develop quantum "everything". Not talking about "classical" quantum mechanics and quantum field theory, among the fashionable themes are "quantum chaos", "quantum computation", "quantum cryptography", "quantum information", "quantum technology", etc. In such a "quantum atmosphere", the appearance of quantum games had to be expected. The foundation of this theory was laid out in papers [140], [49], [133], see also [75] and [158]. So, this theory is one of the youngest offspring of both game theory and quantum theory. As entering quantum world with only elementary tools is like trying to reach the Moon using a ladder (as baron Münchhausen did) we shall just try to draw the attention of the readers to this exciting new development by discussing very cautiously the conceptual peculiarity of quantum theory and its relevance to conflict interaction.

An elementary classical system representing the basic unit of classical information is a bit, which is the system having two states, usually denoted 1 and 0. The quantum analog of this system is the quantum bit or *qubit*, which describes the quantum spin, i.e. the quantum particle having two states: spin up or down. However, in quantum theory the expression "two states" has a different meaning. It means that there are two *basic states*, but also certain intermediate states. Thus a *qubit* can be represented by the surface of the usual (two-dimensional) sphere. For example, you can imagine a qubit as the surface of the Globe. In this model, the two basic states (spin up and down) can be represented by some chosen opposite points N and S of the sphere, say the North and the South poles of the Earth, other points of the sphere being considered as intermediate. Just comparing the complexity of the globe's surface (qubit) with that of the two point set $\{0, 1\}$ (classical bit) one can imagine what progress can be achieved, as compared with classical calculations, if one would be able to "tame" this quantum bit. This is what quantum computation is about, see e.g. [85].

The difference between quantum and classical systems is much deeper than the difference in formal complexity of the state spaces. In particular, it is reflected in quantum measurements. To measure a state of a classical bit means simply to look, whether it is zero or one. Similarly, if you would have a classical system with the state space being the surface of the sphere, you could just look where your point is, e.g. where is the ship on the surface of the Globe. But you cannot measure the position of a qubit directly; only certain averaged characteristics of its position can be estimated. For instance, you can prepare an experiment to measure whether the spin of your qubit is up. If the spin is precisely down, you get the answer "no", as you could expect. But the remarkable thing is that if the spin is not precisely down, then there is always a positive probability to find that it is actually up! In our metaphor with the ship on the Earth surface, imagine that just by looking for this "quantum ship" which is floating somewhere near the South pole, you can actually find her out in the North pole! Even more, by looking at her you effectively transfer her momentarily to the North or South pole. Welcome to the quantum world! Moreover, there is no way to predict whether you will find her in (transfer her to!) the South or the North pole, only some probabilities can be calculated: the nearer the ship is to the North pole, the higher probability to find her in the North pole.

Furthermore, suppose in your experiment you found that your "quantum" ship is in the North pole. If you would repeat your experiment the next moment, you would find her in the North pole again, as one could expect. But if you now choose any other two opposite points on your sphere N' and S' (as two new basic states) and will measure whether your particle is in N' or S', then almost surely you will find it in either N' or S' thus momentarily transferring it from N to N' or S'. As you can maybe already guess there is, of course, no way to predict whether it will be N' or S'. Thus choosing various opposite points of the sphere and carrying out the corresponding observations, you can drive your "quantum" ship around the Globe along a curious discontinuous random trajectory just by "looking at it". Of course, "looking at" is not a precise description of what is going on, it is not like moving a matchbox by a glance, as some tricksters can do.

The point is that any quantum measurement can be carried out only by an interaction of the particle with a measuring apparatus that inevitably changes the position of the particle.

The remarkable features of quantum measurement are illustratively encapsulated in the famous *Schrödinger cat paradox* invented by E. Schrödinger, one of the fathers of the modern quantum theory. This paradox comes from an imaginary experiment, where a quantum particle (qubit) and a cat are placed together in an isolated room, and where one of the two basic states of the quantum particle is connected with a diabolic mechanism that would kill the cat whenever the particle would enter this state. Our previous discussion makes it clear that only by opening the door of this room, i.e. by looking at the system, can an observer fix the state of the particle and hence the state of the cat (dead or alive). Before this observation the cat is therefore in some intermediate state between life and death. How to make sense of this? Are you killing the cat by looking at it?

New problems and new difficulties arise when trying to build the theory of continuous quantum measurement, i.e. when describing what is going on with a quantum particle, if it is under permanently observation. Logical difficulties which appear when passing from discrete to continuous time were well under discussion in ancient Greece (*Zeno paradoxes of motion*). In quantum case, the transition from discrete to continuous observations is essentially more subtle than in classical mechanics, arising difficulties being usually called *quantum Zeno paradox*. One of the performance of this paradox is the so called *watch dog effect*, which roughly speaking means that if you watch a quantum particle, it can never move; if you watch a "quantum pot", it can never boil, etc. The amount of literature devoted to this subject (foundation of quantum mechanics) is tremendous, see e.g. [181], [17], [86] and references therein for some aspects of this issue.

7.2 Lattices and quantum logic

To describe some peculiarity of quantum logic, we need to talk about logical structures in general. The key structure is the *order*. For the set of real numbers, the *order structure* is given by usual relations $<, >, \leq, \geq$. As is common in mathematics, in order to introduce some general structure, one should first summarize the main features of this structure from the basic example. One can easily observe that the relation $\leq$ on the real numbers enjoy the following basic properties.

L1: $x \leq x$ for all x;

L2: $x \leq y$ and $y \leq z$ imply $x \leq z$ (this property is called *transitivity*);

L3: $x \leq y$ and $y \leq x$ imply $x = y$ (*anti-commutativity*);

L4: for any pairs x, y there exists an element $x \vee y$ (called the maximum of x and y) such that $x \leq x \vee y$, $y \leq x \vee y$, and $x \vee y$ is the least element with these properties, i.e. if $x \leq z$ and $y \leq z$ for some element z, then necessarily $x \vee y \leq z$;

L5: for any pairs x, y there is an element $x \wedge y$ (called the minimum of x and y) such that $x \wedge y \leq x$, $x \wedge y \leq y$, and $x \wedge y$ is the greatest element with these properties, i.e. if $z \leq x$ and $z \leq y$ for some element z, then necessarily $z \leq x \wedge y$.

One can now give the following definition. An arbitrary set L is called a *lattice*, if there is a relation $\leq$ between some pairs of its elements (called the *order relation* or the *order structure*) that satisfies the above properties L1-L5. In general lattices, the elements $\vee, \wedge$ given by L4, L5 are usually called respectively *the least upper bound* and *the greatest lower bound* of x and y. The major difference between general lattices and the order on usual numbers is represented by the fact that for a general lattice not all pairs should be at all comparable, i.e. there can be pairs x, y such that neither $x \geq y$ nor $x \leq y$ holds.

Remark. If only L1-L3 hold for a relation $\leq$, then this relation is called a *partial order*.

An element M of a lattice L is called its *maximum element*, if $x \leq M$ for all x from L. Similarly, an element m is called its *minimum element*, if $m \leq x$ for all x from L. The set of real numbers has neither a maximum nor a minimum element. But, for the set of all non-negative numbers, zero is of course the minimal element.

Examples of the lattices.

1 (Set theory). The set of all subsets of a given set S is ordered by inclusion, i.e. $M \leq N$ means that M is a subset (or a part) of N. The least upper bound of two sets M and N is its union $M \cup N$ (the set of elements that belong either to M or to N), and their greatest lower bound is the intersection $M \cap N$ (the set of elements that belong to both M and N). Properties $L1–L5$ are obvious for this order relation. The whole set S and the empty set $\emptyset$ represent respectively the maximum and the minimum elements for this lattice.

2 (Logic). Propositions (about some group of objects or notions) can be ordered by implication, i.e. $P \leq Q$ for two propositions P and Q means that P implies Q. For instance, "Nick is two years old" implies "Nick is not an old man". For propositions $P \vee Q$ means P or Q, and $P \wedge Q$ means P and Q. The tautology (the class of always correct propositions like "$1 = 1$") is the maximum element for this lattice, and the contradiction (the class of always wrong propositions like "$1 \neq 1$") is the minimum element.

3 (Qubit). Consider the set of all points of the sphere Q (qubit) together with an empty set and the whole sphere Q itself and the order structure induced from the set theory, i.e. defined by inclusion. Then $\emptyset$ and Q are the minimum and maximum elements respectively, so that $\emptyset \leq x \leq Q$ for any point x on Q, and any two different points of Q are not comparable, i.e. neither $x \leq y$, nor $y \leq x$ holds. In particular, $x \wedge y = \emptyset$ and $x \vee y = Q$ for any such points.

Remark. The order structure is one of the basic structures in mathematics and in real life modelling. In particular, it is central to the problem of optimization, and hence to game they, independently of any quantum context. Bringing the order structure to algebraic level leads even to a new branch of mathematical analysis, so called *idempotent or tropical analysis,* see [102], [103].

One can see that Examples 1 and 2 are closely related to each other. For instance, choosing an element x from the whole set S of Example 1, we can make a correspondence between subsets M of S and the propositions "x is an element of M". In this way the whole structure of Example 1 can be interpreted as a part of logical structure of Example 2. Similarly, the order structure of Example 3 has its logical counterpart. With each point x on the sphere Q one can connect a proposition "our qubit is in the state x" or better "if measured, there is a chance (with nonzero probability) to find that it is in x". This leads to quantum logic. Are there any essential differences between quantum and classical logic?

For the characterization of quantum logic the following definition is fundamental. A lattice L is called *distributive* if

$$x \wedge (y \vee z) = (x \wedge y) \vee (x \wedge z) \tag{7.1}$$

for all x, y, z from L.

It is easy to check that the lattices of Examples 1 and 2 of the previous section are distributive, but the lattice of Example 3 is not. To see the latter, pick up any three different points x, y, z of a qubit (sphere) Q. Then the l.h.s. of equation (7.1) equals $x \wedge Q = x$, but its r.h.s. equals $\emptyset \vee \emptyset = \emptyset$.

Thus the main point of the quantum logic is that it is not distributive. And the main reason why the distributivity breaks down is connected with peculiarity of quantum measurement considered in Section 7.1. Namely, an arbitrary two-point set $\{x, y\}$ "covers" the whole qubit, in the sense that for any point z on Q there is always a non-vanishing probability to find it either in x or in y. Returning to our earlier metaphor, for any quantum ship on Earth there is always a non-vanishing probability to find it either in Paris or in Moscow! Again we refer to the textbooks on quantum theory like [181] for a deeper discussion.

7.3 Rendezvous of Bob and Alice

This section is inspired by paper [60] and can be considered as an introduction to that paper. However, we suggest a slightly different model than that of [60] just to make the connection with quantum bit even more apparent. The idea is to model some conflict interaction, where certain lack of information leads to a non-distributive, i.e. quantum like, logic of the players.

Suppose Alice and Bob decided to meet at a corner of a square shaped house, but did not fix a particular corner. Let us numerate the corners by the numbers 1, 2, 3, 4.

Bob arrived at the meeting 1st. If Alice then went to a corner, where she can see Bob (for instance, she comes to the corner 3, and Bob is in the corner 2, 3 or 4), then she is happy. Otherwise she goes away, feeling offended ("Bob did not come again"), and the next day Bob has to regain the favour in her eyes by a certain gift (payoff). Denoting by a_j, $j = 1, 2, 3, 4$, the payoffs that Alice demands from Bob, if she waited him in vain at the corner j, yields the following table representation for this game:

Bob

	1	2	3	4
1	0	0	a_1	0
2	0	0	0	a_2
3	a_3	0	0	0
4	0	a_4	0	0

Alice (label for rows)

where only the payoffs to Alice are inserted. As this is a zero sum game, the payoffs of Bob are just negations of the corresponding payoffs of Alice. The game is quite classical and its analysis causes no difficulties (finding mixed strategy Nash equilibria, etc.).

We can now modify the rules assuming, that both Bob and Alice are short-sighted leading to the following game (which can be called *rendezvous of shortsighted*). When Alice arrives, she just asks loudly: "Are you here, Bob?" If Bob is in the same corner everything is fine. If he is in the opposite corner, he cannot hear her anyway, and she leaves offended as in the previous game. Now, if Bob is in a neighboring corner, he hears her, but there is some positive probability that he would not understand correctly from which side the sound of her voice came, and consequently would come to an opposite corner with the same sorrowful result as before. You can see that Alice's questions are quite similar to quantum measurements of a "quantum ship" on the Globe. Namely, compare the pairs of corners $1, 3$ and $2, 4$ with different pairs of opposite points on the qubit. If Alice is in corner 1, then to her question (measurement), she gets a positive answer with some probability whenever Bob is not in the opposite corner. Moreover, if he is not in the opposite corner, her question would certainly transfer him to one of the opposite corners 1 or 3, precisely as the quantum measurement of a spin does. By the same reasons Alice's assumptions like "Can I find him in corner j?" are subject to the quantum logic, and not to the classical distributive one, as for any two corners, there is a possibility to find him in one of them.

It does not seem difficult to find similar interactions in various economical and sociological contexts. The appearance of quantum logic in suchlike interactions, makes it natural to introduce quantum tools to deal with them. We stop here. For an appropriate application of these tools (groups of unitary operators in separable complex Hilbert spaces) in game theoretical context we refer to [60] and other papers cited at the beginning of the chapter.

Chapter 8

It's party time!

8.1 Combinatorial games

In the previous chapter we touched on the youngest branch in the theory of games. Here we are going to touch on the oldest one. Combinatorial games constitute quite a special class of games whose analysis was started long before the game theory itself, as many games of these type has been played for centuries. We shall just give a flavor of the theory, which is a marvelous illustration of the basic idea of modern mathematics that in order to analyze a large class of complicated objects one can introduce various algebraic operations in the set of these objects, for example, binary compositions of elements that can be considered as a sort of generalized addition or multiplication, and then investigate how the properties of the objects would be transformed under these operation. We shall show that one can add and subtract games(!) just like numbers and that there is a natural order structure on the set of combinatorial games. Though the theory is not trivial, one can go deep inside it using more or fewer elementary tools (unlike quantum games mentioned in the previous chapter). For a complete discussion we refer to monographs [42], [21], [22], and for more recent advances to Proceedings [64], where one can find an exciting variety of interesting "party" games for all tastes. It could be also not a bad idea for the reader to take some encyclopedia of party games, like [63], and to use their examples to test the skills obtained by reading this chapter.

By *combinatorial games* one usually understands the games of two players, called Left and Right (which has nothing to do with politics), of the following type. The games are specified by the set of all possible positions and by the rules that to any position P prescribe the set of positions P^L where Left can move to from P, and the set of positions P^R where Right

can move to from P. The game starts at some position and then Left and Right alternate their moves. We shall restrict our attention to a important subclass of these games, where (i) there are only two outcomes of the game: either Left or Right win (thus excluding chess or Go, where draw can occur as well), (ii) the number of positions is finite (so called *short* games), (iii) the loser is the player who is unable to move when called upon to do so (so called *normal play* convention (which has nothing to do with the normal form of Chapter 1)).

In this chapter all *games* will be the combinatorial games of the type described.

The following standard examples will be used to test our general concepts.

Example: Domineering. On a rectangular board ruled into squares, the players alternately place dominoes. Each domino should cover two adjacent squares and they should not overlap. The difference between Left and Right moves is that Left should place the dominoes vertically, and Right horizontally.

Example: The Game of Nim. A starting position can be an arbitrary number of heaps of matchsticks (each heap of arbitrary size). The legal move is to take any number of sticks out of any heap (but you can not touch two heaps in one go). Of course, the game is trivial in case of only one heap (take it all away, and you win).

Example: Tac-tix (Deleting circles). This game is played on a rectangular filled with circles, the size of the rectangular being arbitrary. For instance, 3×6 initial position looks like

$$\begin{matrix} \circ & \circ & \circ & \circ & \circ & \circ \\ \circ & \circ & \circ & \circ & \circ & \circ \\ \circ & \circ & \circ & \circ & \circ & \circ \end{matrix}$$

Figure 8.1

By a move a player is allowed to cross out one or several neighboring circles in a row or column. But it is forbidden to cross a circle more than once. The following figure shows the result of possible three successive moves from the starting position above:

Figure 8.2

An important particular class of combinatorial games consists of so-called *impartial games*, where the set of possible moves of Left is the same as for the Right in any position. For example, Nim and Tac-tix are impartial and Domineering is not.

In abstract form, any combinatorial game can be identified with its initial position, and any position P is uniquely specified by the collection of possible moves (or *options*) P^L of Left and P^R of Right, i.e. $P = \{P^L | P^R\}$. By the first (the second) player one understands the player (which could be Right or Left) who is (respectively is not) to make the first move in a given position. Let us stress for clearness that unlike being first or second, the property of being Right or Left remains with one and the same player through the whole game.

Example: The simplest games. The simplest games are defined by the positions, where at most one move is available. Clearly there are 4 such games, which have special notations 0, 1, -1, $\star$ (the reason for the numerical notations will be clarified later) They are given by the following game trees:

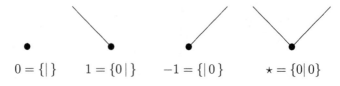

$$0 = \{ | \} \qquad 1 = \{0 | \} \qquad -1 = \{ | 0 \} \qquad \star = \{0 | 0\}$$

Figure 8.3

These pictures designate that in the first game there are no legal moves at all, in the second game there is one legal move for Left, in the third game there is one legal move for Right, and in the last game there is one legal move for Right and one legal move for Left. The reason for the given numerical values for these games will be clear later. These games appear (in some disguise, maybe), as end points of other, more complicated games. Say, for the rectangular boards

 and

in the domineering game, there are no legal moves for Left, and just one legal move for Right. Thus the domineering game that starts with these position is $-1 = \{\,|\,0\}$.

When playing a game, the main question is, of course, who is going to win, and what is the winning strategy (if any). Hence, the following classification is central for the theory.

The game G is called *positive* (notation $G > 0$) if there is a winning strategy for Left. The game G is called *negative* (notation $G < 0$) if there is a winning strategy for Right. G is *zero* or *zero game* (notation $G = 0$) if there is a winning strategy for the second player. The game G is called *fuzzy* (notation $G\|0$) if there is a winning strategy for the first player (irrespectively of whether he/she is Right or Left).

Exercise 8.1. Show that each game belong to one of these four classes.

One combines these notions in a usual way: $G \geq 0$ means $G > 0$ or $G = 0$, $G \leq 0$ means $G < 0$ or $G = 0$, and $G|\!\rhd 0$ means $G > 0$ or $G\|0$, $G \lhd |0$ means $G < 0$ or $G\|0$.

Exercise 8.2. Check that $G \geq 0$ means that there is a winning strategy for Left, if Right starts. Similarly, $G|\!\rhd 0$ means that there is a winning strategy for Left, if Left starts.

Example. The simplest game 0 given above is obviously a zero game according to above definition, games 1 and -1 are positive and negative respectively, game $\star$ is fuzzy.

8.2 Addition and subtraction of games, order structure

Suppose Left and Right have to play n (combinatorial, normal play convention) games $G_1, G_2,..., G_n$ simultaneously according to the following rule. If a player is to move, he/she selects one of these n games (component games) and makes any legal move in that game. A player loses, of course, when there is no legal move in any of the component game. These rules defines a new game G that is called the *disjunctive sum* (or simply the *sum*) of the component games and is denoted $G = G_1 + \dots + G_n$.

Some games have a direct natural representation as a sum of trivial games. For instance, it is clear that game Nim with n heaps is precisely the sum of n trivial 1-heap Nim games. Thus, to learn the wining ways in Nim, you just have to understand how to play sums, if you know how to play the components.

Generally, a position of a game breaks up into a sum, when a move in one part of the position does not affect the other parts. Such situation can often happen in domineering, when the vacant spaces of the board are placed in several separated regions.

The second basic algebraic operation is that of the *negation*. A game obtained by reversing the roles of Left and Right throughout a game G, is called the *negative* of G, and is denoted $-G$. Inductively, if $G = \{G^L | G^R\}$, then $-G = \{-G^R | -G^L\}$, where $-G^L$, say, means the collection of all negations of the positions from G^L. Clearly, if G is positive, then $-G$ is negative, and vice versa. At the same time, if G is zero or fuzzy, the same holds for $-G$.

We can now define the *subtraction* of games by the rule: $G - H = G + (-H)$.

Exercise 8.3. Check that $1 - 1 = 0$ and $\star + \star = 0$, where the games 1 and $\star$ are defined on Figure 8.3. It shows, in particular, that the notations 1 and -1 for games on Figure 8.3 fits to the general definition of negation.

Exercise 8.4. Check that $G - G = 0$ for an arbitrary game, i.e. in any game of the form $G - G$ there is a winning strategy for the second player.

As a hint for the latter exercise one can recall a story of a child who played Chess simultaneously against two Grandmasters and managed to win one of the games. How could this happen? The first Grandmaster played white, and the second played Black. When the first Grandmaster made a move, the child copied this move on the second board, and then copied the reply of the second Grandmaster on the first board, etc.

At last, one can now define the order-relation on the set of all (combinatorial) games in the following way. For any two games G and H one writes $G > H$ whenever $G - H > 0$, $G < H$ whenever $G - H < 0$, $G = H$ if $G - H = 0$, and $G \| H$ if $G - H \| 0$.

One can show that all these definitions are consistent (for instance, that if H is a zero game then $G + H$ always belong to the same class as G) and the order structure has usual connections with addition (say, $G \geq 0$ and $H \geq 0$ imply $G + H \geq 0$). Though these facts are not obvious, the proof is not difficult. It is worth noting that when $G = H$ one usually says G

equals H. However, this does not mean that G and H are identically the same games, but just that they are equivalent in the sense that the game $G - H$ belongs to the class of zero games.

Exercise 8.5. Show that the Nim game with two heaps is a zero game if and only if the sizes of the heaps are equal. (Hint: if sizes are equal than this game has the form $G + G = 0$, where G is the game corresponding to one of the heap. Practically, to win the game with two equal heaps, the second player should just copy the actions of the first player on the different heap.)

Exercise 8.6. Show that if the initial rectangular of Tac-tix has size $m \times n$ with at least one of numbers m and n being odd, then this game is fuzzy, i.e. the first player has a winning strategy. (Hint: under the assumptions on m and n there is a move of the first player breaking the position into two equal separated parts, which represent a zero game.)

8.3 Impartial games and Nim numbers

The theory of Nim was developed by Grundy and Sprague on the Eve of the Second World War, see [62] and [176]. We shall give a short sketch of their theory.

Recall that a game is called impartial if the sets of Right and Left options are always the same. Hence, talking about impartial games, one can simplify the general notation writing just $G = \{A, B, C, ...\}$ instead of $G = \{A, B, C, ...|A, B, C, ...\}$ (where $A, B, C, ...$ is any collection of impartial games). It is clear that the operation of negation does not change an impartial game, so that we have a remarkable relation $-G = G$ or $G+G = 0$ for all impartial game. The following simple notation is fundamental: the Nim game started from one heap of n matchsticks is denoted $\star n$ and is called *impartial number* n or *Nim number* n. Inductively, these numbers are defined as

$$\star n = \{\star 0, \star 1, ..., \star(n-1)\}$$

To reconcile with our previous notations of Figure 8.1 let us observe that $\star 0 = 0$ (so that zero is everywhere the same) and $\star 1 = \{0\} = \{0|0\} = \star$. The main result of Grundy-Sprague theory that we are going to explain states that any impartial game is equivalent (or equal in the sense of the definition of the previous section) to the game $\star n$ with some non-negative integer n. In particular, the explicit construction of this n allows for the explicit recipe for solving Nim.

The proof is very short, but conceptually rather rich. Its main idea is to extend the class of Nim type games by allowing occasional increase in the size of heaps. To catch this idea, let us consider the following "coins on the semi-infinite string" game. This game is played on a semi-infinite string of squares. A position is specified when a finite number of squares are occupied by coins (one coin on a square only). Typical position looks like

Figure 8.4

A legal move is to move any of the coins to the left onto any unoccupied square but not passing over any other coins. As always, game ends when there are no legal moves left. For instance, the end position for the game specified by position of Figure 8.4 will be

It turns out that this game is a disguised version of Nim. To see this let us first alternate the positions of coins from the right to the left: Head, Tail, Head, Tail, etc. Such a marking of the position of Figure 8.4 yields

$$\text{H} \quad \text{T} \qquad \text{H T H} \qquad \text{T} \quad \text{H}$$

Figure 8.5

Now, any such position uniquely defines a Nim game position by the following rule: the number of heaps equals the number of Heads, and the sizes of these heaps are the numbers of free squares between each Head and the nearest Tail to the left. Thus for position on Figure 8.4 or 8.5 the corresponding Nim position is $2, 0, 3, 2$. The main observation is that the Nim game with this position is equivalent to the original game. To see this, one observes first that the legal moves of Nim are in one-to-one correspondence with the moves of Heads of the coin game. The difference is in the possibility of moving Tails, which increases some Nim heap and hence is forbidden in Nim. The key observation then is that the latter moves can not change the outcome of the game, as they are reversible. Namely, if say, I am winning the corresponding Nim, I do not need to touch Tails at all: as long as you do not touch Tails, I just use my winning Nim strategy, and whenever you touch a Tail, I shall return to the same Nim position by a similar move of the corresponding neighboring Head.

Precisely the same argument proves the following general result.

Theorem 8.1. *Let G be any game played with a finite collections of non-negative integer numbers such that any move should change one and only one of this numbers. Any decrease of a number is always a legal move. The rules can allow for some increase of a number, but in such a way that the game always terminates. Then the outcome of such G is the same as in Nim with the same starting position.*

To prove the Grundy theorem it is convenient to introduce the following notation. For any collection $\{a, b, ...\}$ of non-negative integer numbers let $mex(a, b, ...)$ be the least non-negative integer number which does not belong to this collection. This number is called the *minimal excludent*.

Theorem 8.2. (Grundy's theorem) *Each impartial game (as always in this chapter, we mean combinatorial short game with the normal play convention) is equivalent to some Nim heap $\star n$.*

Proof. The proof is conducted by induction. Namely, recall that each impartial game can be represented as $G = \{A, B, ...\}$ with some collection of possible next moves $A, B, ...$ Suppose the required result holds for all options $A, B, ...$, so that they are equivalent to some Nim sizes, say $a, b, ...$ Let us show that then the result holds for G as well. In fact, we claim that G is then equivalent to a single Nim heap of size $n = mex(a, b, ...)$. In fact, all numbers strictly less than n should belong to the collection $a, b, ...$, and hence any decrease of n is allowed. Moreover, some strictly larger numbers than n (but not n) could be allowed. Hence, the result follows from the previous theorem.

If G is equivalent to $\star n$, the number n is called the *Nim number*, or the *Nim value*, or the *Grundy number* of G.

Let us apply this theory to the Nim itself. For any collection of sizes, say $a, b, c,$ of Nim heaps, one knows by Grundy's theorem that there exists a one heap game $\star n$ that is equivalent to the original one. But by definition of the addition of games, $\star n = \star a + \star b + \star c$. So one just need to know how to calculate sums. And of course, to perform addition of arbitrary number of terms, it is enough to understand how to calculate the sums of two elements. Thus one needs to understand how to calculate n from a, b to have the equation $\star n = \star a + \star b$.

Before doing it, let us mention that instead of talking about adding heaps it is often more convenient to talk about *Nim addition* $+_\star$ *of numbers*, the connection being given by the equation

$$\star n + \star m = \star(n +_\star m),$$

i.e. Nim-adding two integer numbers m and n means to find the size of a heap that is equivalent to two heaps with sizes m and n. This new addition defines a remarkable algebraic structure on the set of integer numbers.

To find the rules for Nim-addition, one can do it inductively starting from small a and b and using the constructive procedure for assigning Nim numbers from the proof of Grundy's theorem given above. For instance, we know already that $\star a + \star a = 0$ for all a. Let us find $\star 1 + \star 2$. The legal moves from two heaps with sizes 1 and 2 clearly can lead to positions $\star 1$, $\star 2$ and $\star 1 + \star 1 = 0$. As $mex(0, 1, 2) = 3$ we conclude that $\star 1 + \star 2 = \star 3$. The general rule that can be obtained is the following (we omit the proof): the Nim-sum of different powers of 2 coincides with the usual sum, e.g.

$$1 +_\star 2 +_\star 8 = 1 +_\star 2 +_\star 2^3 = 1 + 2 + 8 = 11.$$

With this rule in mind (and also using $m +_\star m = 0$ for all m), one can easily calculate any Nim sum, e.g.

$$9 +_\star 13 +_\star 2 = 8 +_\star 1 +_\star 8 +_\star 4 +_\star 1 +_\star 2 = 4 +_\star 2 = 4 + 2 = 6.$$

Remark. Nim-addition can be easily performed on numbers in the binary representation (used in all computers). Namely, Nim-addition of two binary numbers is done by adding corresponding digits according to the rule: $1 + 0 = 0 + 1 = 1$ and $1 + 1 = 0 + 0 = 0$ (i.e. by modulus 2 addition).

Well, now you know how to play Nim with any number of arbitrary heaps. If the Nim-value of a position you are starting with is a positive number, you have to make a move bringing it to zero. If the Nim-value of a position you are starting with is zero, then you are in a losing position and can only hope that your opponent does not understand it.

Example. Let $9, 13, 2$ be a Nim position and it is your go. Can you win? And what can be a good move? Solution: by the calculations above the Nim value of this game is 6, so you can win. Taking 2 out of 13 brings the value to zero.

Of course, the theory can be applied to other impartial games (however, one should not expect that this should be a simple task in any particular game).

Exercise 8.7. Show that $1 \times m$ rectangle for the Tac-tix has Nim value m for any m. (Hint: show by induction that for any m only the options $0, \star 1, ..., \star(m-1)$ are available when starting from rectangular $1 \times m$.)

Exercise 8.8. Calculate the Nim-values for the simple positions

$$
\begin{array}{cccccc}
\circ\,\circ\,\phi & \phi\,\circ\,\circ & \phi\,\circ\,\phi & \circ\,\circ\,\circ & \circ\text{-}\!\ominus\!\text{-}\ominus & \circ\,\circ\,\circ \\
\circ\,\circ\,\phi & \circ\,\circ\,\phi & \circ\,\circ\,\circ & \circ\,\circ\,\phi & \circ\,\circ\,\circ & \circ\,\circ\,\circ
\end{array}
$$

Figure 8.6

of Tac-tix. Answer: $0, 1, 2, 5, 4, 2$. (Hint: make the strategic analysis of all possible options and use the constructive proof of Grundy's theorem. For instance, for the first picture, there are two legitimate moves leading to games $\star 2, \star 3$ (check this!), and hence the Nim-value is $0 = mex(2,3)$; for the second picture there are three options $0, \star 2, \star 3$, and hence the Nim-value is $1 = mex(0,2,3)$; for the third picture there are options $0, \star 1, \star 3$; for the last picture there are options $0, \star 1, \star 3, \star 4, \star 5$.)

8.4   Games as numbers and numbers as games

Impartial games of the previous section describe fuzzy games. Here we shall say a few words about positive and negative games showing their connection with usual numbers. To this end, let us return to the simplest games of Figure 8.3. Having the game 1 and the rule for addition one can naturally define the games corresponding to the integer numbers by

$$2 = 1+1, \ 3 = 1+1+1, \ 4 = 1+1+1+1, ..., \ -2 = -1-1, \ -3 = -1-1-1, ...$$

Consequently, all integer numbers have representations as positive or negative games, and vise versa: there is a class of games that can be represented by integer numbers. You can easily invent the positions in domineering that yield examples for all these numbers.

What about other, non-integer numbers? Can we find a game that corresponds to number $1/2$, say? Well, it turns out that the game $1/2$ should be defined as $\{0|1\}$. One can check that this definition implies the expected relation $1/2 + 1/2 = 1$. Afterwards, one can define all numbers of the form $n/2$ with integer n as the sum of n games $1/2$. The game $1/4$

is then defined as $\{0|1/2\}$, etc. Going on by this procedure we define the games corresponding to binary rational, i.e. to the numbers of the form $n/(2^m)$. Recall that addition and subtraction of such games correspond to usual addition and subtraction of real numbers. An appropriate development lead to the representation of all real numbers by games and thus to a large class of games behaving like numbers. For a full discussion, as well as for applications to concrete games, we refer to the monographs mentioned at the beginning of the chapter.

PART II
Armed with mathematics

Chapter 9

A rapid course in mathematical game theory

9.1 Three classical examples of Nash equilibria in economics

Definition of the Nash equilibria for games in normal form was given in Section 2.1. This section is devoted to three examples of games with infinite set of strategies, which were analyzed long before the appearance of the game theory itself.

Example. Cournot's duopoly model.

Ruth and Charlie are directors of two firms, R and C, that produce and sell a product on the same market. The price of the product is supposed to decrease proportionally to the supply, i.e. if Q_R and Q_C are the quantities of the product produced by R and C, the market price for the unit of the product becomes

$$P(Q) = \begin{cases} P_0(1 - Q/Q_0) & \text{if} \quad Q < Q_0 \\ 0 & \text{if} \quad Q \geq Q_0 \end{cases} \tag{9.1}$$

where $Q = Q_R + Q_C$ is the aggregate amount produced, and positive constants P_0, Q_0 denote the highest possible price and the highest reasonable production level. If the marginal cost of the production is c for both firms, the payoffs clearly are

$$\Pi_R(Q_R, Q_C) = Q_R P(Q) - cQ_R, \quad \Pi_C(Q_R, Q_C) = Q_C P(Q) - cQ_C. \tag{9.2}$$

As for $P_0 \leq c$ the model is meaningless (no profit is available), one always assumes that $P_0 > c$. Thus we have a two-player symmetric game with an infinite set of strategies Q_R, Q_C that are numbers from the interval $[0, Q_0]$ (clearly it makes no sense to produce anything outside this interval).

Exercise 9.1. Analyze this model by the following scheme.

(a) Given a strategy Q_C find the best response of R, i.e. the amount $\hat{Q}_R = \hat{Q}_R(Q_C)$ that maximises the profit for R. Answer:

$$\hat{Q}_R(Q_C) = \frac{Q_0}{2}\left(1 - \frac{Q_C}{Q_0} - \frac{c}{P_0}\right). \tag{9.3}$$

(Hint: $\hat{Q}_R$ above is found from the condition

$$\frac{\partial \Pi_R}{\partial Q_R}(\hat{Q}_R) = P_0\left(1 - \frac{\hat{Q}_R + Q_C}{Q_0}\right) - \hat{Q}_R\frac{P_0}{Q_0} - c = 0;$$

then one has to check that the second derivative of Π_R is negative at $\hat{Q}_R$; at last one has to check that $\hat{Q}_R(Q_C) + Q_C \le Q_0$ for any $Q_C \le Q_0$. It is still possible that $\hat{Q}_R$ from (9.3) is negative, so that speaking rigorously, the best response is given by the amount that is the maximum of zero and (9.3).)

(b) Similarly the best response of C to a given strategy Q_R is given by

$$\hat{Q}_C(Q_R) = \frac{Q_0}{2}\left(1 - \frac{Q_R}{Q_0} - \frac{c}{P_0}\right). \tag{9.4}$$

As Nash equilibrium is a pair $Q_R^\star, Q_C^\star$ of strategies, each of which is the best response to another one, this pair have to satisfy the system of equations

$$Q_R^\star = \frac{Q_0}{2}\left(1 - \frac{Q_C^\star}{Q_0} - \frac{c}{P_0}\right), \quad Q_C^\star = \frac{Q_0}{2}\left(1 - \frac{Q_R^\star}{Q_0} - \frac{c}{P_0}\right).$$

Find that the only solution to this system is

$$Q_R^\star = Q_C^\star = Q^\star = \frac{Q_0}{3}\left(1 - \frac{c}{P_0}\right).$$

(Hint: assuming that $Q_C^\star = Q_R^\star$ (which is obvious from the symmetry) simplifies the calculations essentially.) Observe that $Q^\star > 0$ due to the assumption $P_0 > c$.

(c) Calculate the equilibrium payoff

$$\Pi_R(Q^\star, Q^\star) = \Pi_C(Q^\star, Q^\star) = \frac{Q_0 P_0}{9}\left(1 - \frac{c}{P_0}\right)^2.$$

(d) Show that for a monopolist, who acts on the market alone and hence has to maximize the payoff $\Pi_m(Q) = QP(Q) - cQ$ over all $Q \in [0, Q_0]$, the optimal production and the corresponding payoff are

$$Q_m = \frac{Q_0}{2}\left(1 - \frac{c}{P_0}\right), \quad \Pi_m(Q_m) = \frac{Q_0 P_0}{4}\left(1 - \frac{c}{P_0}\right)^2,$$

the latter payoff being of course higher than obtained in duopoly.

(e) Suppose now that the two firms R and C form a cartel, i.e. they agree to produce an equal amount of product $Q_R = Q_C$. Show that the optimal production would be $Q_R = Q_C = Q_m/2$, which gives better payoff for both of them, than under the Nash equilibrium of the initial game. Note that this is a performance of the prisoner's dilemma: Nash equilibrium yields worse payoffs than what can be achieved by cooperation (cartel), but the cooperation is unstable under unilateral deviation from it.

Analysis of this model was published by Cournot in 1838 and represents historically the first formal realization of the idea of the Nash equilibrium. Therefore some authors use the term Cournot-Nash equilibrium instead of the Nash equilibrium.

Example. Bertrand's duopoly model.

Two firms R and C produce and sell a product on the same market as above. But their strategies now are the prices, P_R and P_C, that they assign to the product. The firm "captures the market" if it assigns a lower price. This firm sells the whole product, and the second one sells nothing. In case of the equal prices, the firms share the market equally. The payoffs are

$$
\Pi_R(P_R, P_C) = \begin{cases} (P_R - c)Q(P_R) & \text{if } P_R < P_C \\ (P_R - c)Q(P_R)/2 & \text{if } P_R = P_C \\ 0 & \text{if } P_R > P_C \end{cases} \tag{9.5}
$$

$$
\Pi_C(P_R, P_C) = \begin{cases} (P_C - c)Q(P_C) & \text{if } P_C < P_R \\ (P_C - c)Q(P_R)/2 & \text{if } P_C = P_R \\ 0 & \text{if } P_C > P_R \end{cases} \tag{9.6}
$$

where c is the marginal cost of production, the demand function is $Q(P) = Q_0(1 - P/P_0)$ for $P < P_0$, and a constant $P_0 > c$ denotes the highest possible reasonable price for this product (notice that under assumption $c \geq P_0$ the model becomes meaningless: no profit is available). Thus we have a two-player symmetric game with an infinite set of strategies P_R, P_C that are numbers from the interval $[c, P_0]$ (clearly it makes no sense to choose a price outside of this interval).

Let us prove a remarkable feature of this model that the only Nash equilibrium is given by the prices $P_R = P_C = c$, which gives to both players the vanishing payoff: $\Pi_R(c, c) = \Pi_C(c, c) = 0$. Let (P_R, P_C) be a Nash equilibrium. Obviously $P_R > P_C$ can not be the best response to P_C whenever $P_C > c$, and similarly $P_C > P_R$ can not be the best response to P_R whenever $P_R > c$. This leads to four remaining possibilities:

(i) $c = P_C < P_R \leq P_0$,

(ii) $c = P_R < P_C \leq P_0$,

(iii) $c < P_C = P_R \leq P_0$,

(iv) $c = P_C = P_R$.

Now, in case (i), P_C is not the best response to P_R, as choosing anything between c and P_R yields a better payoff for C. Thus case (i) does not give a Nash equilibrium. By symmetry the same holds for case (ii). It remains to observe that case (iii) does not yield Nash equilibrium either, for if P_R is given, a slight decrease in price for C would give him better payoff (he would capture the whole market instead of having half of it). Thus only (iv) remains, as we claimed.

Exercise 9.2. Show that for a monopolist in Bertrand's model, who maximizes the profit $(P - c)Q(p)$, the optimal price would be $P_m = (P_0 + c)/2$ with the profit

$$\Pi_m(P_m) = (P_m - c)Q_0 \left(1 - \frac{P_m}{P_0} \right).$$

Example. The Stackelberg duopoly model. As in the Cournot model above, two firms R and C produce and sell a product on the same market. Prices and payoffs are given by the same formulas (9.1), (9.2). But now the decisions of R and C are not simultaneous. The game starts with firm R (called the market leader) making a decision by choosing Q_R. Then C (called the market follower) observes the decision of R and then makes his decision. Let us find the subgame perfect Nash equilibrium (see Section 3.6 for subgame perfection). Method of backward induction leads to the following procedure: for any strategy Q_R find the best response strategy of C (this is given by (9.4) as we know), and then find the best choice for Q_R assuming that C would use his best response (common knowledge of rationality!). Thus we need to find Q_R that maximises the payoff

$$\Pi_R(Q_R, \hat{Q}_C(Q_R)) = Q_R[P(Q_R + \hat{Q}_C(Q_R)) - c],$$

which equals to

$$\Pi_R(Q_R, \hat{Q}_C(Q_R)) = Q_R \frac{P_0}{2} \left(1 - \frac{Q_R}{Q_0} - \frac{c}{P_0} \right),$$

due to (9.1) and (9.4). By differentiation one obtains that the maximum is obtained at

$$Q_R^\star = \frac{Q_0}{2} \left(1 - \frac{c}{P_0} \right)$$

leading to the subgame perfect Nash equilibrium

$$Q_R^\star = \frac{Q_0}{2} \left(1 - \frac{c}{P_0} \right), \quad Q_C^\star = \hat{Q}_C(Q_R^\star) = \frac{Q_0}{4} \left(1 - \frac{c}{P_0} \right).$$

Exercise 9.3. Show that under the equilibrium the profits of the players are

$$\Pi_R^\star = \frac{Q_0 P_0}{8}\left(1 - \frac{c}{P_0}\right)^2, \quad \Pi_C^\star = \frac{Q_0 P_0}{16}\left(1 - \frac{c}{P_0}\right)^2,$$

so that $\Pi_R^\star > \Pi_C^\star$ and $Q_R^\star > Q_C^\star$. Show that firm R makes a larger profit in the Stackelberg model, than under the corresponding conditions of the Cournot model, so that the first move leads to an advantage.

9.2 Mixed strategies for finite games

We introduced mixed strategies in Chapter 4. Here we embark on a more systematic study. Consider a game of m players $i = 1, 2, ..., m$ with finite sets of (pure) strategies $S_1 = \{s_1^1, ..., s_1^{n_1}\}$, $S_2 = \{s_2^1, ..., s_2^{n_2}\}$,..., $S_m = \{s_m^1, ..., s_m^{n_m}\}$. A *mixed strategy* for player i is a probability distribution on the set of his/her pure strategies, i.e. a vector $\sigma = (p_1, ..., p_{n_i})$ of dimension n_i with non-negative coordinates that sum up to 1, i.e. $p_1 + ... + p_{n_i} = 1$. Original pure strategies are, of course, given by vectors with one coordinate being 1 and other being zero. A collection of mixed strategies is called a *profile or a situation* (in mixed strategies). The payoff for the player i under the profile $\sigma_1 = (p_1^1, ..., p_{n_i}^1)$, $\sigma_2 = (p_1^2, ..., p_{n_i}^2)$,...,$\sigma_m = (p_1^m, ..., p_{n_i}^m)$ is given by the formula

$$\Pi_i(\sigma_1, \sigma_2, ..., \sigma_m) = \sum_{j_1=1}^{n_1} \sum_{j_2=1}^{n_2} ... \sum_{j_m=1}^{n_m} p_{j_1}^1 p_{j_2}^2 ... p_{j_m}^m \Pi_i(s_1^{j_1}, s_2^{j_2}, ..., s_m^{j_m}), \quad (9.7)$$

where $\Pi_i(s_1^{j_1}, s_2^{j_2}, ..., s_m^{j_m})$ are the payoffs for the initial game in pure strategies. This formula yields, in fact, the average expected payoff assuming that all players choose their strategies independently. The fundamental theorem of Nash states that any finite game has at least one Nash equilibrium in mixed strategies. Proof of this result can be found in many books, and is omitted.

We shall say that a Nash equilibrium is a *pure strategy* (respectively *mixed strategy*) Nash equilibrium, if all strategies in its profile are pure (respectively if at least one strategy is strictly mixed).

We shall discuss now the technique of the calculation of the Nash equilibria starting with two-player two-action games. These calculations are based on the following result.

Theorem 9.1. *(Equality of payoffs lemma I)* Let $\sigma_R^\star = (p^\star, 1 - p^\star)$ and $\sigma_C^\star = (q^\star, 1 - q^\star)$ be a profile in a two-player game with two actions s_R^1, s_R^2 of Ruth and s_C^1, s_C^2 of Charlie.

(i) If $0 < p^\star < 1$, then $\sigma_R^\star$ is the best response to $\sigma_C^\star$, i.e.

$$\Pi_R(\sigma_R^\star, \sigma_C^\star) \geq \Pi_R(\sigma_R, \sigma_C^\star) \tag{9.8}$$

for all σ_R, if and only if

$$\Pi_R(s_R^1, \sigma_C^\star) = \Pi_R(s_R^2, \sigma_C^\star) \tag{9.9}$$

and consequently

$$\Pi_R(\sigma_R^\star, \sigma_C^\star) = \Pi_R(s_R^1, \sigma_C^\star) = \Pi_R(s_R^2, \sigma_C^\star) = \Pi_R(\sigma_R, \sigma_C^\star) \tag{9.10}$$

for all σ_R;

(ii) If $0 < q^\star < 1$, then $\sigma_C^\star$ is the best response to $\sigma_R^\star$, i.e.

$$\Pi_R(\sigma_R^\star, \sigma_C^\star) \geq \Pi_R(\sigma_R^\star, \sigma_C) \tag{9.11}$$

for all σ_C, if and only if

$$\Pi_C(\sigma_R^\star, s_C^1) = \Pi_C(\sigma_R^\star, s_C^2) \tag{9.12}$$

and consequently

$$\Pi_C(\sigma_R^\star, \sigma_C^\star) = \Pi_C(\sigma_R^\star, s_C^1) = \Pi_C(\sigma_R^\star, s_C^2) = \Pi_C(\sigma_R^\star, \sigma_C) \tag{9.13}$$

for all σ_C;

(iii) if $0 < p^\star < 1$ and $0 < q^\star < 1$, then $(\sigma_R^\star, \sigma_C^\star)$ is a Nash equilibrium if and only if both (9.9) and (9.12) hold. Thus for strictly mixed strategies inequalities (9.8) and (9.11) defining the Nash equilibrium are equivalent to the corresponding equality (9.10) and (9.13).

Proof. As (ii) is similar to (i), and (iii) follows from (i) and (ii), it suffice to prove (i). Suppose (9.9) does not hold. Assume, for example, that $\Pi_R(s_R^1, \sigma_C^\star) > \Pi_R(s_R^2, \sigma_C^\star)$. Then

$$\Pi_R(s_R^1, \sigma_C^\star) = (p^\star + 1 - p^\star)\Pi_R(s_R^1, \sigma_C^\star)$$

$$> p^\star\Pi_R(s_R^1, \sigma_C^\star) + (1 - p^\star)\Pi_R(s_R^2, \sigma_C^\star) = \Pi_R(\sigma_R^\star, \sigma_C^\star),$$

and consequently $\sigma_R^\star$ is not the best response to $\sigma_C^\star$, as condition (9.8) breaks down for $\sigma = (1, 0)$. This contradiction proves (9.9). $\quad\blacksquare$

Corollary 9.1. *Suppose a two-player game is given by the table*

Table 9.1

C

	1	2
R 1	a,b	c,d
2	e,f	g,h

with arbitrary numbers a, b, c, d, e, f, g, h, and suppose $(\sigma_R^\star, \sigma_C^\star)$ is a Nash equilibrium with $\sigma_R^\star = (p^\star, 1 - p^\star)$ and $\sigma_C^\star = (q^\star, 1 - q^\star)$. If $0 < p^\star < 1$, then

$$q^\star[(c - g) + (e - a)] = c - g, \qquad (9.14)$$

and if $0 < q^\star < 1$, then

$$p^\star[(h - f) + (b - d)] = h - f. \qquad (9.15)$$

Proof. As one sees by inspection, equation (9.14) is equivalent to (9.9), and equation (9.15) is equivalent to (9.12).

Corollary 9.2. *Suppose the "generic conditions" $a \neq e$, $c \neq g$, $b \neq d$, $h \neq f$ hold for a two-player game of Table 9.1.*

(i) If at least one of the numbers

$$q^\star = \frac{c - g}{(c - g) + (e - a)}, \quad p^\star = \frac{h - f}{(h - f) + (b - d)} \qquad (9.16)$$

does not belong to the open interval $(0, 1)$ (in particular, if it is $\pm\infty$, which may happen if the corresponding denominator vanishes), then there are no mixed strategy Nash equilibria.

(ii) If both these numbers belong to the interval $(0, 1)$, then there is a unique mixed strategy Nash equilibrium $((p^\star, 1 - p^\star), (q^\star, 1 - q^\star))$.

(iii) The number of pure strategy Nash equilibria under "generic conditions" can not exceed two.

Proof. Let $(\sigma_R = (p, 1 - p), \sigma_C = (q, 1 - q))$ be a Nash equilibrium with $p \in (0, 1)$ (the case with $q \in (0, 1)$ is considered similarly). Then $q = q^\star$ by Corollary 9.1. This would be impossible, if this number would not belong to $[0, 1]$. As by "generic condition" it can be neither 0 nor 1, it follows that it belongs to the open interval $(0, 1)$. Then again by Corollary 9.1, it follows that $p = p^\star$, which is again impossible if this number does not belong to $[0, 1]$. Hence one concludes again that $p = p^\star \in (0, 1)$. And in this case (σ_R, σ_C) is mixed strategy Nash by Theorem 9.1, as required. Statement (iii) follows from a simple observation that "generic conditions" exclude the possibility of two neighboring cells in the table to represent a Nash equilibrium.

Thus under generic condition the analysis of Nash equilibria becomes trivial. For other cases, a bit longer calculations are required.

Example. For the table

Table 9.2

C

		1	2
R	1	3,2	1,1
	2	0,0	2,3

("battle of the sexes" game) one finds the two pure Nash equilibria given by outcomes $(3, 2)$, $(2, 3)$ by inspection, and the unique mixed strategy Nash equilibrium $(3/4, 1/4), (1/4, 3/4)$ by above Corollary 9.2.

Example. Let us find all Nash equilibria for the table

Table 9.3

C

		1	2
R	1	5,0	3,1
	2	5,2	3,0

This game does not satisfy the generic conditions. Let us calculate the best responses explicitly. Let $\sigma_R = (p, 1 - p)$, $\sigma_C = (q, 1 - q)$. As

$$\Pi(\sigma_R, \sigma_C) = 5pq + 3p(1 - q) + 5(1 - p)q + 3(1 - p)(1 - q) = 3 + 2q$$

does not depend on σ_R, any σ_R is the best response to arbitrary σ_C, so one only needs to find when σ_C is the best response to σ_R. As the function

$$\Pi_C(\sigma_R, \sigma_C) = p(1 - q) + 2(1 - p)q = p + q(2 - 3p)$$

increases (respectively decreases) in q for $p < 2/3$ (respectively for $p > 2/3$) and does not depend on q for $p = 2/3$, one concludes that the Nash equilibria are $((p, 1 - p), (1, 0))$ for $p < 2/3$, $((p, 1 - p), (0, 1))$ for $p > 2/3$, and $((2/3, 1/3), (q, 1 - q))$ for arbitrary $q \in [0, 1]$.

The extension of the Equality of Payoffs Lemma to arbitrary finite games is the following.

Theorem 9.2. (Equality of payoffs lemma II) *Suppose a profile (σ_i, σ) is a Nash equilibrium for a finite game (in normal form, with payoffs given* by (9.7)*)*, *where* $\sigma_i = (p^1, ..., p^m)$ *is a mixed strategy of player i that has a collection* $\{s^1, ..., s^m\}$ *of pure strategies, and where* σ *denotes the collection of the strategies of all other players in this profile. Suppose* $p^j \neq 0$, $p^k \neq 0$ *for some* $j \neq k$. *Then*

$$\Pi_i(s^j, \sigma) = \Pi_i(s^k, \sigma).$$

Proof. This is the same as for two players. Namely, assuming $\Pi_i(s^j, \sigma) > \Pi_i(s^k, \sigma)$ one sees that the strategy $\hat{\sigma}_i$ obtained from σ_i by changing p_j to $p_j + p_k$ and p_k to zero, yields a better payoff to i than σ_i.

Practical calculations of Nash equilibria for concrete large games leads to complicated numerical problems of linear algebra and linear programming. Some simplifications are available for symmetric and zero-sum games that we shall discuss now.

Clearly any finite two-player game can be specified by two matrices $A = (a_{ij})$, $B = (b_{ij})$, where a_{ij} and b_{ij} define the payoffs of Ruth and Charlie (first and second players) if they use their strategies i and j respectively. *Symmetric games* have the property that matrices A and B are square matrices such that $a_{ij} = b_{ji}$ for all i, j, in other words that $B = A^T$ (i.e. B is transpose of A).

Theorem 9.3. (**Mixed strategy symmetric Nash equilibria**) *A symmetric profile* (σ, σ) *with* $\sigma = (p_1, ..., p_m)$ *and all* p_j *non-vanishing is a symmetric Nash equilibrium for a symmetric game given by* $m \times m$-*matrix A if and only if*

$$A\sigma = \lambda \begin{pmatrix} 1 \\ 1 \\ 1 \end{pmatrix}, \quad \lambda = \Pi(\sigma, \sigma). \tag{9.17}$$

In particular, if A is non-degenerate, then

$$\sigma = \lambda A^{-1} \begin{pmatrix} 1 \\ 1 \\ 1 \end{pmatrix}, \tag{9.18}$$

and consequently there can be at most one such equilibrium and it exists if and only if all coordinates of the right hand side of (9.18) *are positive.*

Proof. The first equation of (9.17) with some constant λ is obvious, since the equality of payoffs lemma means that the coordinates of the vector $A\sigma$ should be equal. The second equation in (9.17) is obtained by a scalar multiplication of the first one by the vector σ (and using the fact that the sum of the coordinates of σ should be one).

Example. Let us find all symmetric Nash equilibria in the version of the Scissors-Rock-Paper game with fines or credits for a tie, i.e. for the symmetric two-player game specified by the matrix

$$A = \begin{pmatrix} -a & 1 & -1 \\ -1 & -a & 1 \\ 1 & -1 & -a \end{pmatrix} \tag{9.19}$$

with some $a \neq 0$.

(i) First let us look for equilibria (σ, σ) with $\sigma = (p_1, p_2, p_3)$ and all p_j being positive. As the determinant of A equals $-a(a^2 + 3)$ and does not vanish (here the assumption $a \neq 0$ is used), it follows that A is non-degenerate. Calculating its inverse yields

$$A^{-1} = -\frac{1}{a(a^2 + 3)} \begin{pmatrix} 1 + a^2 & 1 + a & 1 - a \\ 1 - a & 1 + a^2 & 1 + a \\ 1 + a & 1 - a & 1 + a^2 \end{pmatrix},$$

and consequently

$$A^{-1} \begin{pmatrix} 1 \\ 1 \\ 1 \end{pmatrix} = -\frac{1}{a} \begin{pmatrix} 1 \\ 1 \\ 1 \end{pmatrix} \iff A \begin{pmatrix} 1 \\ 1 \\ 1 \end{pmatrix} = -a \begin{pmatrix} 1 \\ 1 \\ 1 \end{pmatrix}. \tag{9.20}$$

Consequently $\sigma^* = (1/3, 1/3, 1/3)$ is the unique strategy with positive coordinates that defines a symmetric Nash equilibrium.

(ii) Clearly all symmetric pure strategy profiles define Nash equilibria if and only if $a \leq -1$. It remains to look for strategies with precisely two positive probabilities.

(iii) If a strategy of the form $\sigma = (p, 1 - p, 0)$ defines a symmetric Nash equilibrium, it should also define a Nash equilibrium in the two-action game with the matrix

$$\begin{pmatrix} -a & 1 \\ -1 & -a \end{pmatrix},$$

which by shifting is equivalent to the matrix

$$\begin{pmatrix} 1 - a & 2 \\ 0 & 1 - a \end{pmatrix},$$

and which has (by two-action game theory) a unique strictly mixed Nash equilibrium

$$\left(\frac{a + 1}{2a}, \frac{a - 1}{2a} \right)$$

if $|a| > 1$. Hence the only candidate for a strategy of the form $(p, 1 - p, 0)$ defining a symmetric Nash equilibrium in our game is the strategy

$$\sigma = \left(\frac{a+1}{2a}, \frac{a-1}{2a}, 0 \right).$$

Now

$$\Pi(\sigma, \sigma) = -\frac{1}{2a}(a^2 + 1), \quad \Pi(3, \sigma) = \frac{1}{a}$$

and consequently (σ, σ) is a Nash equilibrium only if

$$-\frac{1}{2a}(a^2 + 1) \geq \frac{1}{a},$$

which holds for $a < -1$ (recall that only $|a| > 1$ were under considerations). Hence the strategy σ and similarly the strategies

$$\left(\frac{a-1}{2a}, 0, \frac{a+1}{2a} \right), \quad \left(0, \frac{a+1}{2a}, \frac{a-1}{2a} \right)$$

define Nash equilibria with two non-vanishing probabilities if and only if $a < -1$.

Example. *Rock-Paper-Scissors or Uta Stansburiana lizard game.* This is a symmetric game specified by the matrix (9.19) with $a = 0$. The arguments of the previous example show that the only candidates for a Nash equilibrium are of the form (p, q, r) with all p, q, r being positive. The right hand side equation from (9.20) implies that $\sigma^* = (1/3, 1/3, 1/3)$ corresponds to a Nash equilibrium. Simple algebra (that we omit) is required to show that this is in fact the unique equilibrium in this game.

Now let us discuss the finite zero-sum games of two players. These games are specified by the condition $B = -A$, where A and B be rectangular matrices of the payoffs to the first and to the second players.

Theorem 9.4. (Nash equilibria for zero-sum games, minimax theorem). *For any finite zero-sum game*

$$\min_{\sigma_C} \max_{\sigma_R} \Pi_R(\sigma_R, \sigma_C) = \max_{\sigma_R} \min_{\sigma_C} \Pi_R(\sigma_R, \sigma_C) \tag{9.21}$$

(minimax equation) holds. The common value of the right and left hand sides of this equation is called the value V of the game. Moreover, $\Pi_R(\sigma_R^\star, \sigma_C^\star) = V$ for any Nash equilibrium $(\sigma_R^\star, \sigma_C^\star)$ of this game, and for any two other strategies σ_R, σ_C the condition

$$\Pi_R(\sigma_R, \sigma_C^\star) \leq V = \Pi_R(\sigma_R^\star, \sigma_C^\star) \leq \Pi_R(\sigma_R^\star, \sigma_C) \tag{9.22}$$

holds, which is called the saddle-point condition for the function Π_R. Equivalently (9.22) can be written as the inequality

$$\Pi_R(s_R^j, \sigma_C^\star) \leq V = \Pi_R(\sigma_R^\star, \sigma_C^\star) \leq \Pi_R(\sigma_R^\star, s_C^i) \qquad (9.23)$$

that holds for all pure strategies s_R^j and s_C^i of the first and second players. Finally

$$\min_{\sigma_C} \max_{j} \Pi_R(s_R^j, \sigma_C) = V = \Pi_R(\sigma_R^\star, \sigma_C^\star) = \max_{\sigma_R} \min_{i} \Pi_R(\sigma_R, s_C^i). \qquad (9.24)$$

Proof. Step 1. Clearly

$$\Pi_R(\sigma_R, \sigma_C) \leq \max_{\sigma_R} \Pi_R(\sigma_R, \sigma_C)$$

for any pair of strategies σ_R, σ_C, and consequently

$$\min_{\sigma_C} \Pi_R(\sigma_R, \sigma_C) \leq \min_{\sigma_C} \max_{\sigma_R} \Pi_R(\sigma_R, \sigma_C)$$

for any strategy σ_R. It implies that the inequality

$$\max_{\sigma_R} \min_{\sigma_C} \Pi_R(\sigma_R, \sigma_C) \leq \min_{\sigma_C} \max_{\sigma_R} \Pi_R(\sigma_R, \sigma_C) \qquad (9.25)$$

holds trivially (without any assumptions on a game).

Step 2. Suppose now that $(\sigma_R^\star, \sigma_C^\star)$ form a Nash equilibrium. Observe that such an equilibrium exists according to the Nash theorem, and it is essential for our proof. By the definition of the Nash equilibrium it follows that

$$\Pi_R(\sigma_R^\star, \sigma_C^\star) \geq \Pi_R(\sigma_R, \sigma_C^\star), \quad \Pi_C(\sigma_R^\star, \sigma_C^\star) \geq \Pi_C(\sigma_R^\star, \sigma_C) \qquad (9.26)$$

for all σ_R, σ_C. As by zero-sum condition $\Pi_R(\sigma_R, \sigma_C) = -\Pi_C(\sigma_R, \sigma_C)$ for all σ_R, σ_C, the second inequality in (9.26) can be written as

$$\Pi_R(\sigma_R^\star, \sigma_C^\star) \leq \Pi_R(\sigma_R^\star, \sigma_C),$$

which together with the first inequality of (9.26) implies the saddle-point condition (9.22).

Step 3. From the right inequality of (9.22) it follows that

$$\Pi_R(\sigma_R^\star, \sigma_C^\star) = \min_{\sigma_C} \Pi_R(\sigma_R^\star, \sigma_C) \leq \max_{\sigma_R} \min_{\sigma_C} \Pi_R(\sigma_R, \sigma_C)$$

and from the left inequality of (9.22) it follows that

$$\Pi_R(\sigma_R^\star, \sigma_C^\star) = \max_{\sigma_R} \Pi_R(\sigma_R, \sigma_C) \geq \min_{\sigma_C} \max_{\sigma_R} \Pi_R(\sigma_R, \sigma_C).$$

The last two inequalities together with (9.25) imply

$$\min_{\sigma_C} \max_{\sigma_R} \Pi_R(\sigma_R, \sigma_C) = \Pi_R(\sigma_R^\star, \sigma_C^\star) = \max_{\sigma_R} \min_{\sigma_C} \Pi_R(\sigma_R, \sigma_C).$$

Equivalence of (9.22) and (9.23) is evident. At last, as for all σ_R

$$\min_{\sigma_C} \Pi_R(\sigma_R, \sigma_C) = \min_i \Pi_R(\sigma_R, s_C^i),$$

it follows that

$$\max_{\sigma_R} \min_{\sigma_C} \Pi_R(\sigma_R, \sigma_C) = \max_{\sigma_R} \min_i \Pi_R(\sigma_R, s_C^i),$$

which implies the l.h.s. of (9.24). Of course, the r.h.s. is obtained similarly, which complete the proof of the Theorem.

It follows, in particular, that any Nash equilibrium in an antagonistic game is a saddle-point, and that the payoffs at all Nash equilibria are all the same and coincide with the guaranteed payoff.

Exercise 9.4. Show that the values of the Scissors-Rock-Paper and the Matching pennies (Head and Tail) games are zero.

9.3 Evolutionary stable strategies

Evolutionary stable strategies were introduced in Chapter 4 for two-action games. We shall give a more general and a more systematic presentation here.

Suppose a symmetric two-player game with a finite set of strategies (of course, the same for each player) $S = \{s_1, s_2, ...\}$ is given. A *population profile* is a probability distribution $\nu = (p_1, p_2, ...)$ on the set S (i.e. a vector $\nu = (p_1, p_2, ...)$ with non-negative co-ordinates that sum up to one), where each p_j denotes the probability with which the strategy s_i is played in this population. Such a profile can be realized by a variety of ways, for example, by (i) *monomorphic population*: all members of the population use the same mixed strategy $\sigma = \nu$, or by (ii) *polymorphic population*: each individual of the population plays only some pure strategy and p_j denotes the probability to meet an individual playing the pure strategy s_j.

Dealing with payoffs we shall adhere to the convention used in the theory of symmetric games, i.e. $\Pi(s_i, s_j)$ will mean the payoff to the player playing the strategy s_i against the strategies s_j.

Suppose all individuals are playing a strategy $\sigma^\star = (p_1^\star, p_2^\star, ...)$, and hence the population is monomorphic with the profile $\nu = \sigma^\star$. Suppose a mutation has occurred that caused a small proportion ϵ of individuals to change their strategy to some mutant strategy σ thus changing the profile ν to the new (so called *post-entry*) profile

$$\nu_\epsilon = \epsilon\sigma + (1 - \epsilon)\sigma^\star. \tag{9.27}$$

A mixed strategy σ^* is called *evolutionary stable* (ESS) if there exists $\tilde{\epsilon} > 0$ such that

$$\Pi(\sigma^*, \nu_\epsilon) > \Pi(\sigma, \nu_\epsilon) \tag{9.28}$$

for all $\sigma \neq \sigma^*$ and all ϵ such that $0 < \epsilon < \tilde{\epsilon}$. In other words, if the proportion of mutants is sufficiently small, ESS does strictly better (e.g. produce more offspring) against the post-entry population than the mutant strategy. This implies that any mutants (if appeared in a small proportion) should eventually die out. We shall obtain now the main criterion for ESS, whose particular case was mentioned in Chapter 4 without a proof.

Theorem 9.5. *The strategy σ^* is ESS if and only if for all strategies $\sigma \neq \sigma^*$ either*

(i) $\Pi(\sigma^*, \sigma^*) > \Pi(\sigma, \sigma^*)$

or

(ii) $\Pi(\sigma^*, \sigma^*) = \Pi(\sigma, \sigma^*)$ and $\Pi(\sigma^*, \sigma) > \Pi(\sigma, \sigma)$.

Proof. From (9.27) it follows that inequality (9.28) reads as

$$\Pi(\sigma^*, \epsilon\sigma + (1 - \epsilon)\sigma^*) > \Pi(\sigma, \epsilon\sigma + (1 - \epsilon)\sigma^*). \tag{9.29}$$

Using the obvious linearity of the function Π leads to the equivalent inequality

$$\epsilon\Pi(\sigma^*, \sigma) + (1 - \epsilon)\Pi(\sigma^*, \sigma^*) > \epsilon\Pi(\sigma, \sigma) + (1 - \epsilon)\Pi(\sigma, \sigma^*)$$

or

$$(\Pi(\sigma^*, \sigma^*) - \Pi(\sigma, \sigma^*)) + \epsilon[(\Pi(\sigma^*, \sigma) - \Pi(\sigma, \sigma)) + (\Pi(\sigma^*, \sigma^*) - \Pi(\sigma, \sigma^*)] > 0. \tag{9.30}$$

This clearly implies the statement of the Theorem, since a linear function of ϵ is positive in a neighborhood of zero if either it is positive at $\epsilon = 0$ (condition (i) of the Theorem) or it vanishes at $\epsilon = 0$ but has a positive slope (derivative) at $\epsilon = 0$ (condition (ii) of the Theorem).

Let us use the Theorem for the analysis of the general Hawk-Dove game

Table 9.4

	hawk	dove
hawk	H	V
dove	0	D

with $V \geq 0$ (recall from Chapter 4 that any symmetric 2-action game can be represented in this form up to a common shift in payoffs). We know from

Chapter 4 that under generic conditions a strictly mixed Nash equilibrium for this game exists if either

(i) $D > V, H > 0$, or

(ii) $D < V, H < 0$,

and in both cases this equilibrium $\sigma^* = (p^*, 1 - p^*)$ is unique and is given by formula

$$p^* = \frac{D - V}{H + D - V}. \tag{9.31}$$

Corollary 9.3. *(i) In the case $D > V, H > 0$ the mixed equilibrium σ^* given by (9.31) is not ESS; (ii) in the case $D < V, H < 0$ the mixed equilibrium given by (9.31) is ESS; (iii) if $H > 0$, the pure strategy profile (hawk, hawk) is a Nash equilibrium and the strategy hawk (h) is ESS; (iv) if $D > V$, the pure strategy profile (dove, dove) is a Nash equilibrium and the strategy dove (d) is ESS.*

Proof. (i)-(ii) From the equality of payoffs lemma $\Pi(\sigma^*, \sigma^*) = \Pi(\sigma, \sigma^*)$ for all σ and hence the ESS condition becomes

$$\Pi(\sigma^*, \sigma) - \Pi(\sigma, \sigma) > 0$$

for all $\sigma \neq \sigma^*$. But

$$\Pi(\sigma^*, \sigma) - \Pi(\sigma, \sigma)$$
$$= Hp^*p + Vp^*(1 - p) + D(1 - p^*)(1 - p) - Hp^2 - Vp(1 - p) - D(1 - p)^2$$
$$= (V - H - D)p^2 + (D - V + p^*(D + H - V))p + (V - D)p^*$$
$$= (V - H - D)(p - p^*)^2.$$

This is positive for $p \neq p^*$ if $V > D + H$, which holds in case $V > D$, $H < 0$, and hence σ^* is ESS in case (ii). In case (i), the latter expression becomes negative and hence σ^* is not ESS in this case.

(iii) $\Pi(h, h) - \Pi(\sigma, h) = (1 - p)H$, which is positive for $H > 0$ and $p \neq 1$, hence strategy h is ESS in this case. Statement (iv) is proved similarly.

9.4 Replicator dynamics, Nash's fields and stability

Replicator dynamics (RD) is designed to model a process by which behavioral patterns of populations (created by some mutation, say) are occasionally transferred to an equilibrium (ESS). Simple arguments leading to the equations of RD are the following. Let N denote the size of a population consisting of individuals playing one of m possible strategies $s_1, ..., s_m$ in a two-player game specified by a payoff function Π, where $\Pi(s_i, s_j)$ denotes the payoff to a player playing s_i against s_j. As usual, the payoff expresses the production rate. Denoting the number of individuals (respectively, its proportion) playing the strategy s_i, $i = 1, ..., m$, by n_i (respectively by $x_i = n_i/N$), the corresponding population profile by $\nu = (x_1, ..., x_m)$, and the background production rate by c, leads to the system

$$\dot{n}_i = (c + \Pi(s_i, \nu))n_i, \quad i = 1, ..., m, \tag{9.32}$$

where

$$\Pi(s_i, \nu) = \Pi(s_i, s_1)x_1 + ... + \Pi(s_i, s_m)x_m$$

is the (average) payoff to strategy s_i against the population profile ν. System (9.32) is the *RD in terms of absolute sizes*. Usually one rewrites it in terms of frequencies $x_i = n_i/N$.

Exercise 9.5. Show that (9.32) implies the dynamics

$$\dot{N} = (c + \Pi(\nu, \nu))N,$$

where

$$\Pi(\nu, \nu) = x_1\Pi(s_1, \nu) + ... + x_m\Pi(s_m, \nu) = \sum_{i,j=1}^{m} x_i x_j \Pi(s_i, s_j)$$

is the average fitness of the population (it represents, of course, the average payoff for playing ν against itself), and deduce further that

$$\dot{x}_i = (\Pi(s_i, \nu) - \Pi(\nu, \nu))x_i, \quad i = 1, ..., m, \tag{9.33}$$

which is the required *RD system of equations in the standard form*.

Observe that equations (9.33) do not depend on the background rate c. This is a performance of a more general fact. Namely, a natural generalization of (9.32) is given by the system

$$\dot{n}_i = (c(N) + \Pi(s_i, \nu))n_i, \quad i = 1, ..., m, \tag{9.34}$$

with the background rate $c(N)$ being a function of the size of the population. For example, a model with $c(N) = -kN$ with some positive constant k is called the Lotka-Volterra model.

Exercise 9.6. Show that (9.34) implies the same RD equation (9.33) as in the case of a constant rate c.

The RD equations (9.33) enjoy the following remarkable property.

Theorem 9.6. *(The fundamental theorem of natural selection) If the matrix $\Pi(s_i, s_j)$ is symmetric, the average fitness $\Pi(\nu, \nu)$ of the population does not decrease along any trajectory of RD system (9.33) and satisfies the equation*

$$\frac{d}{dt}\Pi(\nu, \nu) = 2\sum_{i=1}^{m} x_i(\Pi(s_i, \nu) - \Pi(\nu, \nu))^2. \tag{9.35}$$

Proof. By (9.33) (and using symmetry of Π in the first equation)

$$\frac{d}{dt}\Pi(\nu,\nu) = 2\sum_{i,j=1}^{m} x_j(\Pi(s_i,s_j)\dot{x}_i = 2\sum_{i=1}^{m} \Pi(s_i,\nu)x_i(\Pi(s_i,\nu) - \Pi(\nu,\nu))$$

$$= 2\sum_{i=1}^{m} \Pi(s_i,\nu)x_i(\Pi(s_i,\nu) - \Pi(\nu,\nu)) - 2\sum_{i=1}^{m} \Pi(\nu,\nu)x_i(\Pi(s_i,\nu) - \Pi(\nu,\nu))$$

$$= 2\sum_{i=1}^{m} x_i(\Pi(s_i,\nu) - \Pi(\nu,\nu))^2,$$

where the obvious equation $\sum_{i=1}^{m} x_i(\Pi(s_i,\nu) - \Pi(\nu,\nu)) = 0$ was used.

The connection between RD and Nash equilibria are given by the following result.

Theorem 9.7. *If σ is a fixed point of RD equations for a symmetric two-person game G given by a matrix A, then σ defines a symmetric Nash equilibrium either for G or for a symmetric game G' obtained from G by excluding some of its strategies (i.e. given by a matrix that is a major minor of A). And vice versa, any symmetric Nash equilibrium of either G or a game G' obtained from G by excluding some of its strategies, is a fixed point of RD equations for G.*

Proof. This follows directly from Theorems 9.2, 9.3.

Recall that a rest point (or a fixed point) y_0 of an ordinary (vector-valued) differential equation $\dot{y} = f(y)$ is called *asymptotically stable (in the Lyapunov sense)*, if there exists an $\epsilon > 0$ such that its solution with any initial point from the ϵ-neighborhood of y_0 tends to y_0 as $t \to \infty$, *neutrally stable (in the Lyapunov sense)*, if for $\epsilon > 0$ there exists a $\delta > 0$ such that any solution starting at a point from the δ-neighborhood of y_0 stays forever in the ϵ-neighborhood of y_0, and *unstable*, if it is not neutrally stable. Let us recall also the main criterion of stability (*stability through linear approximation*): let $0 \in \mathbf{R}^d$ be a single point of the system $\dot{y} = f(y)$ and let $F = \frac{\partial f}{\partial y}(0)$ denote its Jacobian (matrix of first derivatives that specifies the linear part of system $\dot{y} = f(y)$); then 0 is asymptotically stable whenever all eigenvalues of F have negative real parts and unstable if there exists at least one eigenvalue with a positive real part. It is important to stress that this theorem can not be applied in case of eigenvalues with vanishing real parts.

There is a connection between ESS and rest points of RD. For instance, one can show that if σ is ESS, then it is a rest point of the corresponding RD, see e.g. [72] for a (not very difficult) proof.

Example. *RD in genetics and sexual reproduction.* Suppose a population is characterized by m types of genes (called *alleles*) $g_1, ..., g_m$, and an individual is characterized by a pair (g_i, g_j) of such genes that are placed in a *genetic locus*. An individual with a pair (g_i, g_j) has fitness f_{ij}. Assume the following schematic procedure of sexual reproduction: all genes of all sexually matured individuals are just put in a common pool (all parents are supposed to die), are arbitrary paired, and any pair (g_i, g_j) forms a new individual that will survive to the next stage of sexual reproduction with probability f_{ij}. Let $n_i = n_i(t)$ (respectively $x_i = x_i(t)$) denote the number (respectively the frequency) of the allele g_i in the population (at the time t). The fitness of allele i is then given by $f_i = \sum_{j=1}^{m} f_{ij} x_j$ and the average fitness of the population is given by

$$f = \sum_{i=1}^{m} x_i f_i = \sum_{i,j=1}^{m} f_{ij} x_i x_j.$$

The rate of production being proportional to average fitness, leads to the dynamics

$$\dot{n}_i = n_i \sum_{j=1}^{n} f_{ij} x_j, \quad i = 1, ..., m.$$

But this is the system (9.32) with $c = 0$ and $\Pi(s_i, s_j) = f_{ij}$. Hence the frequencies x_i enjoy the corresponding RD equation (9.33). It is worth noting that the fundamental theorem of natural selection was first discovered by R.A. Fisher precisely in this context.

As $x_1 + x_2 = 1$ for a profile $\nu = (x_1, x_2)$ of a two-action game, system (9.33) reduces to just one equation for $x = x_1$ of the form

$$\dot{x} = (\Pi(s_1, \nu) - (x\Pi(s_1, \nu) + (1 - x)\Pi(s_2, \nu))x$$

or better

$$\dot{x} = x(1 - x)(\Pi(s_1, \nu) - \Pi(s_2, \nu)). \tag{9.36}$$

This RD equation has always two boundary fixed points $x = 0$ and $x = 1$. Mixed strategy fixed points are found from equation $\Pi(s_1, \nu) = \Pi(s_2, \nu)$.

It is easy to investigate stability of the fixed points of (9.36). Let us analyze the game given by Table 9.4 in case $H < 0$, $D < V$. This game has only one symmetric Nash equilibrium $\sigma^\star = (p^\star, 1 - p^\star)$ with $p^\star$ given by

(9.31), and this equilibrium is ESS. Consequently it follows from a general fact mentioned above that σ^* is a stable fixed point of RD equation (9.36) that now takes the form

$$\dot{x} = x(1-x)[x(H+D-V)-(D-V)].$$

To check stability directly one rewrites this equation in terms of $y = x - p^*$:

$$\dot{y} = \left(\frac{D-V}{H+D-V}+y\right)\left(\frac{H}{H+D-V}-y\right)y(H+D-V) = y\frac{H(D-V)+O(y)}{H+D-V}.$$

As the coefficient at y is negative, one concludes that the fixed point p^* is stable.

Example. The RD system for generalized Rock-Paper-Scissors game given by the matrix (9.19) has the form

$$\begin{cases} \dot{x} = [-ax+y-z+a(x^2+y^2+z^2)]x \\ \dot{y} = [-x-ay+z+a(x^2+y^2+z^2)]y \\ \dot{z} = [x-y-az+a(x^2+y^2+z^2)]z \end{cases} \qquad (9.37)$$

since $\Pi(\nu,\nu) = -a(x^2+y^2+z^2)$ for $\nu = (x,y,z)$. Writing this equation around the equilibrium $\sigma^* = (1/3,1/3,1/3)$ in terms of $u = x - 1/3$, $v = y - 1/3$ yields

$$\begin{cases} \dot{u} = \frac{1}{3}[(1-a)u+2v]+\ldots \\ \dot{v} = \frac{1}{3}[-2u-(1+a)v]+\ldots \end{cases},$$

whereby the dots designate the higher terms (quadratic, etc.) in u, v. Thus the linearized equation in a neighborhood of the fixed point is specified by the matrix

$$\frac{1}{3}\begin{pmatrix} 1-a & 2 \\ -2 & -(1+a) \end{pmatrix}$$

with eigenvalues $-a \pm i\sqrt{3}$. Consequently σ^* is a stable (hyperbolic) point for $a > 0$, and an unstable hyperbolic point for $a < 0$. For $a = 0$ (i.e. for classical Scissors-Rock-Paper game) the eigenvalues are purely imaginary, and hence a fixed point of a linearized system is a center with closed orbits around it. As the point is not hyperbolic, one can not deduce from this directly the same property of the full non-linear system. However, one can check that the function $V(x,y,z) = \ln x + \ln y + \ln z$ is the integral of motion for system (9.37), and this allows us to deduce that this σ^* is actually a center for this system, and it has closed orbits around it. Thus in case $a = 0$, the replicator dynamics generates oscillation, and they are in fact observed by field researchers of Uta Stansburiana lizards, whose interaction is described by suchlike game (see Chapter 4).

The next result is devoted to a possible extension of the theory of Rock-Paper-Scissors to n-person setting. This extension is based on the observation that the Rock-Paper-Scissors is both symmetric and zero-sum, which means that the matrix A of the payoffs to the first player is antisymmetric ($A = -A^T$, where A^T is the transpose of A), and the payoffs to the second player are given by $-A = A^T$. (A different extension of Rock-Scissors-Paper is analyzed in [73].)

Theorem 9.8. *Let Γ be a symmetric zero-sum game specified by an antisymmetric $n \times n$-matrix A and assume it has a zero vector v, i.e. such that $Av = 0$, with all co-ordinates being (strictly) positive. Then the strategy $\sigma = v/(v_1 + ... + v_n)$ specifies a neutrally stable symmetric equilibrium.*

Proof. $\sigma = (p_1, ..., p_n)$ is a symmetric equilibrium according to the Equality of Payoffs Lemma. The main additional observation (linked with the antisymmetry of A) is the following: the function

$$L(\eta) = x_1^{p_1} x_2^{p_2} ... x_n^{p_n}$$

is an integral of the replicator dynamics on the strategies $\eta = (x_1, ..., x_n)$, i.e. it does not change along its trajectories. In fact, clearly L is strictly positive on the simplex of pure mixed strategies η and vanishes on its boundary. Next, differentiation along the RD system yields

$$\frac{d}{dt}(\ln L) = \sum_{i=1}^{n} p_i((Ax)_i - (x, Ax)) = (\sigma, Ax) - (x, Ax) = 0.$$

The latter equality holds, because $(\sigma, Ax) = -(A\sigma, x) = (0, x) = 0$ and $(x, Ax) = 0$ due to the anti-symmetry of A. In order to see that starting near σ the trajectory of RD can not go far away from it, it remains to observe that the positive function $-\ln L$ has a unique minimum in the simplex of mixed strategies η attained at $\eta = \sigma$. This follows from the classical *Gibbs inequality* (which is not difficult to prove): for any two sets of probabilities $\eta = (x_1, ..., x_n)$ and $\sigma = (p_1, ..., p_n)$

$$-\sum_{j=1}^{n} p_j \ln p_j \leq -\sum_{j=1}^{n} p_j \ln x_j$$

with equality only when $\eta = \sigma$. The expression on the left hand side of this inequality is called the *entropy* of the probability law σ.

A natural extension of RD for a general non-symmetric game Γ defined at the beginning of Section 9.2 is the system

$$\dot{x}_i^j = (\Pi_j(s_j^i, \sigma) - \Pi_j(\sigma_j, \sigma))x_i^j, \quad i = 1, ..., n_j, \, j = 1, ..., m, \qquad (9.38)$$

where

$$\sigma_1 = (x_1^1, ..., x_{n_1}^1), \quad \sigma_2 = (x_1^2, ..., x_{n_2}^2), \quad ..., \quad \sigma_m = (x_1^m, ..., x_{n_m}^m)$$

is a profile of (mixed) strategies and σ in (9.38) denotes shortly the collection of the strategies of all players others than j. This system describes an evolution of the behavior of the players applying a try-and-error method of shifting the strategies in the direction of a better payoff. Clearly (9.38) reduces to (9.33) for the symmetric two-person setting. As in Theorem 9.7 it follows from the Equality of payoffs Lemma that a Nash equilibrium is a rest point of system (9.38). The r.h.s. of (9.38) is sometimes called the Nash vector field of the game.

A Nash equilibrium for a game Γ is called *asymptotically stable, neutrally stable or unstable in the Lyapunov sense (or dynamically)* if it so for the corresponding dynamics (9.38). (This means roughly speaking that if starting with strategies near the equilibrium, players would adjust their strategies in the direction of better payoffs, their strategies would converge to this equilibrium.)

Let us write down (9.38) more explicitly for 2×2 bi-matrix games, i.e. for two-person two-action games with payoffs given by a pair of matrices (a_{ij}), (b_{ij}), $i, j = 1, 2$. The payoffs are thus given by formulae

$$\Pi_1(X, Y) = x_1 a_{11} y_1 + x_1 a_{12} y_2 + x_2 a_{21} y_1 + x_2 a_{22} y_2,$$

$$\Pi_2(X, Y) = x_1 b_{11} y_1 + x_1 b_{12} y_2 + x_2 b_{21} y_1 + x_2 b_{22} y_2,$$

where $X = (x_1, x_2)$, $Y = (y_1, y_2)$ denote the strategies of the two players. The evolution (9.38) along the Nash vector field takes the form

$$\begin{cases} \dot{x}_1 = x_1(a_{11} y_1 + a_{12} y_2 - \Pi_1(X, Y)) \\ \dot{x}_2 = x_2(a_{21} y_1 + a_{22} y_2 - \Pi_1(X, Y)) \\ \dot{y}_1 = y_1(b_{11} x_1 + b_{21} x_2 - \Pi_2(X, Y)) \\ \dot{y}_2 = y_2(a_{12} x_1 + b_{22} x_2 - \Pi_2(X, Y)) \end{cases} .$$

As $x_1 + x_2 = y_1 + y_2 = 1$, this can be rewritten in terms of just two variables x_1, y_1 as

$$\begin{cases} \dot{x}_1 = x_1(1 - x_1)(Ay_1 - a) \\ \dot{y}_1 = y_1(1 - y_1)(Bx_1 - b) \end{cases} , \tag{9.39}$$

where

$$A = a_{11} + a_{22} - a_{12} - a_{21}, \quad a = a_{22} - a_{12},$$

$$B = b_{11} + b_{22} - b_{12} - b_{21}, \quad b = b_{22} - b_{21}.$$

Let us analyze the stability of equilibria of (9.39) subject to the assumption of "general position", i.e. if $A, B, a, b, A - a, B - b$ do not vanish. In this case system (9.39) have either four or five rest points. Namely it has always four pure strategy rest points $(0,0), (0,1), (1,0), (1,1)$, and it has the fifth rest point

$$x^* = \frac{b}{B}, y^* = \frac{a}{A} \tag{9.40}$$

whenever

$$0 < a/A < 1, \quad 0 < b/B < 1. \tag{9.41}$$

Exercise 9.7. Show that $(0,0)$ is asymptotically stable for $a, b > 0$, $(0,1)$ is asymptotically stable for $a > 0, b < 0$, $(1,0)$ is asymptotically stable for $a < 0, b > 0$ and $(1,1)$ is asymptotically stable for $a, b < 0$. Hint: say for $(0,0)$ one sees that $\dot{x}_1, \dot{y}_1$ are negative in a neighborhood of $(0,0)$ whenever $a, b > 0$.

Exercise 9.8. Show that if $a = b = 0$, then $(0,0)$ is stable if and only if $A \leq 0$, $B \leq 0$.

For the analysis of the fifth single point (9.40), write (9.39) in terms of $\xi = x_1 - x^*, \eta = y_1 - y^*$:

$$\begin{cases} \dot{\xi} = \frac{Ab}{B}(1 - \frac{b}{B})\eta + A(1 - \frac{2b}{B})\xi\eta - A\xi^2\eta, \\ \dot{\eta} = \frac{aB}{A}(1 - \frac{a}{A})\xi + B(1 - \frac{2a}{A})\xi\eta - B\xi\eta^2. \end{cases} \tag{9.42}$$

Exercise 9.9. If $AB > 0$ (or $ab > 0$), the equilibrium (x^*, y^*) is unstable in the Liapunov sense. Hint: by (9.40) conditions $AB > 0$ and $ab > 0$ are equivalent. In case they hold the Jacobian (matrix of linear approximation) has a positive eigenvalue.

If $AB < 0$, the eigenvalues of the Jacobian of the r.h.s. of (9.42) have vanishing real parts so that the stability by linear approximation theorem can not be applied and more subtle methods are required.

Theorem 9.9. *If $AB < 0$ and (9.41) hold, the point (9.40) is neutrally stable, but not asymptotically stable.*

Proof. Proof is based on the observation that the function

$$(B - b)\ln(1 - x_1) + b\ln x_1 - (A - a)\ln(1 - y_1) - a\ln y_1$$

is a first integral of system (9.39), i.e. its derivative vanishes along the solutions of (9.39). This can be checked by straightforward direct calculations.

(In order to see, where this formula comes from, one rewrites system (9.39) as a single equation

$$\frac{dx_1}{dy_1} = \frac{x_1(1 - x_1)(Ay_1 - a)}{y_1(1 - y_1)(Bx_1 - b)}$$

that is solvable by means of the separation of variables.) Assuming, say, that $B > 0$ and $A < 0$, one deduces that the function

$$(1 - x_1)^{B-b} x_1^b (1 - y_1)^{|A-a|} y_1^{|a|}$$

is a constant on each trajectory. This function is a product of two positive functions on a unit segment that vanish at the end points, have maxima at x^* and y^* respectively and are monotone outside these points. It follows that any trajectory starting at $x_1 \in (0,1)$, $y_1 \in (0,1)$ can not approach the boundary points and moreover, for any neighborhood U of point (9.40) there is a smaller neighborhood V such that starting from V one can not exit from U, i.e., stability holds. (Similar arguments show that any trajectory starting outside (9.40) can not approach it infinitely close, i.e. it is not an asymptotically stable point.)

Exercise 9.10. This is meant to outline another approach to the analysis of (9.42) under the assumptions of the above theorem. (i) Show that by the change of the variables $\eta = \mu\eta'$ with

$$\mu = \sqrt{-\frac{B^3 a(A - a)}{A^3 b(B - b)}}$$

system (9.42) takes the form

$$\dot{\xi} = -\lambda\eta' - \beta_1\xi\eta' + \gamma_1\xi^2\eta',$$

$$\dot{\eta}' = \lambda\xi + \beta_2\xi\eta' - \gamma_2\xi\eta'^2$$

with an asymmetric matrix of linear approximation, where

$$\lambda = \sqrt{-\frac{ab(A-a)(B-b)}{AB}}, \beta_1 = (B - 2b)\sqrt{-\frac{aB(A-a)}{Ab(B-b}}, \beta_2 = \frac{B(A - 2a)}{A},$$

$$\gamma_1 = |A|\mu = \sqrt{-\frac{B^3 a(A-a)}{Ab(B-b)}}, \quad \gamma_2 = |b|\mu = \sqrt{-\frac{B^5 a(A-a)}{A^3 b(B-b)}}.$$

(ii) Show that introducing polar coordinates $\xi = r\cos\phi, \eta' = r\sin\phi$ yields

$$\dot{r} = r^2 \frac{\sin 2\phi}{2}(-\beta_1\cos\phi + \beta_2\sin\phi) + r^3\frac{\sin 2\phi}{2}(\gamma_1\cos^2\phi - \gamma_2\sin^2\phi),$$

$$\dot{\phi} = \lambda + [r\frac{\sin 2\phi}{2}(\beta_2 \cos \phi + \beta_1 \sin \phi) - r^2\frac{\sin^2 2\phi}{4}(\gamma_2 + \gamma_1)]$$

so that for small r, $\dot{\phi} \approx \lambda > 0$, and this system can be equivalently described by the equation

$$\frac{dr}{d\phi} = \frac{r^2 \sin 2\phi[-\beta_1 \cos \phi + \beta_2 \sin \phi + r(\gamma_1 \cos^2 -\gamma_2 \sin^2)]}{2\lambda + r \sin 2\phi(\beta_2 \cos \phi + \beta_1 \sin \phi) - r^2 \sin^2 2\phi(\gamma_1 + \gamma_2)/2}. \quad (9.43)$$

Stability of (9.42) is thus reduced to the question of the existence of periodic solutions to (9.43). (iii) Show that if either $B = 2b$ or $A = 2a$, the point (9.40) is a center for (9.42), i.e. the trajectories of (9.42) are closed in a neighborhood of (9.40) and hence the point (9.40) is neutrally stable. (Hint: Suppose $A = 2a$, i.e. $\beta_2 = 0$. Then the r.h.s. of (9.43) is an odd function of ϕ so that if $r = R(\phi)$ is a solution of (9.43), then $R(-\phi)$ is again a solution. As $R(\phi)$ and $R(-\phi)$ coincide for $\phi = 0$, they coincide identically (by the uniqueness of solutions to ordinary differential equations). In particular, $R(\pi) = R(-\pi)$, and hence $R(\phi)$ is 2π-periodic. And this means that solutions to (9.42) are closed. Similarly, if $B = 2b$, i.e. $\beta_1 = 0$, one shows that for a solution $R(\phi)$ the functions $R(\pi - \phi)$ and $R(-\pi - \phi)$ are solutions as well and again by uniqueness one concludes that $R(\phi) = R(\pi - \phi) = R(-\pi - \phi)$ for all ϕ, which again leads to periodicity.)

Exercise 9.11. By definition a profile $(s_1^{i_1}, ..., s_m^{i_m})$ of pure strategies in the mixed strategies extension of a general finite game described at the beginning of Section 10.2 is a Nash equilibrium if and only if for all $j = 1, ..., m$ and all $s \in S_j$

$$\Pi_j(s_1^{i_1}, ..., s_m^{i_m}) \geq \Pi_j(s_1^{i_1}, ..., s_{j-1}^{i_{j-1}}, s, s_{j+1}^{i_{j+1}}, ..., s_m^{i_m}).$$

Show that if all these inequalities are strict, then this equilibrium is asymptotically stable. Hint: writing system (9.38) in terms of independent variables

$$x_1^k, k \in \{1, ..., n\} \setminus i_1, \quad x_2^k, k \in \{1, ..., n\} \setminus i_2, ..., \quad x_m^k, k \in \{1, ..., n\} \setminus i_m,$$

yields

$$\begin{cases} \dot{x}_1^k = x_1^k(\Pi_1(s_1^k, s_2^{i_2}, ..., s_m^{i_m}) - \Pi_1(s_1^{i_1}, ..., s_m^{i_m})) + ..., k \in \{1, ..., n_1\} \setminus i_1, \\ \qquad\qquad\qquad\qquad\quad \cdot \quad \cdot \quad \cdot \quad \cdot \quad \cdot \\ \dot{x}_m^k = x_m^k(\Pi_m(s_1^{i_1}, ..., s_{m-1}^{i_{m-1}}, s_m^k)) - \Pi_1(s_1^{i_1}, ..., s_m^{i_m}) + ..., k \in \{1, ..., n_m\} \setminus i_m, \end{cases} \quad (9.44)$$

where dots in the first and the last lines denote nonlinear terms. Single point $0 \in \mathbf{R}^{n_1+...+n_m-m}$ stands for the equilibrium $(s_1^{i_1}, ..., s_m^{i_m})$. Apply stability by linear approximation theorem.

Remarks. RD and Nash fields are not the only useful dynamics describing approach to an equilibrium. A comprehensive review on these dynamics for evolutionary games can be found in [73]. There exists another important notion of stability for an equilibrium, different from dynamic or Lyapunov stability, namely the so called *structural stability*. This is a very general concept in mathematics meaning roughly speaking the continuous dependence of some object or a collection of objects on the parameters of the setting. This is crucial for practical application, as usually one can not specify the parameters precisely (everything is measured only with some degree of precision) and structural stability means that if the parameters are chosen close enough to the exact ones, then the object under consideration would be also near the exact one.

Applied to the game equilibria this concept reads as follows. A Nash equilibrium $(\sigma_1, ..., \sigma_m)$ in a general finite game Γ described at the beginning of Section 9.2 is called *structurally stable* if for arbitrary $\epsilon > 0$ there exists a $\delta > 0$ such that for all games $\tilde{\Gamma}$ with the same number of players and pure strategies and with payoffs $\tilde{\Pi}$ that differ from Π no more than by δ, i.e. such that

$$|\Pi_i(s_1^{j_1}, s_2^{j_2}, ..., s_m^{j_m}) - \tilde{\Pi}_i(s_1^{j_1}, s_2^{j_2}, ..., s_m^{j_m})| < \delta \qquad (9.45)$$

for all $i, s_1^{j_1}, s_2^{j_2}, ..., s_m^{j_m}$, there exists a Nash equilibrium $(\tilde{\sigma}_1, ..., \tilde{\sigma}_m)$ for the game $\tilde{\Gamma}$ such that $|\tilde{\sigma}_j - \sigma_j| < \epsilon$ for all $j = 1, ..., m$.

Examples. 1. For the game given by Table 9.4 with $D > V, H > 0$ the mixed equilibrium given by (9.31) is dynamically unstable, but structurally stable, and for $D < V, H < 0$ it is both dynamically and structurally stable. 2. For a two-person two-action game given by 2×2-matrices A and B such that $a_{11} \geq a_{21}$ and $b_{11} = b_{12}$ the pure strategy equilibrium $(0, 0)$ is structurally unstable (as it vanishes by an arbitrary small increase of b_{12}), though it can be dynamically stable (as in Exercise 9.7).

It makes sense also to speak about structural stability of dynamically stable or unstable equilibria, i.e. a dynamically stable (or unstable) Nash equilibrium $(\sigma_1, ..., \sigma_m)$ in a game Γ is called *structurally stable* if for arbitrary $\epsilon > 0$ there exists a $\delta > 0$ such that for all games $\tilde{\Gamma}$ with the same number of players and pure strategies and with payoffs $\tilde{\Pi}$ that differ from Π no more than by δ (i.e. (9.45) holds), there exists a dynamically stable (respectively unstable) Nash equilibrium $(\tilde{\sigma}_1, ..., \tilde{\sigma}_m)$ for the game $\tilde{\Gamma}$ such that $|\tilde{\sigma}_j - \sigma_j| < \epsilon$ for all $j = 1, ..., m$.

The notion of stability is closely related to another important notion of a *generic property*: a property (object or characteristics) in a class of structures parametrized by a collection of real numbers s from a given subset S of a Euclidean space is called *generic* if it holds for s from a subset $\tilde{S} \subset S$ that is both open (which means that if $s_0 \in \tilde{S}$, then all $s \in S$ that are closed enough to s_0 belong to $\tilde{S}$ as well, i.e. the property of being in $\tilde{S}$ is structurally stable in any point $s \in \tilde{S}$) and dense (which means that for any s_0 there exists an $s \in \tilde{S}$ that is arbitrary close to s_0, i.e. the negation of being in $\tilde{S}$ is nowhere structurally stable).

Example. As an instructive example consider a class Γ_n^2 of mixed strategy extensions of games of n players each having only two strategies. Let $A_{j_1,...,j_n}^i$ denote the payoff to i under pure profile $\{j_1, ..., j_n\}$, $j_k = 1, 2$. A mixed strategy profile can be described by families

$$\sigma_1 = (x_1, 1 - x_1), \sigma_2 = (x_2, 1 - x_2), ..., \sigma_n = (x_n, 1 - x_n).$$

Equations (9.38) can be written in terms of $x_1, ..., x_n$ yielding (check it!)

$$\dot{x}_i = x_i(1 - x_i) \sum_{I \in \{1,...,n\} \setminus i} \tilde{A}_I^i \prod_{k \in I} x_k \prod_{k \notin I} (1 - x_k), \quad i = 1, ..., n, \qquad (9.46)$$

where

$$\tilde{A}_I^i = A_{j_1...j_{i-1}1j_{i+1}...j_n}^i - A_{j_1...j_{i-1}2j_{i+1}...j_n}^i$$

with $j_k = 1$ whenever $k \in I$ and $j_k = 2$ otherwise. Hence pure mixed (i.e. with all probabilities being positive) Nash equilibria for a game in Γ_n^2 are given by vectors $x^\star = (x_1^\star, ..., x_n^\star)$ with coordinates from $(0, 1)$ solving the following system of n equations

$$\sum_{I \in \{1,...,n\} \setminus i} \tilde{A}_I^i \prod_{k \in I} x_k \prod_{k \notin I} (1 - x_k) = 0, \quad i = 1, ..., n. \qquad (9.47)$$

In particular for $n = 3$, denoting x_1, x_2, x_3 by x, y, z and arrows of payoffs A^1, A^2, A^3 by A, B, C yields for system (9.46) the following explicit form

$$\begin{cases} \dot{x} = x(1 - x)(a + A_2 y + A_3 z + Ayz), & 0 \le x \le 1 \\ \dot{y} = y(1 - y)(b + B_1 x + B_3 z + Bxz), & 0 \le y \le 1 \\ \dot{z} = z(1 - z)(c + C_1 x + C_2 y + Cxy), & 0 \le z \le 1 \end{cases} \qquad (9.48)$$

as well as the form

$$\begin{cases} a + A_2 y + A_3 z + Ayz = 0 \\ b + B_1 x + B_3 z + Bxz = 0 \\ c + C_1 x + C_2 y + Cxy = 0 \end{cases} \tag{9.49}$$

for system (9.47), where

$$a = A_{122} - A_{222}, \quad A_2 = A_{112} - A_{212} - a, \ A_3 = A_{121} - A_{221} - a,$$

$$A = A_{111} - A_{211} - a - A_2 - A_3,$$

and the coefficients in other two lines are defined analogously.

Exercise 9.12. Assuming $x^\star = (x_1^\star, ..., x_n^\star)$ solves (9.47), write down system (9.46) in terms of the deviations from the equilibrium $\xi_i = x_i - x_i^\star$ and check that the matrix of linear approximation (the Jacobian matrix) $J^\star$ has the entries

$$J_{ij}^\star = x_i^\star (1 - x_i^\star) \sum_{I \in \{1,...,n\} \setminus \{i,j\}} (\tilde{A}_{I \cup j}^i - \tilde{A}_I^i) \prod_{k \in I} x_k \prod_{k \notin (I \cup \{i,j\})} (1 - x_k) \tag{9.50}$$

for $i \neq j$ and $J_{ii}^\star = 0$ for all i.

Theorem 9.10. *Let $J^\star$ be the Jacobian matrix (described in Exercise 9.12) of a pure mixed equilibrium $x^\star = (x_1^\star, ..., x_n^\star)$ solving (9.47). If at least one of the eigenvalues of $J^\star$ has a non-vanishing real part, then $x^\star$ is unstable in Liapunov sense (i.e. dynamically). In particular, if n is an odd number, then a necessary condition for the Liapunov stability of $x^\star$ is the degeneracy of $J^\star$, that is $\det J^\star = 0$.*

Proof. From Exercise 9.12 one deduces that $J^\star$ has zeros on the main diagonal. Hence the sum of its eigenvalues vanishes. Consequently if there exists an eigenvalue with a non-vanishing real part there should necessarily exist also an eigenvalue with a positive real part, which implies instability. As eigenvalues with vanishing real parts appear as pairs of conjugate imaginary numbers, it follows that in case of odd n the fact that all real parts vanish implies that zero should be an eigenvalue, i.e. the degeneracy of $J^\star$.

Exercise 9.13. Write down the condition $\det J^\star = 0$ explicitly (i.e. in terms of the payoff coefficients) for the case $n = 3$. Hint and answer: (i) Using the notations of Exercise 9.12 the condition $\det J^\star = 0$ writes down as

$$(A_2 + Az^\star)(B_3 + Bx^\star)(C_1 + Cy^\star) + (B_1 + Bz^\star)(C_2 + Cx^\star)(A_3 + Ay^\star) = 0.$$
$$(9.51)$$

(ii) Solving (9.49) by expressing y, z, in terms of x and putting this in the first equation leads to the quadratic equation:

$$v(x^\star)^2 + ux^\star + w = 0, \tag{9.52}$$

where

$$w = aC_2B_3 + cbA - bA_3C_2 - cA_2B_3, v = abC + AB_1C_1 - BA_2C_1 - CA_3B_1,$$

$$u = a(BC_2 + CB_3) + b(AC_1 - CA_3) + c(AB_1 - BA_2) - A_2B_3C_1 - A_3B_1C_2.$$

(iii) Using the system that is solved by $x^\star, y^\star, z^\star$ to express $y^\star z^\star$ as a linear function of $y^\star, z^\star$ allows to rewrite (9.51) as a quadratic equation (in $x^\star, y^\star, z^\star$). Expressing the quadratic terms of this equation again via linear terms leads (after a series of remarkable cancelations) to the equation on $x^\star$ only:

$$u + 2x^\star v = 0.$$

Comparing this with (9.52) leads to the conclusion that (9.51) is equivalent to the equation

$$u^2 - 4vw = 0, \tag{9.53}$$

which is the required polynomial homogeneous equation of the sixth order in coefficients $a, A, A_2, A_3, b, B, B_1, B_3, c, C, C_1, C_2$.

Corollary. The property to have a dynamically unstable pure mixed equilibrium is generic among the games of type Γ_3^2 that have pure mixed equilibria. More precisely, apart from the games from the (algebraic of the sixth order) manifold M described by equation (9.53) pure mixed equilibria are always dynamically unstable and structurally stable (as dynamically unstable equilibria).

Exercise 9.14. Show that under the assumption (9.53) (or equivalently (9.51)) there exist α, β, γ such that a function V of the (relative entropy) form

$$\alpha[x^\star \ln x + (1 - x^\star) \ln(1 - x)] + \beta[y^\star \ln y + (1 - y^\star) \ln(1 - y)]$$
$$+ \gamma[z^\star \ln z + (1 - z^\star) \ln(1 - z)] \tag{9.54}$$

is a first integral for (9.48), i.e. it does not change along the trajectories of (9.48) $(dV/dt = 0)$ if and only if the condition

$$AB_1C_1 + aBC = BA_2C_1 + CA_3B_1 \tag{9.55}$$

holds, in which case

$$V = (B_1 + Bz^\star)(C_1 + Cy^\star)[x^\star \ln x + (1 - x^\star)\ln(1 - x)]$$

$$-(A_2 + Az^\star)(C_1 + Cy^\star)[y^\star \ln y + (1 - y^\star)\ln(1 - y)]$$

$$-(A_3 + Ay^\star)(B_1 + Bz^\star)[z^\star \ln z + (1 - z^\star)\ln(1 - z)]$$

is an integral. Use this fact to give a sufficient condition for the equilibrium $(x^\star, y^\star, z^\star)$ to be (neutrally) stable in the Liapunov sense. Hint: show by substituting (9.54) in the equation $dV/dt = 0$ that (9.54) is an integral of motion (9.48) if and only if

$$\begin{cases} \alpha(A_2 + Az) + \beta(B_1 + Bz) = 0 \\ \alpha(A_3 + Ay) + \gamma(C_1 + Cy) = 0 \\ \beta(B_3 + Bx) + \gamma(C_2 + Cy) = 0 \\ \alpha A + \beta B + \gamma C = 0. \end{cases}$$

Expressing β, γ in terms of α from the first two equations one observes that the third equation is then automatically satisfied due to (9.51). Solving the last equation leads to (9.55).

9.5   Iterative method of solving matrix games

In this section we shall briefly describe a discrete version of RD or Nash field evolution that can be used in calculating equilibria (this method is often referred to as the *Brown-Robinson method*).

Imagine two people playing a matrix game repeatedly many times. Each player chooses the next move on the basis of the previously observed moves of the opponent. Namely, suppose k games were completed to a given moment of time. During this period the opponent played the first strategy k_1 times, the second strategy — k_2 times, etc. Hence the first player can reasonably estimate the probability of player 2 choosing j-th strategy as k_j/k. Equivalently it means that player 2 uses the mixed strategy $x^k = (k_1/k, \ldots, k_n/k)$. Assuming these actions of the opponent, player 1 should choose for the next move the strategy that maximizes the gain against x^k. This is a simple model of learning by experience. Such a model would be useful, if it would be known that the opponent really follows a certain mixed strategy. One could imagine that in some games with the nature such an assumption would be reasonable.

Developing this approach assume now that both players decided to play the game using these (learning by experience) strategies. In this case, the first moves of the players clearly determine uniquely the whole dynamics of the conflict becomes determined. For example, consider a two-person zero-sum game given by the table (a version of a baseball game):

Table 9.5

Player 2

		t_1	t_2	t_3
	s_1	2	-1	-1
Player 1	s_2	-1	2	-1
	s_3	-1	-1	1

Exercise 9.15. Show that the value of this game is $-1/7$ and the corresponding optimal (minimax) strategies are $(2/7, 2/7, 3/7)$, $(2/7, 2/7, 3/7)$.

To have a start, assume that in the first round both players chose their first strategies. Let us write it down in the following table form:

		Player 1			Player 2		
1	s_1	2	-1	-1	t_1	2	-1	-1

The number 1 on the left is placed to designate the first game. The number s_1 on the second place designates the strategy chosen by player 1. In the third column we placed the row of the matrix of payoffs corresponding to this choice of player 1. The last two columns describe the strategy, chosen by 2, namely t_1, and the corresponding column of the payoff matrix.

We can now find the next move of 1 according to the assumed strategy of learning. Namely, player 1 observes that the opponent used the strategy t_1 and chooses s_1 that maximizes the gain against t_1 giving him the payoff 2. On the other hand, player 2 should choose t_2 or t_3 in order to maximise his/her payoff against s_1. Let us assume that in case of several equivalent moves the preference is given to the move with the lower number, i.e. t_2 in our case. Hence the second game could be described by the following table:

		Player 1			Player 2		
2	s_1	2	-1	-1	t_2	-1	2	-1

Let us summarize the course of these two games in the table

		Player 1				Plater 2		
1	s_1	2	-1	-1	t_1	2	-1	-1
2	s_2	4	-2	-2	t_2	1	1	-2

Note that in the second and third columns we put not the row corresponding to s_1, but the sum of the rows corresponding to the strategies, chosen in the first two games. In both cases s_1 was used. For the second player in the last column of the second row the sum of the first and second columns of the payoff matrix is placed. To see that rational behind this choice of tabular description, let us look for the choice of player 1 in the third game.

According to the learning procedure, he/she has to assume that the strategy of 2 is the mixture of t_1 and t_2 with equal probabilities. The expected payoff for each of his/her three strategies can be defined as the half of the values of the each element of the last row and column: $(1/2, 1/2, -1)$. To find the next move, it is necessary the maximum component of this vector and to apply the corresponding strategy. Note that in order to do this, it is not necessary to divide by 2. Player 1 just has to observe, that the first and the second component of the vector $(1, 1, -2)$ are maximal, and hence to choose s_1. Similarly, player 2 observes, that the second and the third coordinates in the vector $(4, -2, -2)$ (second row, third column) are minimal, and hence uses t_2 in the third game:

		Player 1				Player 2		
3	s_1	6	-3	-3	t_2	0	3	-3

Here again the vectors of payoffs of players 1 and 2 are obtained by adding the first and the second row of the payoff matrix to the corresponding rows of the previous table. The procedure of the construction of the table is thus as follows. On the k-th step we have the vector u^k of player 1 and the vector v^k of player 2. A new vector u^{k+1} of the player 1 is obtained by summing together u^k and the i-th row a_i of the payoff matrix, where i corresponds to the maximal component of the vector u^k. Similarly $v^{k+1} = v^k + a^j$, where j corresponds to the minimal component of the vector u^k. Below is the table obtained by playing 21st rounds of the baseball game.

		Player 1				Player 2		
1	s_1	2	-1	-1	t_1	2	-1	-1
2	s_1	4	-2	-2	t_2	1	1	-2
3	s_1	6	-3	-3	t_2	0	3	-3
4	s_2	5	-1	-4	t_2	-1	5	-4
5	s_2	4	1	-5	t_3	-2	4	-3
6	s_2	3	3	-6	t_3	-3	3	-2
7	s_2	2	5	-7	t_3	-4	2	-1
8	s_2	1	7	-8	t_3	-5	1	0
9	s_2	0	9	-9	t_3	-6	0	1
10	s_3	-1	8	-8	t_3	-7	-1	2
11	s_3	-2	7	-7	t_3	-8	-2	3
12	s_3	-3	6	-6	t_3	-9	-3	4
13	s_3	-4	5	-5	t_3	-10	-4	5
14	s_3	-5	4	-4	t_1	-8	-5	4
15	s_3	-6	3	-3	t_1	-6	-6	3
16	s_3	-7	2	-2	t_1	-4	-7	2
17	s_3	-8	1	-1	t_1	-2	-8	1
18	s_3	-9	0	0	t_1	0	-9	0
19	s_1	-7	-1	-1	t_1	2	-10	-1
20	s_1	-5	-2	-3	t_1	4	-11	-2
21	s_1	-3	-3	-3	t_1	6	-12	-3

Remarkably enough, it turns out (see e.g. a rather lengthy proof in [82]) that if the players keep playing in this way, the average income (of player 1) will tend to the value of the game!

Let us convince ourselves that such a conclusion is plausible from contemplating the above table. During the 21 steps player 1 used 6 times the strategy s_1, 6 times the strategy s_2 and 9 times the strategy s_3, which corresponds to the mixed strategy $(2/7, 2/7, 3/7)$ (which is, in fact, a solution) with the payoff vector $(-1/7, -1/7, -1/7)$. The same vector is obtained also by dividing u^{21} by 21.

However, the rate of convergence of this method is not fast, and one can not recommend the above method for practical calculations of the optimal strategy. So this convergence has mostly a theoretical value, as another performance of stability discussed earlier. Results of this kind could be often observed on economics models.

9.6 Zero-sum games and linear programming

A general problem of linear optimization or programming is the problem

(P) Find a vector $x = (x_1, ..., x_n)$ that maximizes $(c, x) = x_1 c_1 + ... + x_n c_n$ on the polyhedron

$$x \in P = \{x \geq 0 : Ax \leq b\}$$

(i.e. all co-ordinates of x are non-negative and $A x_i \leq b_i$ for all $i = 1, ..., m$), where c, b are given vectors and A is a given $m \times n$-matrix.

We presented in Chapter 6 some practical examples, where such problems naturally arise. In fact, the reason for the inclusion of this subject in the book is an outstanding applicability in practice of these problems, whose solution was based on game theoretic reasoning.

The main progress in solving (P) was achieved by introducing the following *dual problem*

(P') Find vector $y = (y_1, ..., y_m)$ that minimizes (b, y) on the polyhedron

$$y \in P' = \{y \geq 0 : A^T y \geq c\},$$

which turns out to satisfy the following remarkable *duality theorem*: $(c, x) \leq (b, y)$ for all $x \in P$, $y \in P'$. Moreover, if the polyhedrons P and P' not empty, then the optimal value for problems (P) and (P') coincide and there exist optimal solutions $\hat{x} \in P$, $\hat{y} \in P'$ so that

$$(b, \hat{y}) = \inf\{(b, y) : y \in P'\} = \sup\{(c, x) : x \in P\} = (c, \hat{x}).$$

We shall not prove this theorem which is widely presented in the literature, but will stress the precise connection with games making it clear that one can use game theory (e.g. minimax theorem) to analyze linear programming problems and vice versa. The following simple result shows that finding a solution to a game amounts to solving certain pair of dual linear programs.

Theorem 9.11. *Let an $m \times n$ matrix A denote a matrix of a zero-sum game of two players with a value $v(A) > 0$. Then $p = (p_1, ..., p_n)$ is an optimal (minimax) strategy of the first player if and only if the vector $\hat{y} = p/v(A)$ is an optimal solution to the linear program*

$$\inf\{y_1 + ... + y_n : y_i \geq 0, (A^T y)_j \geq 1 \text{ for all } i, j\}.$$

And $q = (q_1, ..., q_m)$ is an optimal (minimax) strategy of the second player if and only if the vector $\hat{x} = q/v(A)$ is an optimal solution to the dual linear program

$$\sup\{x_1 + ... + x_n : x_j \geq 0, (Ax)_i \leq 1 \text{ for all } i, j\}.$$

Proof. According to (9.23) optimality of p and q in the game specified by the matrix A rewrites as

$$(A\hat{x})_i \leq 1 \leq (A^T\hat{y})_j,$$

which implies the statement of the theorem.

Now return to the general problems (P) and (P') and show how it can be given a game theoretic formulation. To this end one introduces the following *Lagrange function* associated with (P):

$$L(x,y) = (c,x) + (y,b) - (y,Ax).$$

Theorem 9.12. *If $\hat{x} \geq 0$, $\hat{y} \geq 0$ and $(\hat{x}, \hat{y})$ is a saddle point for L, i.e.*

$$L(x,\hat{y}) \leq L(\hat{x},\hat{y}) \leq L(\hat{x},y), \quad x \geq 0, y \geq 0,$$

then $\hat{x}$ is a solution to (P) and $\hat{y}$ is a solution to (P').

Proof. We shall prove only the statement concerning $\hat{x}$. The right inequality of the saddle point condition rewrites as

$$(\hat{y}, b - A\hat{x}) \leq (y, b - A\hat{x})$$

for all $y \geq 0$. On one hand, this implies $b - A\hat{x} \geq 0$ (by choosing y large enough) and on the other hand, it implies $(\hat{y}, b - A\hat{x}) \leq 0$ (by choosing $y = 0$). Hence

$$(\hat{y}, b - A\hat{x}) = 0.$$

Taking this equation into account allows us to rewrite the left inequality in the saddle point condition as

$$(c,x) + (\hat{y}, b - Ax) \leq (c,\hat{x}), \quad x \geq 0.$$

Hence $(c,x) \leq (c,\hat{x})$ for any $x \geq 0$ with $b - Ax \geq 0$. And thus $\hat{x}$ is a solution to (P).

9.7 Backward induction and dynamic programming

Dynamic programming is a mathematical (algorithmic) realization of the idea (and procedure) of backward induction that was illustrated on many examples in Chapter 3. Here we prove a general result underlying the application of dynamic programming in games.

Assume X and Y are arbitrary sets, h is a real function on $X \times Y$ and $\mathcal{H} : X \times Y \mapsto \Gamma$ is a mapping with values in the set Γ of zero-sum games of two players. More precisely, for any x, y, $\mathcal{H}(x, y)$ is a game $\Gamma_{A,B,H}(x,y)$

with $A = A(x, y)$, $B = B(x, y)$ being the sets of strategies of the first and second player respectively and $H = H(x, y)$ being the payoff function to the second player, so that $H(x, y; a, b)$ denotes his payoff in the profile (a, b), $a \in A, b \in B$ (the first player gets $-H(x, y; a, b)$). Consider now the following two-step game $G_{h,\mathcal{H}}$. First the players I and II choose (independently) their strategies $x \in X, y \in Y$. Then the second player obtains $h(x, y)$ and afterwards the game $\mathcal{H}(x, y)$ is played. The strategies of the first (resp. second) player are clearly described by the pairs (x, α) (resp. (y, β)), where $x \in X, y \in Y$ and α, β are functions from $X \times Y$ to A and B respectively.

Theorem 9.13. *(Dynamic programming for zero-sum games)* *Suppose for any x, y the game $\mathcal{H}(x, y)$ has a value $H(x, y, \alpha_0(x, y), \beta_0(x, y))$ given by certain minimax strategies $\alpha_0(x, y), \beta_0(x, y)$, i.e. $H(x, y, \alpha_0(x, y), \beta_0(x, y))$ is a saddle-point of $H(x, y, a, b)$, i.e.*

$$H(x, y, \alpha_0(x, y), b) \leq H(x, y, \alpha_0(x, y), \beta_0(x, y)) \leq H(x, y, a, \beta_0(x, y))$$
$$(9.56)$$

for $a \in A(x, y), b \in B(x, y)$. Suppose also that (x_0, y_0) is a saddle-point of the function $h(x, y) + H(x, y, \alpha_0(x, y), \beta_0(x, y))$:

$$h(x_0, y) + H(x_0, y, \alpha_0(x_0, y), \beta_0(x_0, y))$$

$$\leq h(x_0, y_0) + H(x_0, y_0, \alpha_0(x_0, y_0), \beta_0(x_0, y_0))$$

$$\leq h(x, y_0) + H(x, y_0, \alpha_0(x, y_0), \beta_0(x, y_0)), \quad x \in X, y \in Y. \quad (9.57)$$

Then the game $G_{h,\mathcal{H}}$ has a value that equals

$$h(x_0, y_0) + H(x_0, y_0, \alpha_0(x_0, y_0), \beta_0(x_0, y_0))$$

with minimax strategies being (x_0, α_0) and (y_0, β_0) respectively.

Proof. It is straightforward. Namely, one has to show that

$$h(x_0, y) + H(x_0, y, \alpha_0(x_0, y), \beta(x_0, y))$$

$$\leq h(x_0, y_0) + H(x_0, y_0, \alpha_0(x_0, y_0), \beta_0(x_0, y_0))$$

$$\leq h(x, y_0) + H(x, y_0, \alpha(x, y_0), \beta_0(x, y_0))$$

for all strategies (x, α) and (y, β), and this follows directly from (9.56), (9.57).

The method of dynamic programming consists in using this result for solving k steps games by solving recursively the auxiliary one step games.

In particular, it implies the existence of (minimax) solutions in multi-step finite games with alternating moves like chess.

The following characteristic example is classical.

Example. Inspection game.

Player I (offender) wants to carry out an unlawful action (a crime). There are N time periods, in which such an action could be carried out. Player II (inspector), who is interested in preventing the crime, is able to make only one inspection (in any of these periods). The gain equals 1, if the crime is committed and undiscovered, and equals -1, if the offender is found and prosecuted (this happens if he chooses to commit a crime at the same time as the inspector chooses to carry out the inspection. The gain equals zero, if the offender does not act at all.

In the first period (first step of the game) each player has two alternatives. Player I can act (commit a crime) or not. Player II can carry out an inspection or not doing so. If player I acts and player II carries out the inspection, the game is over and the gain (of the first player) equals -1. If player I acts and player II does not inspect, the game is over as well and the gain equals 1. If player I does not act and player II carries out the inspection, then player I can without fear commit the crime in the next period (whenever $N > 1$ of course) and the gain again equals 1 (it is assumed that after each step player I becomes aware of any inspection that has been carried out). If player I does not act and player II does not inspect, then the game moves to the second step that differs from the previous one only by the remaining number of periods. Consequently the matrix of the (zero-sum game of the) first step can be symbolically expressed as follows:

$$\begin{pmatrix} -1 & 1 \\ 1 & \Gamma_{N-1} \end{pmatrix}. \tag{9.58}$$

Here Γ_{N-1} denotes the necessity to play this game again. If the values of the games Γ_i are equal respectively v_i, the perspective to play these games is equivalent (in the sense of the expectations) to their values (Theorem 9.13). Hence the matrix (9.58) can be written as

$$\begin{pmatrix} -1 & 1 \\ 1 & v_{N-1} \end{pmatrix}. \tag{9.59}$$

This yields the following recursive equation (by $Val A$ we denote the value of a the game with the matrix A):

$$v_N = Val \begin{pmatrix} -1 & 1 \\ 1 & v_{N-1} \end{pmatrix}.$$

Assuming $v_{N-1} \leq 1$ one deduces (see e.g. Revision Exercise 10) the following difference equation:

$$v_N = \frac{v_{N-1} + 1}{3 - v_{N-1}}.$$

Solving this equation combined with the initial condition $v_1 = 0$ yields

$$v_N = \frac{N - 1}{N + 1}.$$

Thus one gets the value of the game on each step. Then one can calculate also the corresponding equilibrium strategies. Namely, as the matrix takes the form

$$\begin{pmatrix} -1 & 1 \\ 1 & (N\text{-}2)/N \end{pmatrix}$$

the equilibrium strategies for $N \geq 2$ are

$$\left\{ \begin{array}{l} x^N = (\frac{1}{N+1}, \frac{N}{N+1}) \\ y^N = (\frac{1}{N+1}, \frac{N}{N+1}) \end{array} \right\}.$$

Theorem 9.13 has a natural extension to n-person games that we shall present now.

Assume X_i, $i = 1, ..., n$, are arbitrary sets, $h = (h_1, ..., h_n)$ is a collection of real functions on $X = X_1 \times ... \times X_n$ and $\mathcal{H} : X \mapsto \Gamma$ is a mapping with values in the set Γ_n of n person games. More precisely, for any $x = (x^1, ..., x^n)$, $\mathcal{H}(x)$ is a game $\Gamma_{A,H}(x)$, where $A = A_1 \times ... \times A_n$, $A_i = A_i(x)$ are the sets of the strategies of the player i, and $H = (H_1, ..., H_n)(x)$ is the collection of payoff functions, so that $H(x; a_1, ..., a_n)$ denotes the payoff to i in the profile $(a_1, ..., a_n)$. Consider now the following two step game $G_{h,\mathcal{H}}$. First the players choose (independently) their strategies $x_i \in X_i$ forming the first step profile $x = (x^1, ..., x^n)$. Then each player obtains $h_i(x)$ and afterwards the game $\mathcal{H}(x)$ is played. The strategies of the player i are clearly described by the pairs (x^i, α^i), where $x^i \in X_i$ and α^i is a functions from X to A_i.

For a profile $\alpha = (\alpha^1, ..., \alpha^n)$ in a game $\Gamma_{A,H}(x)$ we shall denote by $\hat{\alpha}^i$ the collection of the $(n - 1)$ strategies of the players $j \neq i$ in the profile α. Similarly for $x = (x^1, ..., x^n)$ we denote by $\hat{x}^i$ the collection of x^j with $j \neq i$.

Theorem 9.14. *(Dynamic programming for n-person games)* *Suppose for any x the game $\mathcal{H}(x)$ has a Nash equilibrium given by the profile $\alpha_0(x) = (\alpha_0^1(x), ..., \alpha_0^n(x))$, i.e.*

$$H_i(x; \alpha(x)) \geq H_i(x; a_i, \hat{\alpha}_0^i(x)) \tag{9.60}$$

for any i, $a_i \in A_i(x)$. Suppose also that $x_0 = (x_0^1, ..., x_0^n)$ is an equilibrium profile for the n person games with the strategy spaces X_i and the payoffs of i-th player given by $h_i(x) + H_i(x; \alpha_0(x))$, i.e.

$$h_i(x_0) + H_i(x_0; \alpha_0(x_0)) \geq h_i(x_i, \hat{x}_0^i) + H_i(x_i, \hat{x}_0^i; \alpha_0(x_i, \hat{x}_0^i)) \qquad (9.61)$$

for all i and $x^i \in X_i$. Then the profile (x_0, α_0) is a Nash equilibrium in $G_{h,\mathcal{H}}$.

Proof. As above the required inequality

$$h_i(x_0) + H_i(x_0; \alpha_0(x_0)) \geq h_i(x_i, \hat{x}_0^i) + H_i(x_i, \hat{x}_0^i; \alpha^i(x_i, \hat{x}_0^i), \hat{\alpha}_0^i(x_i, \hat{x}_0^i)) \qquad (9.62)$$

follows from (9.60) and (9.61).

The rest of this section is devoted to a slightly more advanced material on multi-step games with the players alternating their moves (like in chess) in a setting which is general enough to be applied both to finite and infinite games.

Let X be a locally compact metric space with metric (or distance) ρ. To compare the compact subsets, say A, B, of X it is natural to use their *Hausdorf distance (or metric)* defined as

$$D(A, B) = \max\left(\sup\{\rho(x, B) : x \in A\}, \sup\{\rho(x, A) : x \in B\}\right), \qquad (9.63)$$

where

$$\rho(x, A) = \inf\{\rho(x, y) : y \in Y\} = \min\{\rho(x, y) : y \in Y\}$$

for a point $x \in X$ and a compact set $Y \subset X$. From this definition we shall need mainly the following simple observation: for any $a \in A$ (resp. $b \in B$) there exists $b \in B$ (resp. $a \in A$) such that $\rho(b, a) \leq D(A, B)$.

Now let h and f be bounded continuous functions on $X \times X$ and X respectively. Moreover there are given two mappings $x \mapsto K_i(x)$, $i = 1, 2$, which for each x define compact subsets $K_i(x)$ of X. These (multi-valued) mappings are supposed to be continuous in the sense that for any $x \in X$ and any $\epsilon > 0$ there exists a $\delta > 0$ such that if the distance between y and x does not exceed δ, then $D(K_i(x), K_i(y)) < \epsilon$.

Let the (multi-step) zero-sum games $\Gamma_i(x, k)$, $i = 1, 2$ of two players be defined in the following way. In $\Gamma_1(x, k)$ the first player starts by choosing a $x_1 \in K_1(x)$. On the next step the second player chooses $x_2 \in K_2(x_1)$. Then the first player makes the next move by choosing $x_3 \in K_1(x_2)$, etc. After k steps the game ends with the second player getting the payoff

$$h(x, x_1) + h(x_1, x_2) + ... + h(x_{k-1}, x_k) + f(x_k).$$

The game $\Gamma_2(x, k)$ is defined analogously, but the second player starts.

Theorem 9.15. *(Dynamic programming for games with perfect information)* *(i) Games* $\Gamma_1(x,k)$, $\Gamma_2(x,k)$ *have values for all* $x \in X$, *$k = 1,2,...,$ that we denote* $v_1(x,k)$ *and* $v_2(x,k)$ *respectively, which are continuous functions of x satisfying the recursive equations*

$$v_1(x,k) = \min_{y \in K_1(x)}(h(x,y) + v_2(y,k-1)),$$
$$v_2(x,k) = \max_{y \in K_2(x)}(h(x,y) + v_1(y,k-1)). \tag{9.64}$$

In particular,

$$v_1(x,k) = (Bv_1)(x,k-2), \tag{9.65}$$

where the operator B (sometimes called the Bellman operator of the game, in honor of the contribution [19]) is defined as

$$Bf(x) = \min_{y \in K_1(x)} \max_{z \in K_2(y)} (h(x,y) + h(y,z) + f(z)), \tag{9.66}$$

and hence by induction

$$v_1(x,2m) = (B^m f)(x), \tag{9.67}$$

where B^m is of course the m th power of B. (ii) Assume additionally the Lipschitz continuity of the mappings K_i, i.e. that

$$D(K_i(x), K_i(y)) \le L\rho(x,y) \tag{9.68}$$

with some constant L for all x,y and $i = 1,2$. Then

$$\sup_{\rho(x,y) \le \delta} |Bf(x) - Bf(y)| \le \sup_{\rho(x,y) \le \delta L^2} |f(x) - f(y)|$$

$$+ \sup_{\rho(x,y) \le \delta L(1+L)} \sup_{\rho(\tilde{x},\tilde{y}) \le \delta(1+L)} |h(\tilde{x},x) - h(\tilde{y},y)|, \tag{9.69}$$

so that the Bellman operator B preserves the Lipschitz continuity. Moreover, for arbitrary m

$$\sup_{\rho(x,y) \le \delta} |B^m f(x) - B^m f(y)| \le \sup_{\rho(x,y) \le \delta L^{2m}} |f(x) - f(y)|$$

$$+ \sum_{j=1}^{m} \sup_{\rho(x_j,y_j) \le \delta L(1+L)^j} \sup_{\rho(\tilde{x}_j,\tilde{y}_j) \le \delta(1+L)^j} |h(\tilde{x}_j,x_j) - h(\tilde{y}_j,y_j)|. \tag{9.70}$$

Proof. (i) is a straightforward application of Theorem 9.13 combined with the standard induction in k. The continuity assumptions of the theorem ensure that on each step the max or min is well defined and the obtained value function is continuous. To get (ii) one observes that from the first equation of (9.64) and (9.68) it follows that

$$\sup_{\rho(x,y)\le\delta} |v_1(x,k) - v_1(y,k)| \le \sup_{\rho(x,y)\le\delta L} |v_2(x,k-1) - v_2(y,k-1)|$$

$$+ \sup_{\rho(x,y)\le\delta L} \sup_{\rho(\tilde{x},\tilde{y})\le\delta} |h(\tilde{x},x) - h(\tilde{y},y)|.$$

Using similarly the second equation (9.64) and the definition of B yields (9.69). The last inequalities follows by induction.

In the last chapter, when introducing differential games, we shall need also the following corollary of this result concerning a limiting game when the number of steps go to infinity.

Theorem 9.16. *Consider a sequence of the games $\Gamma_i^n(x,k(n))$, $n = 1, 2, ...,$ from the previous theorem, when K_i, k depend on n in such a way that $k(n) \to \infty$ as $n \to \infty$ and for both $i = 1, 2$*

$$\sup_{y\in K_i^n(x)} \rho(x,y) \to 0$$

as $n \to \infty$ uniformly in $x \in X$. Assume now that

$$|h(x,y)| \le C\rho(x,y)$$

with some constant C and that $v_1^n(x,k(n))$ and $v_2^n(x,k(n))$ converge uniformly on compact sets to some (necessarily continuous) limits $v_1(x), v_2(x)$, as $n \to \infty$. Then $v_1(x) = v_2(x)$ for all x.

Proof. One has for any x

$$v_1^n(x,k(n)) = h(x,y(n)) + v_2^n(y(n),k(n)-1) \tag{9.71}$$

with $y(n) \in K_1^n(x)$ (in particular $y(n)$ stays in a compact for all n). Taking into account that $h(x,y(n)) \to 0$ as $n \to \infty$ and the uniform convergence of $v_2^n(y,k(n))$ yields $v_1(x) = v_2(x)$ by passing to the limit as $n \to \infty$.

Thus, as one could expect, in the limit of infinitely many steps the result does not depend on who starts the game. This fact is crucial for the analysis of differential games touched upon in the last chapter.

9.8   Cooperative games: Nucleus and the Shapley vector

To complete our introduction to the mathematical methods of game theory we shall present here briefly two popular approaches to the solutions of cooperative games, namely the notions of a nucleus and of the Shapley vector. First of these notions (together with the concept of co-operative games) was discussed in Chapter 6 for the simplest case of three players. There one can find also some motivating examples. Here we shall give general definitions.

A finite game in coalition form is a pair (N, v), *where* N *is a finite set, called the set of players, and* v *is a real function, defined on the set of all subsets of* N, *called coalitions, such that* $v(\emptyset) = 0$. *For a subset* $S \subset N$ *the number* $v(S)$ *is called the value of the coalition* S *and is interpreted as the total gain, which this coalition can achieve for its members without any agreements with other players.*

By a distribution (more precisely an effective distribution) of gains one means an arbitrary collection of numbers $y_1, ..., y_N$ such that $y_1 + ... + y_N = v(N)$. One says that such a distribution belongs to the *core of the game*, if the *stand-alone principle* is satisfied, i.e. when for any coalition $S \in N$

$$\sum_{j \in S} y_j \geq v(S). \tag{9.72}$$

A serious shortcoming of this notion is connected with the fact that the nucleus can be very large, as well as empty. In many cases another notion of solution turns out to be reasonable, namely the *Shapley vector* that we shall introduce now and that is always uniquely defined.

If S is an arbitrary set, let us call a S-vector any real function on S and denote by E^S the set of all these vectors. This set is obviously a Euclidean space of dimension $|S|$ (the number of elements in S). If N denotes the set of players, then the set of games with this set of players can be clearly identified with the Euclidean space of dimension $2^{|N|} - 1$, which we denote by G^N. By a permutation of N one means an arbitrary one-to-one mapping of N in itself. If θ is a permutation, $v \in G^N$, $x \in E^N$, we define $\theta_* v$ in G^N by the equation $(\theta_* v)(S) = v(\theta S)$, and θ_* in E^N by the equation $(\theta_* x)(i) = x(\theta i)$. For a $v \in G^N$ a *zero player* for the game v is any $i \in N$ such that $v(S \cup \{i\}) = v(S)$ for al $S \subset N$.

The following definition is fundamental. *The value (or the Shapley vector) on* G^N *is a function* $\phi : G^N \mapsto E^N$ *such that* (i) ϕ *is linear;* (ii) $\phi(\theta_* v) = \theta_*(\phi v)$ *for all* $v \in G^N$ *and* θ, *i.e.* ϕ *is invariant under permutations;* (iii) $(\phi v)(i) = 0$ *for any zero player* i; (iv) $\sum_{j=1}^{N} (\phi v)_j = v(N)$, i.e. ϕ yields always effective distributions. And the following result is quite remarkable.

Theorem 9.17. *On G^N there exists a unique value, and it is given by the formula*

$$(\phi v)(i) = \sum_{S \subset N \setminus \{i\}} \gamma_S [v(S \cup \{i\}) - v(S)], \qquad (9.73)$$

where

$$\gamma_S = |S|!(n - |S| - 1)!/n!.$$

We omit the proof of this theorem that is not difficult (can be considered as an exercise) and is presented in all books on cooperative games, see e.g. [183]. The analysis of Shapley value and its connections with other notions of solutions is quite popular in modern game theory, see e.g. [182].

9.9 Revision exercises

Unlike the exercises scattered through the text that are aimed at the further development of the theory, the exercises below are meant just for the revision of the knowledge acquired.

1. *A Duopoly model.* Two firms, labeled 1 and 2, produce a product in quantities Q_1 and Q_2 respectively. The market price depends only on the total supply $Q = Q_1 + Q_2$:

$$P(Q) = \begin{cases} 5 - Q & \text{if} \quad Q < 5 \\ 0 & \text{if} \quad Q \geq 5. \end{cases}$$

The production costs are quadratic: $C(Q_i) = cQ_i^2$, where c is a positive constant. Each firm is aiming to maximise its profit.

(i) Suppose the firms choose the quantities Q_1, Q_2 independently (Cournot type model). Find the Nash equilibrium for this game.

(ii) Suppose the two firms collude by agreeing to produce the amount $Q = Q_1 = Q_2$ and that they have some way of enforcing this agreement (a Cartel model). What is the optimal choice Q^* for Q?

(iii) Choosing $c = 1$ for simplicity, confirm that under the Cartel agreement of (ii) the firms get higher profits than playing the Nash strategies obtained in (i).

2. *Cournot model for arbitrary number of firms.* N firms produce the quantities $Q_1, Q_2, ..., Q_m$ of some product on the same market. The market price is $P(Q) = P_0(1 - Q/Q_0)$, where $Q = Q_1 + \dots + Q_m$, and the profit of firm i is given by

$$\Pi_i(Q_1, \dots, Q_m) = (P(Q) - c)Q_i,$$

$i = 1, \dots, m$, where a positive constant $c < P_0$ denotes the marginal costs of production. Find symmetric Nash equilibria $Q_1^* = \dots = Q_m^* = Q^*$.

3. Consider a game given by the table

$$C$$

		1	2
R	1	a, b	c, d
	2	a, f	c, h

with arbitrary a, b, c, d, f, h. Let $p^\star = (h - f)/(h - f + b - d)$. Show that

(i) if $h > f$ and $b > d$, the Nash equilibria are $((p, 1 - p), (1, 0))$ for $p > p^\star$, $((p, 1 - p), (0, 1))$ for $p < p^\star$ and $((p^\star, 1 - p^\star), (q, 1 - q))$ for all q;

(ii) if $h < f$ and $b < d$, the Nash equilibria are $((p, 1 - p), (0, 1))$ for $p > p^\star$, $((p, 1 - p), (1, 0))$ for $p < p^\star$ and $((p^\star, 1 - p^\star)), (q, 1 - q)$ for all q;

(iii) if $h > f$, $b < d$ and $h - f > (d - b)/2$, the Nash equilibria are $((p, 1 - p), (1, 0))$ for all p.

4. Show that any pair of strategies constitutes a Nash equilibrium for a game given by the table

Table 9.6

$$C$$

		1	2
R	1	a, b	c, b
	2	a, f	c, f

with arbitrary numbers a, b, c, f.

5. Find all symmetric Nash equilibria and ESS for a Hawk-Dove game of the form

Table 9.7

	hawk	dove
hawk	0	V
dove	0	D

for arbitrary numbers D and V.

6. Find all symmetric Nash equilibria and ESS for a pure coordination game with the table

$$C$$

	1	2	3
1	1,1	-1,-2	-1,-3
2	-2,-1	2,2	-1,-2
3	-3,-1	-3,-2	3,3

with R labeling the rows.

and write down the corresponding RD equations. Find the fixed points and check which of them are stable.

7. Consider a Hawk-Dove (or Lion-Lamb) type game G with payoff table

Table 9.8

	A	B
A	-1,-1	5,0
B	0,5	3,3

(i) Show that there is only one symmetric Nash equilibrium $\sigma^\star = (p^\star, 1 - p^\star)$, find this equilibrium and prove that it defines ESS.

(ii) Write down the standard Replicator Dynamics (RD) equation for the proportion x of the players of the game G using action A with probability 1, identify the fixed points of this RD equation and show which of them are asymptotically stable and which are not.

(iii) Consider a game in which the game G is repeated an infinite number of times and payoffs are discounted by a factor δ ($0 < \delta < 1$). Assume the players are limited to selecting strategies from the following 3 options:

(a) σ_A: play A in every stage game,

(b) σ_B: play B in every stage game,

(c) the "trigger strategy" σ_T: begin by playing B and continue to play B until your opponent plays A; once your opponent has played A, play A forever afterwards.

Find out whether $[\sigma_A, \sigma_A]$ and/or $[\sigma_B, \sigma_B]$ represent a Nash equilibrium, find a condition on δ such that $[\sigma_T, \sigma_T]$ is a Nash equilibrium, and find out whether $[\sigma_T, \sigma_T]$ is ESS.

(iv) Consider a modification of the game G, where the first player chooses his/her strategy and then advises his/her opponent about this choice, so that the second player makes his/her move depending on the choice of the first player. Draw the game tree (or extensive form) for this game and find a solution using the method of backward induction. Give the strategic form of this game and find all the pure Nash equilibria. Identify the subgames of this game and show that the solution found by backward induction is subgame perfect.

8. *Sherlock Holmes versus professor Moriarty.* This is an example from Conan Doyle's story "The final Problem" that was subjected to game theoretical analysis by von Neumann and Morgenstern. Holmes is going from London to Dover and then to the Continent in order to escape Moriarty who is going to kill Holmes as an act of revenge. When his train pulls out, Holmes observes Moriarty on the platform and realized that Moriarty might secure a special train to overtake him (reaching Dover earlier). So Holmes can either proceed to Dover or leave train at Canterbury, the only intermediate station. Moriarty is intelligent enough to visualize this possibility and consequently is faced with the same choice. Both leave the trains independently, and if they turn out to be on the same platform, Holmes would be almost certainly killed by his adversary. Assuming that Holmes's chances of survival are 100 per cent, if he escapes via Dover and 50 cent if he escapes via Canterbury (as in the latter case pursuit continues), the game can be represented by the following table

Table 9.9

		Moriarty	
		Canterbury	Dover
Holmes	Canterbury	0	50
	Dover	100	0

where Holmes's payoffs are shown, Moriarty's payoffs being the negatives of these.

Find the minimax strategies of Holmes and Moriarty and hence the value of the game.

9. *Game Morra.* This is a two-player symmetric zero-sum Italian game that is played by the following rules. Two players simultaneously extend one, two or three fingers and at the same time call out a number between one and three. The number called out is a guess of the number of fingers shown by the opponent. If only one player guesses correctly, then the loser pays the winner the amount corresponding to the total number of fingers displayed, otherwise no payoff is due. The matrix of the game is

Table 9.10
C

	1-1	1-2	1-3	2-1	2-2	2-3	3-1	3-2	3-3
1-1	0	2	2	-3	0	0	-4	0	0
1-2	-2	0	0	0	3	3	-4	0	0
1-3	-2	0	0	-3	0	0	0	4	4
2-1	3	0	3	0	-4	0	0	-5	0
2-2	0	-3	0	4	0	4	0	-5	0
2-3	0	-3	0	0	-4	0	5	0	5
3-1	4	4	0	0	0	-5	0	0	-6
3-2	0	0	-4	5	5	0	0	0	-6
3-3	0	0	-4	0	0	-5	6	6	0

R is the row label at the left of the table.

Find the solutions (minimax strategies) of this game.

10. Suppose that a two-player two-action zero sum game is given by a matrix $A = (a_{ij})$, $i, j = 1, 2$. (i) Show that if there do not exist pure strategy Nash equilibria (or saddle points), then either

$$1) \quad \max(a_{11}, a_{22}) \leq \min(a_{12}, a_{21}), \quad a_{11} + a_{22} < a_{12} + a_{21},$$

or

$$2) \quad \max(a_{12}, a_{21}) \leq \min(a_{11}, a_{22}), \quad a_{11} + a_{22} > a_{12} + a_{21}.$$

(ii) Show that in both cases the value of the game is $\det(A)/t(A)$, where $\det(A)$ is the determinant of A and $t(A) = a_{11} + a_{22} - a_{12} - a_{21}$, and

$$\sigma = \left(\frac{a_{22} - a_{21}}{t}, \frac{a_{11} - a_{12}}{t} \right), \quad \eta = \left(\frac{a_{22} - a_{12}}{t}, \frac{a_{11} - a_{21}}{t} \right)$$

is a pair of minimax strategies.

9.10 Solutions to revision exercises

1. (i) The payoff for the firm i is $\Pi_i(Q_1, Q_2) = Q_i P(Q) - cQ_i^2$.

The best response $Q_1 = Q_1(Q_2)$ of the first firm to a strategy Q_2 of the second firm is found from the equation

$$\frac{\partial \Pi_1}{\partial Q_1} = 5 - Q_1 - Q_2 - Q_1 - 2cQ_1 = 0,$$

and is given by

$$Q_1(Q_2) = \frac{5 - Q_2}{2(1 + c)}.$$

To check that this is really a best response one needs to check that

$$\frac{\partial^2 \Pi_1}{\partial Q_1^2} = -2 - 2c < 0.$$

By symmetry, the best response $Q_2 = Q_2(Q_1)$ of the second firm to a strategy Q_1 of the first firm is found as follows:

$$\frac{\partial \Pi_2}{\partial Q_2} = 5 - Q_2 - Q_1 - Q_2 - 2cQ_2 = 0, \quad Q_2 = \frac{5 - Q_1}{2(1 + c)}.$$

A Nash equilibrium can be found by solving the simultaneous equations

$$Q_1 = \frac{5 - Q_2}{2(1 + c)}, \quad Q_2 = \frac{5 - Q_1}{2(1 + c)}.$$

The solution is $Q_1 = Q_2 = Q^\star = 5/(3 + 2c)$. Of course, one needs to check that $2Q^\star < 5$, but this is obvious.

(ii) In a cartel each firm maximises $\Pi(Q, Q) = QP(2Q) - cQ^2$. The optimal choice Q^{car} is found by

$$\frac{\partial \Pi}{\partial Q} = 5 - 4Q - 2cQ = 0, \quad Q^{car} = \frac{5}{2c + 4}.$$

This is actually the best response, because

$$\frac{\partial^2 \Pi}{\partial Q^2} = -4 - 2c < 0,$$

and $2Q^{car} < 5$.

(iii) If $c = 1$, then $\Pi(Q^{car}, Q^{car}) = 65/18 > \Pi(Q^\star, Q^\star) = 2$.

2. By symmetry, a situation $(Q^\star, ..., Q^\star)$ is a symmetric Nash equilibrium, if and only if $Q_1 = Q^\star$ is the best response of the first firm to the situation when all other firms choose $Q^\star$. This best response $\hat{Q}_1$ is found from the condition that the derivative of

$$\Pi_1(Q_1, Q, ..., Q) = Q_1 P_0 \left(1 - \frac{Q_1 + (m - 1)Q}{Q_0} \right) - cQ_1$$

with respect to Q_1 vanishes at $\hat{Q}_1$. This leads to the equation

$$Q^\star = \frac{Q_0}{2} \left(1 - (m - 1)\frac{Q^\star}{Q_0} - \frac{c}{P_0} \right)$$

with a unique solution

$$Q^\star = \frac{Q_0}{m + 1} \left(1 - \frac{c}{P_0} \right).$$

3. For $\sigma_R = (p, 1 - p)$, $\sigma_C = (q, 1 - q)$,

$$\Pi_R(\sigma_R, \sigma_C) = aq + c(1 - q)$$

and hence $\Pi_R(\sigma_R, \sigma_C)$ does not at all depend on σ_R. Consequently, any strategy σ_R of R yields the best response to any strategy σ_C of C in this game. So, the Nash equilibria are arbitrary pairs (σ_R, σ_C) such that σ_C is the best response to σ_R. Now

$$\Pi_C(\sigma_R, \sigma_C) = bpq + f(1-p)q + dp(1-q) + h(1-p)(1-q)$$

$$= q[p(h - f + b - d) - (h - f)] + dp + h(1 - p).$$

In cases (i) and (ii) it can be written as

$$q(h - f + b - d)(p - p^\star) + dp + h(1 - p).$$

In case (i) this function is increasing (respectively decreasing) in q for $p > p^\star$ (respectively $p < p^\star$) and does not depend on q for $p = p^\star$. Hence the best response is given by $q = 1$ (respectively $q = 0$) for $p > p^\star$ (respectively $p < p^\star$), and by arbitrary q for $p = p^\star$ leading to the required Nash equilibria. Similarly case (ii) is considered. In case (iii) one sees by inspection that $\Pi_C(\sigma_R, \sigma_C)$ is an increasing function of q for all p from $[0, 1]$ and hence the best response is always given by $q = 1$.

4. $\Pi_R(\sigma_R, \sigma_C)$ does not depend on σ_R (like in the previous exercise), and $\Pi_C(\sigma_R, \sigma_C)$ does not depend on σ_C. Hence (9.10) and (9.13) hold for all strategies σ_R, σ_C, $\sigma_R^\star$, $\sigma_C^\star$.

5. (i) If $V = D$, then all symmetric profiles are Nash equilibria, and there are no ESS, because

$$\Pi(\sigma, \eta) = \Pi(\eta, \eta)$$

for all pairs of strategies σ and η.

(ii) If $V \neq D$, then there are no strictly mixed Nash equilibria according to Theorem 4.1. Consequently in this case, the only candidates for symmetric Nash equilibria are (h, h) and (d, d), from which (h, h) is always a Nash equilibrium, and (d, d) is a Nash equilibrium only for $D > V$.

(iii) If $V > D$, the unique symmetric Nash equilibrium (h, h) defines ESS, since $\Pi(h, h) - \Pi(\sigma, h) = 0$ and $\Pi(h, \sigma) - \Pi(\sigma, \sigma) = (1 - p)^2(V - D)$ for $\sigma = (p, 1 - p)$.

(iv) If $V < D$, then (h, h) does not define ESS (follows from the same formula as in (iii)), but (d, d) does define it, since $\Pi(d, d) - \Pi(\sigma, d) = p(D - V) > 0$ for $p \neq 0$.

6. (i) This game is specified by the square matrix

$$A = \begin{pmatrix} 1 & -1 & -1 \\ -2 & 2 & -2 \\ -3 & -3 & 3 \end{pmatrix}.$$

The determinant of A is -24, and hence A is non-degenerate. Standard calculations yield

$$A^{-1} = -\frac{1}{12}\begin{pmatrix} 0 & 3 & 2 \\ 6 & 0 & 2 \\ 6 & 3 & 0 \end{pmatrix}, \quad A^{-1}\begin{pmatrix} 1 \\ 1 \\ 1 \end{pmatrix} = -\frac{1}{12}\begin{pmatrix} 5 \\ 8 \\ 9 \end{pmatrix}.$$

Consequently, by Theorem 9.5 $(\sigma^\star, \sigma^\star)$ with

$$\sigma^\star = \left(\frac{5}{22}, \frac{8}{22}, \frac{9}{22}\right)$$

is the only symmetric Nash equilibrium that randomizes over all strategies (i.e. with all coordinates being positive).

(ii) Clearly all three symmetric pure strategy profiles are Nash equilibria. It remains to look for symmetric equilibria that randomize over precisely two strategies. The calculations yield three such equilibria (η, η) with η being $(1/2, 1/2, 0)$, $(1/2, 0, 1/2)$, $(0, 1/2, 1/2)$. Let us only indicate how the first one is found. For this end one looks for equilibria with η of the form $\eta = (p, 1 - p, 0)$. Clearly, to define an equilibrium, it should also define an equilibrium for the two-action symmetric game with the matrix

$$\begin{pmatrix} 1 & -1 \\ -2 & 2 \end{pmatrix},$$

which yields (by the theory of two-action game) the only possibility $p = 1/2$. To confirm that $\eta = (1/2, 1/2, 0)$ defines an equilibrium it remains to check that $\Pi(\eta, \eta) \geq \Pi(3, \eta)$, which is straightforward.

(iii) One easily checks that all three pure strategy profiles are ESS. On the other hand, the strategies $(1/2, 1/2, 0)$, $(1/2, 0, 1/2)$, $(0, 1/2, 1/2)$ are not ESS, as they are not ESS in the corresponding two-action games. It remains to analyze the strategy σ. One needs to check whether

$$\Pi(\sigma^\star, \sigma) - \Pi(\sigma, \sigma) > 0$$

for all σ. Writing $\sigma = (p, q, 1 - p - q)$ one gets

$$\Pi(\sigma, \sigma) = 8p^2 + 10q^2 + 12pq - 10p - 11q + 3$$

and

$$\Pi(\sigma^\star, \sigma) - \Pi(\sigma, \sigma) = -\frac{6}{11} - 3 - 8p^2 - 10q^2 - 12pq + 10p + 11q.$$

As the derivatives of this function with respect to p and q do not vanish at $\sigma^\star$ (check this!), $\sigma^\star$ is not a minimum point of this function, and hence it is not positive in a neighborhood of $\sigma^\star$. Consequently $\sigma^\star$ is not ESS.

At last, one gets

$$A\nu = \begin{pmatrix} 2x - 1 \\ 4y - 2 \\ 3 - 6x - 6y \end{pmatrix}$$

for a profile $\nu = (x, y, 1 - x - y)$ and using the formula for $\Pi(\nu, \nu)$ given above one can write the RD equations (9.33) as

$$\begin{cases} \dot{x} = (12x + 11y - 8x^2 - 10y^2 - 12xy - 4)x \\ \dot{y} = (10x + 15y - 8x^2 - 10y^2 - 12xy - 5)y \end{cases}.$$

One checks directly that $(1,0)$, $(0,1)$, $(0,1/2)$, $(1/2,0)$ and $(5/22, 8, 22)$ are the only fixed points of this dynamics, as follows of course also from Theorem 9.7. In order to check that, say, the first fixed point is stable (as we expect from the theory, because it is ESS), we can put $y = 0$ and $x = 1 - u$ in the RD system leading to the equation

$$\dot{u} = 4u(u - 1)(1 - 2u) = -4u + O(u^2),$$

which implies stability, as the coefficient at u is negative.

7. (i) Obviously there are no symmetric pure strategy Nash equilibria. To find mixed strategy symmetric Nash equilibria $\sigma^\star = (p^\star, 1 - p^\star)$ with $p^\star \in (0, 1)$ one can use "equality of payoffs" lemma

$$\pi(A, \sigma^\star) = \pi(B, \sigma^\star) \iff 5 - 6p^\star = 3(1 - p^\star)$$

(or otherwise), which yields $p^\star = 2/3$. From above ("equality of payoffs" lemma) $\pi(\sigma, \sigma^\star) = \pi(\sigma^\star, \sigma^\star)$ for all σ. Hence the main ESS condition becomes $\pi(\sigma^\star, \sigma) > \pi(\sigma, \sigma)$ for all $\sigma = (p, 1 - p)$ with $p \neq 2/3$, i.e.

$$-p^\star p + 5p^\star(1 - p) + 3(1 - p^\star)(1 - p) > -p^2 + 5p(1 - p) + 3(1 - p)^2,$$

which by tiding up leads to an equivalent inequality $(2 - 3p)^2 > 0$, which obviously holds for $p \neq 2/3$.

(ii) General RD equation for a 2-action 2-player game is

$$\dot{x} = x(1 - x)[\pi(A, \nu) - \pi(B, \nu)], \quad \nu = (x, 1 - x),$$

which in our case yields

$$\dot{x} = x(1 - x)(2 - 3x).$$

Fixed points are $x_1^\star = 0$, $x_2^\star = 1$, $x_3^\star = p^\star = 2/3$.

From the properties of ESS it follows that $x_3^\star$ is stable. This can be also checked directly as for other points below.

Consider $x_1^\star$. Equation in terms of ϵ with $x = x_1^\star + \epsilon$ is

$$\dot{\epsilon} = \epsilon(1 - \epsilon)(2 - 3\epsilon) = 2\epsilon + O(\epsilon^2).$$

Hence x_1^* is unstable, as the signs of ϵ and $\dot{\epsilon}$ coincide. Similarly writing the equation in terms of ϵ with $x = x_2^* - \epsilon$ yields that x_2^* is unstable.

(iii) $[\sigma_A, \sigma_A]$ is not Nash, as $\pi(\sigma_B, \sigma_A) = 0 > \pi(\sigma_A, \sigma_A) = -1/(1-\delta)$. $[\sigma_B, \sigma_B]$ is not Nash, as $\pi(\sigma_A, \sigma_B) = 5/(1-\delta) > \pi(\sigma_B, \sigma_B) = 3/(1-\delta)$. Next, as

$$\pi(\sigma_B, \sigma_B) = \pi(\sigma_B, \sigma_T) = \pi(\sigma_T, \sigma_B) = \pi(\sigma_T, \sigma_T) = \frac{3}{1-\delta},$$

the condition for $[\sigma_T, \sigma_T]$ to be Nash is

$$\pi(\sigma_T, \sigma_T) \geq \pi(\sigma_A, \sigma_T) \Longleftrightarrow \frac{3}{1-\delta} \geq 5 - \frac{\delta}{1-\delta} \Longleftrightarrow \delta \geq \frac{1}{3}.$$

At last, as

$$\pi(\sigma_B, \sigma_B) = \pi(\sigma_B, \sigma_T) = \pi(\sigma_T, \sigma_B) = \pi(\sigma_T, \sigma_T),$$

the main ESS conditions do not hold. Hence $[\sigma_T, \sigma_T]$ is not ESS.

(iv) The game tree is

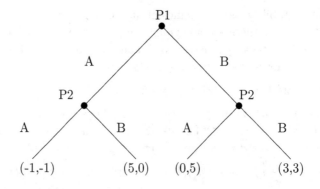

Induction implies that P_1 plays A and then P_2 plays B with the resulting payoff $(5, 0)$. The strategic form is

	AA	AB	BA	BB
A	-1,-1	-1,-1	**5,0**	**5,0**
B	**0,5**	3,3	0,5	3,3

the Nash equilibria being bolded. There are three subgames. Clearly the solution $[A, BA]$ obtained by backward induction is the only subgame perfect equilibrium.

8. The minimax strategies are $(2/3, 1/3)$ for Holmes and $(1/3, 2/3)$ for Moriarty, and the value of the game is $100/3$ meaning that the Holmes's chances of survival are 33.33 percent.

Chapter 10

Examples of game models and methods

In the wide sense, by mathematical modeling one can understand any application of mathematics to the description of real life processes. At the end of the day this leads to control and optimization with often several criteria of performance. In fact, one needs, say, a plane to fly in the desired direction (control), to do it as quick as possible but to remain safe and to use the least amount of fuel possible (optimization with several criteria). In economics one can be interested in the investments with maximum income and minimum risk (again control and optimization). The examples are around everywhere, leading to a large scale interference (permanently increasing) of mathematical methods in basically all domains of science. In this chapter we start with a selection of concrete models, where deep insights can be obtained by the mathematical methods of game theory. The second half of the chapter (starting from Section 10.5) is an introduction to the game-theoretical treatment of option pricing, as developed in [254] and [255] based on the initial ideas from [87], [90]. In this approach risk-neutral probabilities, playing the key role in finances, emerge automatically from the robust-control evaluation. This seems to be especially appealing for incomplete market encompassing extensive, so to say untamed, randomness, when the coexistence of an infinite number of risk-neutral measures precludes one from unified pricing of derivative securities. Sometimes, we can specify even a unique risk-neutral selector. Our method is robust enough to be able to accommodate various market rules and settings including path-dependent payoffs, American options and transaction costs. On the other hand, it leads to rather simple numerical algorithms. The continuous-time limit is described by nonlinear and/or fractional Black-Scholes type equations. Full story and extensive bibliography can be found in monograph [214], and some new developments in [297].

10.1 A static model of strategic investment

As a warming up example consider a very simple static model of the competition of agents in strategic investment (advertising, discount cards etc.). In the pre-final section we shall develop a more sophisticated dynamic model. Assume two agents I and II sell the same product on a market by the prices c_1 and c_2 respectively. To increase the chances of selling it, they invest x_1 and x_2 (e.g. dollars) respectively in advertising respectively. Assume the probabilities $p_1(x_1, x_2)$, $p_2(x_1, x_2)$ for players I and II to sell their product under this level of investment re known.

Remark. Equivalently, instead of the probabilities of selling a single product, one could speak about the expected proportion of some stock of this product to be sold.

The expectation of the total income of the players equal

$$H_i = p_i(x_1, x_2)c_i - x_i, \quad i = 1, 2. \tag{10.1}$$

As clearly it makes no sense for players i to invest above c_i (since probabilities are less than one, this would definitely lead to a loss) we came to the game Γ of two players with infinite sets of strategies $[0, c_1]$, $[0, c_2]$ and payoffs (10.1). We are interested in finding Nash equilibria.

Remark. It is of course possible (and is much easier) to find the compromise point.

The *law of diminishing returns* on the investment states (see e.g. [65]) that there should be a positive effect of extra advertising which diminishes as investment increases. Mathematically, it means that $p_i(x_1, x_2)$ should be an increasing concave function of x_i, $i = 1, 2$. In other words, assuming that p_i are twice continuously differentiable this is equivalent to

$$\frac{\partial p_i(x_1, x_2)}{\partial x_i} > 0, \frac{\partial^2 p_i(x_1, x_2)}{\partial^2 x_i} < 0, \quad \forall x_1, x_2. \tag{10.2}$$

Looking for the maximum of H_i by differentiating one gets straightforward that the best reply x_1 of the first player to the strategy x_2 of the second player should satisfy the equation

$$\frac{\partial p_1(x_1, x_2)}{\partial x_1} = \frac{1}{c_1}, \tag{10.3}$$

and the best reply x_2 of the second player to the strategy x_1 of the first player should satisfy the equation

$$\frac{\partial p_2(x_1, x_2)}{\partial x_2} = \frac{1}{c_2}. \tag{10.4}$$

Clearly under condition (10.2) there can be no more than one solution of each of this equation, and the solutions do yield a maximum. So only the existence question is an issue, a possible solution being given in the following statement.

Proposition 10.1. *Assume* (10.2) *holds. (i) If*

$$\frac{\partial p_1(c_1, y)}{\partial x_1} < \frac{1}{c_1} < \frac{\partial p_1(0, y)}{\partial x_1} \quad \forall y \in [0, c_2], \qquad (10.5)$$

then for any $x_2 \in [0, c_2]$ there exists a unique solution $x_1 = X_1(x_2)$ of (10.3) *from $(0, c_1)$ yielding a unique best reply to x_2. (ii) If*

$$\frac{\partial p_2(y, c_2)}{\partial x_2} < \frac{1}{c_2} < \frac{\partial p_2(y, 0)}{\partial x_2} \quad \forall y \in [0, c_1], \qquad (10.6)$$

then for any $x_1 \in [0, c_1]$ there exists a unique solution $x_2 = X_2(x_1)$ of (10.4) *from $(0, c_2)$ yielding the unique best reply to x_1. (iii) If both* (10.5), (10.6) *hold, then there exists a solution $\hat{x}_1 \in (0, c_1), \hat{x}_2 \in (0, c_2)$ of the system* (10.3), (10.4) *yielding (possibly non-unique) Nash equilibrium to the game Γ.*

Proof. Statements (i), (ii) are obvious from the continuity and monotonicity of the first derivatives of p_i. To get (iii) one observes that the continuous curves $X_1(x_2)$ and $X_2(x_1)$ connect respectively the left and right, the lower and upper, sides of the rectangular $[0, c_1] \times [0, c_2]$ and hence ought to intersect.

Of course, in symmetric case, symmetric equilibria are of particular interest.

Proposition 10.2. *(i) Under the assumption of Proposition 10.1 assume that $p_1 = p_2$, $c_1 = c_2$. Then there exists a symmetric Nash equilibrium. (ii) If additionally*

$$\frac{\partial^2 p_1}{\partial x_1 \partial x_2} < \frac{\partial p_1}{\partial x_1^2} \qquad (10.7)$$

on the diagonal, i.e. for all $x_1 = x_2$, then this symmetric Nash equilibrium is unique.

Proof. A mapping $x_2 \mapsto X_1(x_2)$ is a continuous mapping $[0, c_1]$ in itself. Hence it has a fixed point. Next, assuming (10.7) we conclude by continuity that there exists a $\epsilon > 0$ such that (10.7) holds also for all $|x_1 - x_2| < \epsilon$. Differentiating yields

$$\frac{\partial X_1}{\partial x_2} = -\left(\frac{\partial^2 p_1}{\partial x_1 \partial x_2} \bigg/ \frac{\partial p_1}{\partial x_1^2} \right) (X_1(x_2), x_2)$$

implying (under condition (ii)) that $X_1(x_2) - x_2$ is a decreasing function whenever x_2 and $X_1(x_2)$ are near (in the ϵ-neighborhood of) the diagonal. Hence $X_1(x_2)$ can not cross the diagonal twice, and hence the uniqueness of the solution to the equation $X_1(x_2) = x_2$.

Problems. 1. Obtain reasonable general conditions on p_1, p_2 that ensure the uniqueness of Nash equilibrium. 2. Develop an n-person and/or T period version of such a game.

10.2   Variations on Cournot's theme: Territorial price building

Suppose there is a market that include selling sites $M_1, \ldots, M_m$, products $1, \ldots, K$ and production sites $1, \ldots, L$.

There are N agents that buy the products at the production sites and sell them in $M_1, \ldots, M_m$.

Assume that at the each production site l there is a large stock of all products available at the price p_{kl} for the unit of the product k, and that the transportation cost to M_i of a unit of the product k is ξ_{ikl}.

Let Y_{ikl}^n denote the amount of the product k, which the agent n brought to M_i from the site l. Then the total amount of kth product in M_i is

$$Y_{ik} = \sum_{n=1}^{N} Y_{ik}^n, \quad Y_{ik}^n = \sum_{l=1}^{L} Y_{ikl}^n.$$

The selling price of the product k in M_i is the function of the total supply of the product. Following the standard Cournot model (see Chapter 9) assume that this price is given by the formula

$$R_{ik}(Y_{ik}) = (1 - Y_{ik}/\alpha_{ik})\beta_{ik}, \quad Y_{ik} \in [0, \alpha_{ik}],$$

with some positive constants α_{ik}, β_{ik}.

Then the total income of the player n can be calculated by the formula

$$H_n = \sum_{i=1}^{m} \sum_{k=1}^{K} H_n(ik) \tag{10.8}$$

with

$$H_n(ik) = Y_{ik}^n \left(1 - \frac{Y_{ik}}{\alpha_{ik}}\right)\beta_{ik} - \sum_{l=1}^{L} Y_{ikl}^n(\xi_{ikl} + p_{kl}), \quad Y_{ik} \in [0, \alpha_{ik}]. \tag{10.9}$$

Thus we have defined a symmetric N-person game Γ^N with the payoffs given by (10.8) and with the strategies of each player n being the arrays (Y_{ikl}^n) with positive entries bounded by α_{ik}.

For simplicity we reduce our attention to the case of 2 players.

Proposition 10.3. *The game Γ^2 has a symmetric Nash equilibrium. If for each pair (i, k) there exists a unique $l = q(i, k)$ where the minimum of $\xi_{ikl} + p_{kl}$ is attained, then this equilibrium is unique and the corresponding strategies of each player are given by*

$$\hat{Y}_{ik} = \frac{1}{3}\delta_{q(i,k)}^l \alpha_{ik}\left(1 - \frac{\xi_{ikl} + p_{kl}}{\beta_{ik}}\right), \qquad (10.10)$$

where δ_m^l denotes the usual Dirac symbol that equals 1 for $l = m$ and zero otherwise.

Proof. Clearly looking for a best reply one has to maximize each $H_n(ik)$ separately for each pair (ik). It is also clear from (10.9) that to have a best reply one has to bring the product only from those sites that minimize $\xi_{ikl} + p_{kl}$. Hence given (Y_{ik}^1) to find the best reply of the second player one has to find the maximum of the functions

$$\bar{H}_2^{ik}(Y_{ikq(i,k)}^2) = Y_{ikq(i,k)}^n\left(1 - \frac{Y_{ik}^1 + Y_{ikq(i,k)}^2}{\alpha_{ik}}\right)\beta_{ik} - Y_{ikq(i,k)}^n(\xi_{ikq(i,k)} + p_{kq(i,k)})$$

for each pair (ik). Differentiating one gets

$$\frac{d\bar{H}_2^{ik}(z)}{dz} = \left(1 - \frac{Y_{ik}^1 + z}{\alpha_{ik}}\right)\beta_{ik} - (\xi_{ikq(i,k)} + p_{kq(i,k)}) - z\frac{\beta_{ik}}{\alpha_{ik}} = 0$$

for the best reply $z = Y_{ikq(i,k)}^2$, yielding

$$Y_{ikq(i,k)}^2 = \frac{1}{2}\alpha_{ik}\left(1 - \frac{Y_{ik}^1}{\alpha_{ik}} - \frac{\xi_{ikq(i,k)} + p_{kq(i,k)}}{\beta_{ik}}\right). \qquad (10.11)$$

As the second derivatives of $\bar{H}_2^{ik}$ is negative this point does really define a maximum (and not a minimum). By the symmetry one concludes that symmetric equilibrium strategy solves the equation

$$\hat{Y}_{ik} = \frac{1}{2}\alpha_{ik}\left(1 - \frac{\hat{Y}_{ik}}{\alpha_{ik}} - \frac{\xi_{ikq(i,k)} + p_{kq(i,k)}}{\beta_{ik}}\right) \qquad (10.12)$$

and thus equals (10.10) as required.

Problems. 1. Extend the theorem to the N-person game (this is straightforward). 2. Analyze possible cooperation (agreements) between the

Understanding Game Theory

agents and the Stakelberg version of the game (in the spirit of the classical Cournot model of Section 10.1), calculate the compromise set. 3. Calculate the equilibria in a version of the game when each agent has a fixed capital and his/her purpose is to distributed it in the optimal way between products and realization sites. 4. Extend the model to include the possibility of several production firms on each of the site with the price at the production site being increased by the total demand of the selling agents (as well as decreased by the total supply) yielding a game with players including both the selling agents and production firms.

10.3 Models of inspection

The classical model of inspection was discussed in Section 9.7. Here we shall analyze more advanced models.

Example 10.1. Bi-matrix inspection games.

The game is carried out between a trespasser (player I) and an inspector (player II) in n-periods (or steps). We shall analyze a 'stop to abuses' version of this game when it lasts only till the first time trespassing is discovered by the inspector.

Player I has 2 pure strategies: to break (violate) the law (B) or to refrain from it (R). Player II has also 2 pure strategies: to check the actions of player I (C) or to have a rest (R). If player I chooses (R) he gets the legal income $r > 0$. If he chooses (B), he obtains additionally the illegal surplus $s > 0$. However, if his illegal action is discovered by player II, the player I pays the fine $f > 0$ and the game is over.

If player II chooses (C) he spends the amount $c > 0$ on this procedure and can discover the trespassing of player I with the probability $p(\bar{p} = 1-p)$. If player I breaks the law and this action is not discovered, the inspector loses the amount $l > 0$.

Consequently one step of this game can be described by the table

| | | Inspector | |
		Check (C)	Rest (R)
Trespasser	Break (B)	$-pf + \bar{p}(r+s), -(c+\bar{p}l)$	$r+s, -l$
	Refrain (R)	$r, -c$	$r, 0$

or shortly by the matrix

$$\begin{pmatrix} -pf + \bar{p}(r+s), -(c+\bar{p}l) & r+s, -l \\ r, -c & r, 0 \end{pmatrix} \tag{10.13}$$

It is natural to assume, that $c < pl$, so that the pair (B, R) is not a Nash equilibrium (otherwise the inspector has no reasons at all to conduct checks).

Let us say that a bi-matrix game *has a value*, if payoffs in any of the existing Nash equilibrium are the same, the corresponding pair of payoffs being called the *value of the game*.

Proposition 10.4. *Assume $f, r, s, c, l > 0$, $c < pl$. Then the game with the matrix* (10.13) *has a value $V = (u, v)$. Moreover, 1) if $s < s_1 = \frac{p}{\overline{p}}(f+r)$, the unique Nash equilibrium is given by the pair of mixed strategies $(x, \overline{x})$, $(y, \overline{y})$, where*

$$x = \frac{c}{pl}, y = \frac{s}{p(f+r+s)}; \quad u = r, v = -\frac{c}{p}.$$

2) if $s > s_1$, the unique Nash equilibrium is the profile (B,C) and

$$u = -pf + \overline{p}(r+s), v = -(c+\overline{p}l);$$

3) if $s = s_1$, then $u = r, v = -(c+\overline{p}l)$ and the Nash equilibria are given by all pairs (X, C), where X is any (pure or mixed) strategy of the trespasser.

Proof. Clearly the only candidate for Nash equilibrium in pure strategies is the pair (B, C). This profile is an equilibrium if and only if $s \geq s_1$. Formulae for x, y follow from (9.16). Other statements are checked by a straightforward inspection.

Of course of greater interest is the analysis of a multi-step version of this game. Let us consider the n-step game, where during this time the player I can break the law at most k times and the player II can organize the check at most m times. Assume that after the end of each period (step), the result becomes known to both players. Total payoff in n steps equals the sum of payoffs in each step. It is also assumed that all this information (rules of the game) is available to both players.

Let us denote the game described above by $\Gamma_{k,m}(n)$. Let $(u_{k,m}(n), v_{k,m}(n))$ be the value of this game. We then get the following system of recurrent equations:

$$(u_{k,m}(n), v_{k,m}(n))$$

$$= Val \begin{pmatrix} -pf + \overline{p}(r+s+u_{k-1,m-1}(n-1)), & r+s+u_{k-1,m}(n-1), \\ -(c+\overline{p}l) + \overline{p}v_{k-1,m-1}(n-1) & -l + v_{k-1,m}(n-1) \\ r+u_{k,m-1}(n-1), -c+v_{k,m-1}(n-1) & r+u_{k,m}(n-1), v_{k,m}(n-1) \end{pmatrix}$$

$$(10.14)$$

(if all $\Gamma_{k,m}(n)$ have values, i.e. their equilibrium payoffs are uniquely defined), with the boundary conditions $(m, n, k \geq 0)$:

$$(u_{0,m}(n), v_{0,m}(n)) = (nr, 0); \qquad (10.15)$$

$$(u_{k,0}(n), v_{k,0}(n)) = (nr + ks, -kl); \quad k \leq n, \qquad (10.16)$$

reflecting the following considerations: if the trespasser is unable to break the law, the pair of solutions (R, R) will be repeated over all periods; and if the inspector is unable to check, the trespasser will commit the maximum number of violations available.

Though $k \leq n, m \geq n$, the form of the recurrent equations below is slightly simplified if one allows all non-negative k, m, n together with the agreement

$$(u_{k,m}(n), v_{k,m}(n)) = (u_{k',m'}(n), v_{k',m'}(n)); k' = \min(k,n), m' = \min(m,n) \qquad (10.17)$$

Let us reduce our attention further to the game $\Gamma_{n,n}(n)$. Let us write $U_n = u_{n,n}(n)$ and $V_n = v_{n,n}(n)$. Then (10.14) takes the form:

$$(U_n, V_n) = (U_{n-1}, V_{n-1}) + ValM_n, \qquad (10.18)$$

where $n \geq 0$, $U_0 = V_0 = 0$) and

$$M_n = \begin{pmatrix} \overline{p}(r+s) - p(f + U_{n-1}), -(c + \overline{p}l + pV_{n-1}) & r + s, -l \\ r, -c & r, 0 \end{pmatrix}. \quad (10.19)$$

For $n = 1$ the game becomes the same as the game (10.13), and its solution is given by Proposition 10.4.

Let us find the solution to the game $\Gamma_{2,2}(2)$. Plugging the values U_1 and V_1 into M_2 yields

$$M_2 = \begin{pmatrix} \overline{p}(r+s) - p(f+r), -\overline{p}l & r + s, -l \\ r, -c & r, 0 \end{pmatrix}, \quad 0 < s < s_1;$$

$$M_2 = \begin{pmatrix} \overline{p}(-pf + \overline{p}(r+s)), -\overline{p}(c + \overline{p}l) & r + s, -l \\ r, -c & r, 0 \end{pmatrix}, \quad s > s_1;$$

$$s_2 = \frac{p}{\overline{p}}(f + r) + \frac{p}{\overline{p}^2}r.$$

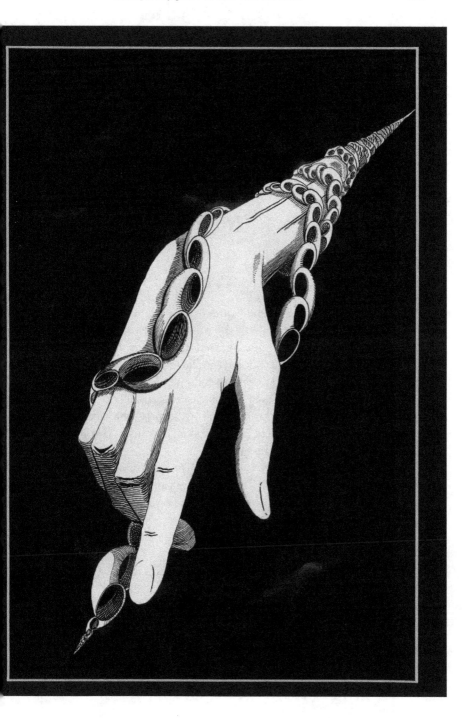

Direct calculations show that under the assumptions of Proposition 10.4 the M_2 also has a value, and one can distinguish three basic cases (equilibrium strategies are again denoted by $(x, \overline{x})$, $(y, \overline{y})$):

1) if $0 < s < s_1$, then

$$x = \frac{c}{pl + c}, y = \frac{s}{p(f + 2r + s)}; \quad U_2 = 2r, V_2 = -\frac{c(2pl + c)}{p(pl + c)}$$

2) if $s_1 < s < s_2$, then

$$x = \frac{c}{p(c + (1 + \overline{p})l)}, \quad y = \frac{s}{p(\overline{p}f + (1 + \overline{p})(r + s))};$$

$$U_2 = -pf + \overline{p}(r + s), \quad V_2 = -(\frac{cl}{p(c + (1 + \overline{p})l)} + c + \overline{p}l)$$

3) if $s > s_2$, then

$$U_2 = (1 + \overline{p})(-pf + \overline{p}(r + s)), \quad V_2 = -(1 + \overline{p})(c + \overline{p}l),$$

and the profile (B,C) is an equilibrium.

Analogously one can calculate the solutions for other $n > 2$.

Example 10.2. Tax payer against the tax man.

This is a game between a tax payer (player I) and the tax police (player II). Player I has 2 pure strategies: to hide part of the taxes (H) or to pay them in full (P). Player II has also 2 strategies: to check player I (C) and to rest (R). Player I gets the income r if he pays the tax in full. If he chooses the action (H), he gets the additional surplus l. But if he is caught by player II, he has to pay the fine f.

In the profile (C,H) player II can discover the unlawful action of player I with the probability $p(\overline{p} = 1 - p)$, so that p can be called the efficiency of the police. Choosing (C), player II spends c on the checking procedure. Of course $l, r, f, c > 0$.

Hence we defined a bi-matrix game given by the table

Player II (Police)

		Check (C)	Rest (R)
Player I	Hide (H)	$r + \overline{p}l - pf, -c + pf - \overline{p}l$	$r + l, -l$
	Pay (P)	$r, -c$	$r, 0$

or shortly by the payoff matrix

$$\begin{pmatrix} r + \overline{p}l - pf, -c + pf - \overline{p}l & r + l, -l \\ r, -c & r, 0 \end{pmatrix}. \tag{10.20}$$

According to formulas (9.16), the candidates to the mixed equilibrium are the strategies $(\beta, \overline{\beta})$, $(\alpha, \overline{\alpha})$, where

$$\alpha = \frac{a_{22} - a_{12}}{a_{11} - a_{12} - a_{21} + a_{22}} = \frac{l}{p(l+f)} > 0$$

$$\beta = \frac{b_{22} - b_{21}}{b_{11} - b_{12} - b_{21} + b_{22}} = \frac{c}{p(l+f)} > 0.$$

In order to have these strategies well defined, it is necessary to have $\alpha < 1$ and $\beta < 1$ respectively. By a direct inspection one gets the following result.

Proposition 10.5. *1) If $c \geq p(f+l)$, the pair (H,R) is an equilibrium, and moreover the strategy (R) is dominant for the police (even strictly, if the previous inequality is strict). 2) If $c < p(f+l)$ and $fp \leq \overline{p}l$, the pair (H,C) is an equilibrium and the strategy (H) is dominant (strictly if the previous inequality is strict). 3) If $c < p(f+l)$, $fp > \overline{p}l$, then the unique Nash equilibrium is the profile of mixed strategies $(\beta, \overline{\beta})$, $(\alpha, \overline{\alpha})$.*

Remarks. 1. Consequently, in cases 1) and 2) the actions of the police are not effective. 2. It is not difficult to show that the equilibrium in case 3) is stable.

It is more interesting to analyze the game obtained by extending the strategy space of player I by allowing him to choose the amount l of tax evasion: $l \in [0, l_M]$, where l_M is the full tax due to player I. For example, we shall assume that the fine is proportional to l, i.e. $f(l) = nl$. Say, in the Russian tax legislation $n = 0.4$. Under these assumptions the key coefficients α, β take the form

$$\alpha = \frac{1}{p(n+1)}, \quad \beta = \frac{c}{l} \frac{1}{p(n+1)}.$$

Let $H_I(l)$ denote the payoff to player I in the equilibrium when l is chosen. One can distinguish two cases:

1) $p > \frac{1}{n+1} \Longleftrightarrow \alpha < 1$. If

$$l > l_1 = \frac{c}{p(n+1)} \Longleftrightarrow \beta < 1,$$

then $(\beta, \overline{\beta})$, $(\alpha, \overline{\alpha})$ is a stable equilibrium. If $l < l_1 \Longleftrightarrow \beta > 1$, then (H, R) is a stable equilibrium. Since

$$H_I(l < l_1) = r + l < r + l_1,$$

$$H_I(l > l_1) = \beta\alpha(r + \overline{p}l - pf) + \beta\overline{\alpha}(r+l) + \overline{\beta}\alpha r + \overline{\beta}\,\overline{\alpha}r = r + \beta l(\alpha\overline{p} + \overline{\alpha} - \alpha pn),$$

it follows that $H_I(l > l_1) > H_I(l < l_1)$ would be possible whenever $\beta l(\alpha \bar{p} + \bar{\alpha} - \alpha pn) \geq l_1$. However $\alpha \bar{p} + \bar{\alpha} - \alpha pn = 0$, hence this is not the case. Consequently $H_I(l > l_1) < H_I(l < l_1)$ and therefore player I will avoid tax on the amount $l = l_1$.

2) $p < \frac{1}{n+1} \iff \alpha > 1$.

If $l > l_1 \iff \beta < 1$, then (H, C) is an equilibrium, and if $l < l_1 \iff \beta > 1$, then (H, R) is an equilibrium. Since

$$H_I(l < l_1) = r + l < r + l_1, \quad H_I(l > l_1) = r + (1 - p)l - pln,$$

the choice $l > l_1$ is reasonable for player I as

$$l(1 - p - pn) \geq l_1.$$

Hence $H_I(l > l_1) > H_I(l < l_1)$ whenever $l \geq \frac{l_1}{1 - p(n+1)}$.
Consequently, if

$$\frac{l_1}{1 - p(n + 1)} \leq l_M, \tag{10.21}$$

the equilibrium strategy for player I is $l = l_M$ and otherwise $l = l_1$.

One can conclude that in both cases it is profitable to avoid tax on the amount l_1, but as the efficiency of tax man increases, it becomes unreasonable to avoid tax on a higher amount.

Let us see which condition in the second case would ensure the inequality (10.21) when the amount of tax avoidance is l_M in the equilibrium. Plugging l_1 in (10.21) yields

$$\frac{c}{p(n + 1)(1 - p(n + 1))} \leq l_M.$$

Denoting $x = p(n + 1) < 1$ one can rewrite it as

$$x^2 - x + \frac{c}{l_M} \leq 0. \tag{10.22}$$

The roots of the corresponding equation are

$$x_{1,2} = \frac{1 \pm \sqrt{1 - \frac{4c}{l_M}}}{2}.$$

Hence for $c > l_M/4$ inequality (10.21) does not hold for any p, and for $c \leq l_M/4$ the solution to (10.22) is

$$x \in \left[\frac{1 - \sqrt{1 - \frac{4c}{l_M}}}{2}; \frac{1 - \sqrt{1 + \frac{4c}{l_M}}}{2} \right].$$

Thus for

$$c \leq \frac{l_M}{4}, \quad p \in \left[\frac{1 - \sqrt{1 - \frac{4c}{l_M}}}{2(n+1)}; \frac{1 - \sqrt{1 + \frac{4c}{l_M}}}{2(n+1)} \right] \tag{10.23}$$

it is profitable to avoid tax payment on the amount l_M.

Let us consider a numeric example with $n = 0.4$ so that $1/(n+1) = 0.714$. If, say, $c = 1000$, then $l_M = 100000$.

1) If $p < 0.714$, the condition $c \leq l_M/4$ holds true, as $1000 < 2500$. Hence, for $p \in [0.007; 0.707]$ it is profitable to avoid tax on the whole amount, i.e. 100000.

2) If $p > 0.714$ it is profitable to avoid tax on the amount $l_1 = 714.29$.

Hence if the efficiency of tax payment checks is $p < 0.707$, it is profitable to avoid tax on the whole amount of 100000, and if $p > 0.707$, then not more than on 1010.

Other models of inspection games can be found in [7], [8], [9], [52].

Problems. 1. In the first example get explicit formulas for the solutions in case $n > 2$. Do there exist explicit formulas for general $\Gamma_{k,m}(n)$? 2. Construct and analyze a multi step analog of the second example.

10.4   A dynamic model of strategic investments

Here we suggest and analyze a dynamic model describing the changes to market shares in a duopoly caused by strategic investments in advertising, where firms aim at maximizing the discounted flow of profits. This section is based on papers [26] and [216].

In usual models of competitive strategic investments (see [173], [187], [78] and references therein) it is assumed that the firms are correcting their rate of investment continuously, which leads to positional differential games. These game are very difficult to analyze, so that the solutions are usually obtained under strong additional assumptions, say, by fixing the rate of investment for a fixed period of time [156], [50].

Here we assume from the beginning that the rate of strategic investment is constant for each firm. This allows for a much simpler analysis avoiding the heavy machinery of differential games. From the practical point of view this approach seems to be quite realistic, as it can be used recursively by fixing the rate for a certain (large or small) time and re-evaluating the strategy each planning period.

Suppose two firms I and II compete with each other in order to increase their market shares, which initially are s_1 and s_2 respectively. The control parameters of the firms are the rate of their investments in advertising x_1 and x_2 respectively. The investment of each firm enlarges the total market and attracts its competitor's customers, so that the evolution of the market shares $v^1(t), v^2(t)$ of the firms can be described by the system

$$\begin{cases} \dot{v}^1 = F\left(x_1, x_2, v^1, v^2\right) + G\left(x_1, x_2, v^1, v^2\right) \\ \dot{v}_2 = F\left(x_1, x_2, v^1, v^2\right) - G\left(x_1, x_2, v^1, v^2\right) \end{cases}, \qquad (10.24)$$

where F describes the impact on the market shares of market enlargement and G accounts for the changes in market shares due to customers swapping allegiance.

Denote by r_i the difference between the unit sale price p_i and the unit cost, c_i, of a product or service provided by the ith firm, that is $r_i = p_i - c_i$. Assuming that the positive inflation over time is described by a discount fact $0 < \beta < 1$ leads to the following form of the *discount flow of profit for each firm i during a time period* $[\tau_1, \tau_2]$:

$$\pi^i\left(\tau_1, \tau_2, x_1, x_2\right) = \int_{\tau_1}^{\tau_2} \beta^t \left(r_i v^i\left(t\right) - x_i \right) dt. \qquad (10.25)$$

And the total *discounted flow of profits* $\pi^i\left(x_1, x_2\right)$ *obtained by the* i^{th} *firm,* $i = 1, 2,$ *if Firm I uses investment* x_1 *and Firm II uses investment* x_2 is the limit

$$\pi^i\left(x_1, x_2\right) = \lim_{T \to \infty} \pi^i\left(0, T, x_1, x_2\right). \qquad (10.26)$$

Thus we have a two-person game with profits (10.26) and infinite number of strategies: x_1 and x_2 are arbitrary positive numbers.

We are aiming at finding equilibria for this game.

Recall that the strategy (in our case the rate of investment) $\hat{x}_1$ is *the best reply to the investment* $\hat{x}_2$ if

$$\pi^1\left(\hat{x}_1, \hat{x}_2\right) \geq \pi^1\left(x_1, \hat{x}_2\right) \quad \forall x_1. \qquad (10.27)$$

Similarly an investment $\hat{x}_2$ is *the best reply to the investment* $\hat{x}_1$ if

$$\pi^2\left(\hat{x}_1, \hat{x}_2\right) \geq \pi^2\left(\hat{x}_1, x_2\right) \quad \forall x_2. \qquad (10.28)$$

A pair $(\hat{x}_1, \hat{x}_2)$ is a *Nash Equilibrium solution* if conditions (10.27) and (10.28) are satisfied simultaneously.

Of course for the analysis of the model the functions F and G have to be specified. A natural form of F is

$$F\left(x_1, x_2, v^1, v^2\right) = \left(a_1\left(x_1\right) + a_2\left(x_2\right)\right) m \left(v^1 + v^2\right), \qquad (10.29)$$

where the functions $a_1(x_1)$ and $a_2(x_2)$ stand for the efficiency of advertising made by firms I and II and the function m specifies the dependence of the speed of market growth on the current market state (i.e. its saturation).

Assuming that the number of customers swapping allegiance is directly proportional to the number of the competitor's customers, the simplest reasonable G can be taken in the form:

$$G\left(x_1, x_2, v^1, v^2\right) = l_2 v^2 b_1(x_1) - l_2 v^1 b_2(x_2), \qquad (10.30)$$

where the coefficients l_1 and l_2 are interpreted as indicators of 'customers' loyalty and the functions $b_1(x_1)$ and $b_2(x_2)$ are also efficiency functions.

Assuming (normalizing) that the total saturated market is 1 and that the speed of market growth decreases as the market approaches its maximum state we choose the function m as

$$m(v) = 1 - v. \qquad (10.31)$$

This leads finally to the following specific form of the general system (10.24):

$$\begin{cases} \dot{v}^1 = \{a_1(x_1) + a_2(x_2)\}\left(1 - v^1 - v^2\right) + l_2 v^2 b_1(x_1) - l_1 v^1 b_2(x_2) \\ \dot{v}^2 = \{a_1(x_1) + a_2(x_2)\}\left(1 - v^1 - v^2\right) + l_1 v^1 b_2(x_2) - l_2 v^2 b_1(x_1) \end{cases}. \qquad (10.32)$$

It turns out that this system is explicitly solvable.

Proposition 10.6. *The solution to the Cauchy problem for system* (10.32) *under initial conditions* $v^1(0) = s_1$ *and* $v^2(0) = s_2$ *(i.e. the market shares of the firms with initial amounts* $s_1\ s_2$ *and the investment rates* x_1 *and* x_2*) are given by*

$$v^1(t) = \frac{l_2 b_1(x_1)}{B^+}\left(1 - e^{-B^+ t}\right)$$

$$+ \frac{(s_1 + s_2 - 1)(l_2 b_1(x_1) - A)}{(B^+ - 2A)}\left(e^{-2At} - e^{-B^+ t}\right) + s_1 e^{-B^+ t}, \qquad (10.33)$$

$$v^2(t) = \frac{l_1 b_2(x_2)}{B^+}\left(1 - e^{-B^+ t}\right)$$

$$+ \frac{(s_1 + s_2 - 1)(l_1 b_2(x_2) - A)}{(B^+ - 2A)}\left(e^{-2At} - e^{-B^+ t}\right) + s_2 e^{-B^+ t}, \qquad (10.34)$$

where

$$A = A(x_1, x_2) = \{a_1(x_1) + a_2(x_2)\}$$

and

$$B^+ = l_2 b_1(x_1) + l_1 b_2(x_2), \quad B^- = l_2 b_1(x_1) - l_1 b_2(x_2).$$

Proof. Notice that by choosing the new coordinates $v = v^1 + v^2$ and $u = v^1 - v^2$, where v represents the total amount of market and u stands for the difference in the market shares the firms hold, leads to the system

$$\begin{cases} \dot{v} = 2\left\{a_1\left(x_1\right) + a_2\left(x_2\right)\right\}\left(1 - v\right) \\ \dot{u} = l_2 b_1\left(x_1\right)\left(v - u\right) - l_1 b_2\left(x_2\right)\left(v + u\right) \end{cases}, \qquad (10.35)$$

which can be solved by first solving the first equation and then substituting the result to the second one. This leads to (we omit the direct computations):

$$\begin{cases} v\left(t\right) = 1 + \left(s_1 + s_2 - 1\right)e^{-2At} \\ u\left(t\right) = \frac{B^-}{B^+} + \frac{B^-}{\left(B^+ - 2A\right)}\left(s_1 + s_2 - 1\right)e^{-2At} + e^{-B^+ t}C_u \end{cases}, \qquad (10.36)$$

where C_u is the constant which depends on the initial conditions. One can now find v^1, v^2 and express C_u in terms of their initial conditions These straightforward (thought rather lengthy) calculations (that we omit) lead to (10.33), (10.34).

Substituting (10.33), (10.34) to (10.26) leads (after simple manipulations) to

Proposition 10.7. *The discounted flows of profits $\pi^i\left(x_1, x_2\right)$ for the i-th firm, $i = 1, 2$, if Firm I uses investment x_1 and Firm II uses investment x_2 are as follows*

$$\pi^1\left(x_1, x_2\right) = \frac{r_1}{\left(B^+ - \ln\beta\right)}\left(-\frac{l_2 b_1\left(x_1\right)}{\ln\beta} + \frac{\left(s_1 + s_2 - 1\right)\left(l_2 b_1\left(x_1\right) - A\right)}{\left(2A - \ln\beta\right)} + s_1\right) + \frac{x_1}{\ln\beta},$$

$$\pi^2\left(x_1, x_2\right) = \frac{r_2}{\left(B^+ - \ln\beta\right)}\left(-\frac{l_1 b_2\left(x_2\right)}{\ln\beta} + \frac{\left(s_1 + s_2 - 1\right)\left(l_1 b_2\left(x_2\right) - A\right)}{\left(2A - \ln\beta\right)} + s_2\right) + \frac{x_2}{\ln\beta}.$$

In order to find best replies and equilibria, we need to specify the form of the functions $a_i\left(x_i\right)$ and $b_i\left(x_i\right)$ describing the efficiency of investments. The *law of diminishing returns* on the investment (already mentioned in Section 10.1) states (see e.g. [65]) that there should be a positive effect of extra advertising which diminishes as investment increases. Mathematically, it can be interpreted as the requirement that the first derivatives of these functions should be positive but the second derivatives should be negative. We will choose the functions $a_i\left(x_i\right)$ and $b_i\left(x_i\right)$ to have the form

$$a_i\left(x_i\right) = \frac{x_i}{x_i + a_i}, \qquad b_i\left(x_i\right) = \frac{x_i}{x_i + b_i}. \qquad (10.37)$$

As required, we have

$$\frac{da_i\left(x_i\right)}{dx_i} = \frac{a_i}{\left(x_i + a_i\right)^2} > 0, \qquad \frac{d^2 a_i\left(x_i\right)}{dx_i^2} = \frac{-2a_i}{\left(x_i + a_i\right)^3} < 0 \qquad \forall x_i \geq 0,$$

and the same is true for functions $b_i(x_i)$. Since the functions $a_i(x_i)$ describe the efficiency of the investment we will assume that the maximum possible efficiency is one. This fact is taken into account since

$$\lim_{x_i \to \infty} a_i(x_i) = 1, \quad \lim_{x_i \to \infty} b_i(x_i) = 1.$$

Finally, coefficients a_i and b_i represent the efficiency rate for the investments. The lower the value of a_i and b_i the more efficient is the investment policy.

Proposition 10.8. *Under the above assumptions*

$$\pi^1(x_1, x_2) = -\frac{hx_1^3 + (b - r_1 d) x_1^2 + (c - r_1 f) x_1 - r_1 g}{\kappa (hx_1^2 + bx_1 + c)} \tag{10.38}$$

where

$$\begin{aligned}
h &= (\kappa + 2\vartheta + 2)(\kappa + l_2 + \theta), \\
b &= (a_1 + b_1)\left(\kappa^2 + (\theta + 2\vartheta)\kappa + 2\theta\vartheta\right) + 2b_1(\kappa + \theta) + l_2 a_1(\kappa + 2\vartheta), \\
c &= a_1 b_1\left(\kappa^2 + (\theta + 2\vartheta)\kappa + 2\vartheta\theta\right), \\
d &= s_1\kappa^2 + ((\vartheta + 1)(1 + s_1 - s_2) + l_2(s_1 + s_2))\kappa + 2l_2(\vartheta + 1), \\
f &= s_1(a_1 + b_1)\kappa^2 + (a_1 l_2(s_1 + s_2) + ((a_1 + b_1)\vartheta + b_1)(1 + s_1 - s_2))\kappa \\
&\quad + 2l_2 a_1 \vartheta, \\
g &= a_1 b_1 \kappa\left(s_1\kappa + \vartheta(1 + s_1 - s_2)\right)
\end{aligned} \tag{10.39}$$

and

$$\theta = l_1\frac{x_2}{x_2 + b_2}, \quad \vartheta = \frac{x_2}{x_2 + a_2}, \quad \kappa = -\ln\beta.$$

Proof. Follows from the previous Proposition by lengthy but direct calculations. (It is convenient to use 'Maple' to perform these calculations.)

Proposition 10.9. *If there exists a best reply $\hat{x}_1$ to the investment x_2, it can be found as a root of the equation $P(x_1) = 0$, where*

$$P(x_1) = h^2 x_1^4 + 2hbx_1^3 + \left(r_1 fh - r_1 db + 2hc + b^2\right) x_1^2$$

$$+ (2bc - 2r_1 dc + 2r_1 gh)x_1 + \left(r_1 gb - r_1 fc + c^2\right).$$

Proof. Follows from the exact formula for the discounted flow of profits $\pi^1(x_1, x_2)$ since the derivative of $\pi^1(x_1, x_2)$ is calculated as

$$\frac{\partial \pi^1(x_1, x_2)}{\partial x_1} = -\frac{P(x_1)}{\kappa(hx_1^2 + bx_1 + c)^2}. \tag{10.40}$$

Thus the analysis of the best reply boils down to the analysis of the roots of the fourth order algebraic equation. This is not an easy task. We give below a sufficient condition for the existence of a unique best reply, based on the *Descartes rule of sign alterations*, stating that a polynomial equation has exactly one positive root, if the number of sign alterations in the sequence of the coefficients of the equation equals one.

Proposition 10.10. *If the discount factor β is close to one and*

$$r_1 > \frac{(l_1 + l_2)}{2l_2} a_1 + b_1 \frac{l_1}{l_2}, \tag{10.41}$$

then there exists a unique best reply $\hat{x}_1$ for any non negative investment $\hat{x}_2$, and $\hat{x}_1$ can be found as a positive root of the equation $P(x_1) = 0$.

Proof. Since the first and the second coefficients of P, h^2 and $2hb$, are positive, to get the uniqueness of a positive it is enough to show that the fourth and the fifth coefficients, $(2bc - 2r_1 dc + 2r_1 gh)$ and $(r_1 gb - r_1 fc + c^2)$, are negative for any $x_2 \geq 0$.

If β is close to one, then $\kappa = -\ln \beta$ is close to zero. Using the above given explicit formulae for all coefficients one obtains after direct manipulations that

$$Q(x_2) = \frac{(x_2 + a_2)^2 (x_2 + b_2)^2}{a_1 b_1} (2bc - 2r_1 dc + 2r_1 gh)$$

is the following fourth order polynomial in x_2: that

$$\begin{aligned}
Q(x_2) = {} & 2a_2^2 \kappa^2 b_2^2 \left(-2r_1 l_2 (1 - s_1) + o(\kappa) \right) \\
& + x_2 \left(2a_2 \kappa b_2 \left(-2l_2 r_1 (b_2 (1 - s_1 + s_2) + a_2 l_1) + o(\kappa) \right) \right) \\
& + x_2^2 \left(-8r_1 l_2 a_2 l_1 b_2 + o(\kappa) \right) \\
& + x_2^3 \left(8l_1 \left(b_1 a_2 l_1 - r_1 l_2 a_2 + l_2 a_1 b_2 - 2r_1 l_2 b_2 \right) + o(\kappa) \right) \\
& + x_2^4 \left(8l_1 \left(2l_1 b_1 + a_1 l_1 + l_2 a_1 - 2r_1 l_2 \right) + o(\kappa) \right),
\end{aligned} \tag{10.42}$$

where as usual o denotes any function such that $\lim_{k \to 0} o(\kappa)/\kappa = 0$. One sees easily that under (10.41) all coefficients of this polynomial are negative for $o(\kappa) = 0$ and hence for small enough $o(\kappa)$. Hence $Q(x_2) < 0$ for all $x_2 \geq 0$.

Analogously for the polynomial

$$R(x_2) = \frac{(x_2 + a_2)^2 (x_2 + b_2)^2}{a_1 b_1} \left(r_1 gb - r_1 fc + c^2 \right)$$

one obtains

$$
\begin{aligned}
R(x_2) &= \kappa^3 a_2^2 b_2^2 \left(-r_1 \left(a_1 l_2 s_2 - b_1 s_2 + b_1 - b_1 s_1\right) + o(\kappa)\right) \\
&+ x_2 \kappa^2 b_2 a_2 \left(-r_1 \left(a_2 b_1 l_1 \left(1 - s_1 - s_2\right) + b_2 l_2 a_1 \left(1 - s_1 + 3 s_2\right)\right. \\
&+ a_1 a_2 l_1 l_2 \left(s_1 + s_2\right)\right) + o(\kappa)) \\
&+ x_2^2 \kappa \left(-2 l_2 a_1 r_1 b_2 \left(b_2 \left(1 - s_1 + s_2\right) + a_2 l_1 \left(1 + s_1 + s_2\right)\right) + o(\kappa)\right) \\
&+ x_2^3 \left(-4 r_1 l_1 a_1 l_2 b_2 + o(\kappa)\right) \\
&+ x_2^4 \left(-4 a_1 l_1 \left(r_1 l_2 - b_1 l_1\right) + o(\kappa)\right)
\end{aligned} \tag{10.43}
$$

and again all coefficients of this polynomial are negative under the assumption of the Proposition so that $Q(x_2) < 0$ for all $x_2 \geq 0$. This completes the proof of the Proposition.

Hence, under the conditions of the previous Proposition, for any non negative constant investment x_2 made by Firm II, there exists a unique best reply investment $x_1 = \hat{X}_1(x_2)$ of Firm I. Under similar assumption one shows that for any non negative constant investment $\hat{x}_1$ made by Firm I, there exists a unique best reply investment $x_2 = \hat{X}_2(x_1)$ of Firm II.

It is possible now to prove the following existence result.

Theorem 10.1. *([26]) If the discount factor β is close to one and*

$$
r_1 > \frac{(l_1 + l_2)}{2 l_2} a_1 + b_1 \frac{l_1}{l_2} \qquad r_2 > \frac{(l_1 + l_2)}{2 l_1} a_2 + b_2 \frac{l_2}{l_1}, \tag{10.44}
$$

then the investment game under consideration has a Nash equilibrium.

Proof. (sketch). One needs to observe that $\lim_{x_2 \to \infty} \hat{X}_1(x_2) < \infty$, which follows from the fact that the limits of all coefficients of $P(x_1)$ are well defined and finite as $x_2 \to \infty$. Similarly $\lim_{x_1 \to \infty} \hat{X}_2(x_1) < \infty$. Consequently the graphs of the continuous curves $\hat{X}_1(x_2)$ and $\hat{X}_1(x_2)$ have to intersect.

Problems. What can be said about best replies and/or Nash equilibria for β away from one? or without assumption (10.41). How robust is the model, i.e. how it depends on parameters and other simplifying assumptions, e.g. on a particular form (10.37) of the efficiency functions?

10.5 Game-theoretic and geometric interpretation of risk-neutral measures

In this section we reveal the geometric meaning for risk-neutral measures, playing key role in finances, and show how they arise naturally when evaluating certain minimax-type Bellman operators. This development is the key ingredient in the game-theoretic treatment of financial markets given in the next section.

In order to explain the ideas clearly, we first develop the theory in dimension $d = 2$, and then extend it to arbitrary dimensions (which requires certain lengthy manipulations with multidimensional determinants).

We shall denote by Int the interior of a closed set. Let a closed convex polyhedron in $\mathbf{R}^2$ contain the origin as an interior point, and let $\xi_1, \ldots, \xi_k$ be its vertices, ordered anticlockwise. We shall denote such a polyhedron by $\Pi = \Pi[\xi_1, \ldots, \xi_k]$. The assumed condition

$$0 \in \text{Int}\Pi[\xi_1, \ldots, \xi_k] \qquad (10.45)$$

implies that all ξ_i do not vanish.

We are interested in the following game-theoretic problem: find

$$\Pi[\xi_1, \ldots, \xi_k](f) = \min_{\gamma \in \mathbf{R}^2} \max_{\xi \in \Pi} [f(\xi) - (\xi, \gamma)] \qquad (10.46)$$

for a convex (possibly non strictly) function f. By convexity, this rewrites as

$$\Pi[\xi_1, \ldots, \xi_k](f) = \min_{\gamma \in \mathbf{R}^2} \max_{\xi_1, \ldots, \xi_k} [f(\xi_i) - (\xi_i, \gamma)]. \qquad (10.47)$$

Having this in mind, we shall analyze a slightly more general problem: for an arbitrary finite collection of non-vanishing vectors $\xi_1, \ldots, \xi_k$ from $\mathbf{R}^2$, ordered anticlockwise, and arbitrary numbers $f(\xi_1), \ldots, f(\xi_k)$, to calculate (10.47) (whenever the minimum exists). The corresponding polyhedron $\Pi[\xi_1, \ldots, \xi_k]$ (obtained by linking together all neighboring vectors ξ_i, ξ_{i+1}, $i = 1, \ldots, k$, and ξ_k, ξ_1 with straight segments) may not be convex anymore.

We shall start with the case of Π being a triangle: $\Pi = \Pi[\xi_1, \xi_2, \xi_3]$. Then condition (10.45) implies that $\xi_i \neq -\alpha\xi_j$ for $\alpha > 0$ and any i, j. Suppose the min in

$$\Pi[\xi_1, \xi_2, \xi_3](f) = \min_{\gamma \in \mathbf{R}^2} \max_{\xi_1, \xi_2, \xi_3} [f(\xi_i) - (\xi_i, \gamma)] \qquad (10.48)$$

is attained on a vector γ_0 and the corresponding max on a certain ξ_i. Suppose this max is unique, so that

$$f(\xi_i) - (\xi_i, \gamma) > f(\xi_j) - (\xi_j, \gamma) \qquad (10.49)$$

for all $j \neq i$. As $\xi_i \neq 0$, by changing γ_0 on a small amount we can reduce the l.h.s. of (10.49) by preserving the inequality (10.49). This possibility contradicts the assumption that γ_0 is a minimal point. Hence, if γ_0 is a minimal point, the corresponding maximum must be attained on at least two vectors. Suppose it is attained on precisely two vectors, that is

$$f(\xi_i) - (\xi_i, \gamma) = f(\xi_j) - (\xi_j, \gamma) > f(\xi_m) - (\xi_m, \gamma) \qquad (10.50)$$

for some different i, j, m. Since the angle between ξ_i, ξ_j is strictly less than π, adding a vector

$$\epsilon(\xi_i/|\xi_j| + \xi_j/|\xi_i|)$$

to γ_0 will reduce simultaneously first two expressions from the l.h.s. of (10.50), but preserve (for small enough ϵ) the inequality on the r.h.s. of (10.50). This again contradicts the assumption that γ_0 is a minimal point. Hence, if γ_0 is a minimal point, it must satisfy the equation

$$f(\xi_1) - (\xi_1, \gamma) = f(\xi_2) - (\xi_2, \gamma) = f(\xi_3) - (\xi_3, \gamma), \tag{10.51}$$

which is equivalent to the system

$$\begin{cases} (\xi_2 - \xi_1, \gamma_0) = f(\xi_2) - f(\xi_1), \\ (\xi_3 - \xi_1, \gamma_0) = f(\xi_3) - f(\xi_1). \end{cases} \tag{10.52}$$

Again by assumption (10.45), the vectors $\xi_2 - \xi_1, \xi_3 - \xi_1$ are independent. Hence system (10.52) has a unique solution γ_0.

For a pair of vectors $u, v \in \mathbf{R}^2$, let $D(u, v)$ denote the oriented area of the parallelogram built on u, v and $R(u)$ the result of the rotation of u on $90°$ anticlockwise. That is, for $u = (u^1, u^2)$, $v = (v^1, v^2)$,

$$D(u, v) = u^1 v^2 - u^2 v^1, \quad R(u) = (u^2, -u^1).$$

Notice that the determinant of system (10.52) is

$$D(\xi_2 - \xi_1, \xi_3 - \xi_1) = D(\xi_2, \xi_3) + D(\xi_3, \xi_1) + D(\xi_1, \xi_2),$$

and by the standard formulas of linear algebra, the unique solution γ_0 is

$$\gamma_0 = \frac{f(\xi_1)R(\xi_2 - \xi_3) + f(\xi_2)R(\xi_3 - \xi_1) + f(\xi_3)R(\xi_1 - \xi_2)}{D(\xi_2, \xi_3) + D(\xi_3, \xi_1) + D(\xi_1, \xi_2)}, \tag{10.53}$$

and the corresponding optimal value

$$\Pi[\xi_1, \xi_2, \xi_3](f) = \frac{f(\xi_1)D(\xi_2, \xi_3) + f(\xi_2)D(\xi_3, \xi_1) + f(\xi_3)D(\xi_1, \xi_2)}{D(\xi_2, \xi_3) + D(\xi_3, \xi_1) + D(\xi_1, \xi_2)}. \tag{10.54}$$

Hence we arrive at the following.

Proposition 10.11. *Let a triangle* $\Pi[\xi_1, \xi_2, \xi_3]$ *satisfy* (10.45), *and let* $f(\xi_1), f(\xi_2), f(\xi_3)$ *be arbitrary numbers. Then expression* (10.48) *is given by* (10.54) *and the minimum is attained on the single* γ_0 *given by* (10.53).

Proof. Our discussion above shows that if γ_0 is a minimum point, then it is unique and is given by (10.53). It remains to show that this γ_0 is in fact the minimal point. But this is straightforward, as any change in γ_0 would necessarily increase one of the expressions $f(\xi_i) - (\xi_i, \gamma)$ (which again follows from (10.45)). Alternatively, the same conclusion can be obtained indirectly from the observation that the minimum exists and is attained on some finite γ, because

$$\max_{\xi_1, \xi_2, \xi_3} [f(\xi_i) - (\xi_i, \gamma)] \to \infty,$$

as $\gamma \to \infty$.

Corollary 10.1. *Expression* (10.54) *can be written equivalently as*

$$\Pi[\xi_1, \xi_2, \xi_3](f) = \mathbf{E}f(\xi),$$

where the expectation is defined with respect to the probability law $\{p_1, p_2, p_3\}$ on ξ_1, ξ_2, ξ_3:

$$p_i = \frac{D(\xi_j, \xi_m)}{D(\xi_2, \xi_3) + D(\xi_3, \xi_1) + D(\xi_1, \xi_2)}$$

((i, j, k) is either (1,2,3) or (2,3,1) or (3,1,2)). Moreover, this distribution is the unique probability on ξ_1, ξ_2, ξ_3 such that

$$\mathbf{E}(\xi) = \sum_{i=1}^{3} p_i \xi_i = 0. \tag{10.55}$$

Proof. The required uniqueness follows from the uniqueness of the expansion of ξ_3 with respect to the basis ξ_1, ξ_2.

We shall call a probability law on ξ_1, ξ_2, ξ_3 *risk-neutral*, if it satisfies (10.55). The reason for this terminology will be seen later. From the point of view of convex analysis this is just a probability on ξ_1, ξ_2, ξ_3 with barycenter in the origin.

We can now calculate (10.46) for arbitrary k.

Theorem 10.2. *Let a polyhedron $\Pi = \Pi[\xi_1, \ldots, \xi_k]$ satisfy the following conditions:*
    (i) No two vectors ξ_i, ξ_j are linearly dependent;
    (ii) The collection $\{\xi_1, \ldots, \xi_k\}$ does not belong to any half-space, i.e. there is no $\omega \in \mathbf{R}^2$ such that $(\omega, \xi_i) > 0$ for all i.
    Then

$$\Pi[\xi_1, \ldots, \xi_k](f) = \max_{i,j,m} \mathbf{E}_{ijm} f(\xi) = \max_{i,j,m} (p_i^{ijm} f(\xi_i) + p_j^{ijm} f(\xi_j) + p_m^{ijm} f(\xi_m)), \tag{10.56}$$

where max *is taken over all triples* $1 \leq i < j < m \leq k$ *such that*

$$0 \in \text{Int}\Pi[\xi_i, \xi_j, \xi_k], \tag{10.57}$$

and $\{p_i^{ijm}, p_j^{ijm}, p_m^{ijm}\}$ *denotes the unique risk-neutral probability on* $\{\xi_i, \xi_j, \xi_m\}$ *(given by Proposition 10.11) with* $\mathbf{E}_{ijm}$ *the corresponding expectation.*

Remark 10.1. Condition (i) is equivalent to the geometrical requirement that the origin does not lie on any diagonal of Π (or its extension), and condition (ii) is equivalent to (10.45).

Proof. For any triple $\{i, j, m\}$ satisfying (10.57),

$$\Pi[\xi_1, \ldots, \xi_k](f) \geq \min_{\gamma \in \mathbf{R}^2} \max_{\xi_i, \xi_j, \xi_m} [f(\xi) - (\xi, \gamma)] = \mathbf{E}_{ijm} f(\xi),$$

where Proposition 10.11 was used for the last equation. Hence

$$\Pi[\xi_1, \ldots, \xi_k](f) \geq \max_{i,j,m} \mathbf{E}_{ijm} f(\xi). \tag{10.58}$$

To get inverse inequality, the key geometrical observation is the following. Conditions (i) and (ii) imply that there exists a subset of the collection $\{\xi_1, \ldots, \xi_k\}$ consisting only of three vectors $\{\xi_i, \xi_j, \xi_m\}$, but still satisfying these conditions (and hence the assumptions of Proposition 10.11). This follows from the Carathéodory theorem (but can be also seen directly, as one can take an arbitrary ξ_i, and then choose, as ξ_m, ξ_j, the vectors with the maximum angle (less than π) with ξ_i when rotating clockwise and anticlockwise respectively). This observation implies that the maximum on the r.h.s. of (10.58) is defined (the set of triples is not empty) and consequently the l.h.s. is bounded from below. Moreover, as for any triple $\{i, j, m\}$ satisfying (10.57),

$$\max_{\xi_i, \xi_j, \xi_m} [f(\xi) - (\xi, \gamma)] \to \infty,$$

as $\gamma \to \infty$, and hence also

$$\max_{i=1,\ldots,k} [f(\xi_i) - (\xi_i, \gamma)] \to \infty,$$

the minimum in (10.47) is attained on some finite γ. Assuming that γ_0 is such a minimum point, we can now argue as above to conclude that

$$f(\xi_i) - (\xi_i, \gamma_0) = f(\xi_j) - (\xi_j, \gamma_0) = f(\xi_m) - (\xi_m, \gamma_0) \tag{10.59}$$

for some triple ξ_i, ξ_j, ξ_m. Moreover, if these triple does not satisfy (i) and (ii), then (by the same argument) the l.h.s. of (10.59) cannot strictly exceed $f(\xi_l) - (\xi_l, \gamma_0)$ for all other ξ_l. Hence we are led to a conclusion that if γ_0

is a minimum point, then there exists a subset $I \subset \{1, \ldots, k\}$ such that the expressions $f(\xi_l) - (\xi_l, \gamma_0)$ coincide for all $l \in I$ and the family $\{\xi_l\}$, $l \in I$, satisfy conditions (i), (ii). But by the above geometrical observation, such a family has to contain a subfamily with three vectors only satisfying (i) and (ii). Consequently, (10.56) holds.

Let us extend the results to higher dimensions d. Let us start with the simplest case of $d + 1$ vectors $\xi_1, \ldots, \xi_{d+1}$ in $\mathbf{R}^d$. Suppose their convex hull $\Pi[\xi_1, \ldots, \xi_{d+1}]$ is such that

$$0 \in \mathrm{Int}\Pi[\xi_1, \ldots, \xi_{d+1}]. \tag{10.60}$$

We are interested in evaluating the expression

$$\Pi[\xi_1, \ldots, \xi_{d+1}](f) = \min_{\gamma \in \mathbf{R}^d} \max_i [f(\xi_i) - (\xi_i, \gamma)]. \tag{10.61}$$

A remarkable fact that we are going to reveal is that this expression depends linearly on f and the minimizing γ is unique and also depends linearly on f.

Assume that $\mathbf{R}^d$ is equipped with the standard basis $e_1, \ldots, e_d$ fixing the orientation. Without loss of generality we shall assume now that the vectors $\xi_1, \ldots, \xi_{d+1}$ are ordered so that the vectors $\{\xi_2, \xi_3, \ldots, \xi_{d+1}\}$ form an oriented basis of $\mathbf{R}^d$. The fact that the vector ξ_1 lies outside any half space containing this basis allows one to identify the orientation of other subsets of $\xi_1, \ldots, \xi_{d+1}$ of size d. Namely, let $\{\hat{\xi}_i\}$ denote the ordered subset of $\xi_1, \ldots, \xi_{d+1}$ obtained by removing ξ_i. The basis $\{\hat{\xi}_i\}$ is oriented if and only if i is odd. For instance, if $d = 3$, the oriented bases form the triples $\{\xi_2, \xi_3, \xi_4\}$, $\{\xi_1, \xi_2, \xi_4\}$, $\{\xi_1, \xi_4, \xi_3\}$ and $\{\xi_1, \xi_3, \xi_2\}$.

The same argument as for $d = 2$ leads us to the conclusion that a minimal point γ_0 must satisfy the equation

$$f(\xi_1) - (\xi_1, \gamma) = \cdots = f(\xi_{d+1}) - (\xi_{d+1}, \gamma), \tag{10.62}$$

which is equivalent to the system

$$(\xi_i - \xi_1, \gamma_0) = f(\xi_i) - f(\xi_1), \quad i = 2, \ldots, d + 1. \tag{10.63}$$

From (10.60) it follows that this system has a unique solution, say γ_0.

To write it down explicitly, we shall use the natural extensions of the notations used above for $d = 2$. For a collection of d vectors $u_1, \ldots, u_d \in \mathbf{R}^d$, let $D(u_1, \ldots, u_d)$ denote the oriented volume of the parallelepiped built on $u_1, \ldots, u_d$ and $R(u_1, \ldots, u_{d-1})$ the rotor of the family $(u_1, \ldots, u_{d-1})$.

That is, denoting by upper scripts the coordinates of vectors,

$$
D(u_1,\dots,u_d) = \det
\begin{pmatrix}
u_1^1 & \cdots & u_1^d \\
u_2^1 & \cdots & u_2^d \\
& \cdots & \\
u_d^1 & \cdots & u_d^d
\end{pmatrix},
$$

$$
R(u_1,\dots,u_{d-1}) = \det
\begin{pmatrix}
e_1 & \cdots & e_d \\
u_1^1 & \cdots & u_1^d \\
& \cdots & \\
u_{d-1}^1 & \cdots & u_{d-1}^d
\end{pmatrix}
$$

$$
= e_1 \det
\begin{pmatrix}
u_1^2 & \cdots & u_1^d \\
& \cdots & \\
u_{d-1}^2 & \cdots & u_{d-1}^d
\end{pmatrix}
- e_2 \det
\begin{pmatrix}
u_1^1 & u_1^3 & \cdots & u_1^d \\
& \cdots & & \\
u_{d-1}^1 & u_{d-1}^3 & \cdots & u_{d-1}^d
\end{pmatrix}
+ \cdots .
$$

Finally, let us define a poly-linear operator $\tilde{R}$ from an ordered collection $\{u_1,\dots,u_d\}$ of d vectors in $\mathbf{R}^d$ to $\mathbf{R}^d$:

$$
\tilde{R}(u_1,\dots,u_d) = R(u_2 - u_1, u_3 - u_1, \dots, u_d - u_1)
$$

$$
= R(u_2,\dots,u_d) - R(u_1,u_3,\dots,u_d) + \cdots + (-1)^{d-1} R(u_1,\dots,u_{d-1}).
$$

Returning to system (10.63) observe that its determinant, which we denote by D, equals

$$
D = D(\xi_2 - \xi_1, \dots, \xi_{d+1} - \xi_1) = \det
\begin{pmatrix}
\xi_2^1 - \xi_1^1 & \xi_2^2 - \xi_1^2 & \cdots & \xi_2^d - \xi_1^d \\
& \cdots & & \\
\xi_{d+1}^1 - \xi_1^1 & \xi_{d+1}^2 - \xi_1^2 & \cdots & \xi_{d+1}^d - \xi_1^d
\end{pmatrix}.
$$

Using the linear dependence of a determinant on columns, this rewrites as

$$
D(\xi_2,\dots,\xi_{d+1}) - \xi_1^1 \det
\begin{pmatrix}
1 & \xi_2^2 & \cdots & \xi_2^d \\
& \cdots & & \\
1 & \xi_{d+1}^2 & \cdots & \xi_{d+1}^d
\end{pmatrix}
$$

$$
- \xi_1^2 \det
\begin{pmatrix}
\xi_2^1 & 1 & \xi_2^3 & \cdots & \xi_2^d \\
& \cdots & & & \\
\xi_{d+1}^1 & 1 & \xi_{d+1}^3 & \cdots & \xi_{d+1}^d
\end{pmatrix}
- \cdots ,
$$

implying that

$$D = D(\xi_2 - \xi_1, \ldots, \xi_{d+1} - \xi_1) = \sum_{i=1}^{d+1} (-1)^{i-1} D(\{\hat{\xi}_i\}). \tag{10.64}$$

Notice that according to the orientation specified above, $D(\{\hat{\xi}_i\})$ are positive (resp. negative) for odd i (resp. even i), implying that all terms in (10.64) are positive, so that the collection of numbers

$$p_i = \frac{1}{D}(-1)^{i-1} D(\{\hat{\xi}_i\}) = \frac{(-1)^{i-1} D(\{\hat{\xi}_i\})}{D(\xi_2 - \xi_1, \ldots, \xi_d - \xi_1)}, \quad i = 1, \ldots, d+1, \tag{10.65}$$

define a probability law on the set $\xi_1, \ldots, \xi_{d+1}$ with a full support.

By linear algebra, the unique solution γ_0 to system (10.63) is given by the formulas

$$\gamma_0^1 = \frac{1}{D} \det \begin{pmatrix} f(\xi_2) - f(\xi_1) & \xi_2^2 - \xi_1^2 & \cdots & \xi_2^d - \xi_1^d \\ & \cdots & & \\ f(\xi_{d+1}) - f(\xi_1) & \xi_{d+1}^2 - \xi_1^2 & \cdots & \xi_{d+1}^d - \xi_1^d \end{pmatrix}, \tag{10.66}$$

$$\gamma_0^2 = \frac{1}{D} \det \begin{pmatrix} \xi_2^1 - \xi_1^1 & f(\xi_2) - f(\xi_1) & \cdots & \xi_2^d - \xi_1^d \\ & \cdots & & \\ \xi_{d+1}^1 - \xi_1^1 & f(\xi_{d+1}) - f(\xi_1) & \cdots & \xi_{d+1}^d - \xi_1^d \end{pmatrix}, \tag{10.67}$$

and similar for other γ_0^i. One sees by inspection that for any i

$$f(\xi_i) - (\gamma_0, \xi_i) = \frac{1}{D} \sum_{i=1}^{d+1} [f(\xi_i)(-1)^{i+1} D(\{\hat{\xi}_i\})], \tag{10.68}$$

and

$$\gamma_0 = \frac{1}{D}(f(\xi_2) - f(\xi_1)) R(\xi_3 - \xi_1, \ldots, \xi_{d+1} - \xi_1)$$

$$- \frac{1}{D}(f(\xi_3) - f(\xi_1)) R(\xi_2 - \xi_1, \xi_4 - \xi_1, \ldots, \xi_{d+1} - \xi_1)$$

$$+ \cdots + \frac{1}{D}(-1)^{d+1}(f(\xi_{d+1}) - f(\xi_1) R(\xi_2 - \xi_1, \ldots, \xi_d - \xi_1),$$

which rewrites as

$$\gamma_0 = -\frac{1}{D} \left[f(\xi_1)\tilde{R}(\{\hat{\xi}_1\}) - f(\xi_2)\tilde{R}(\{\hat{\xi}_2\}) + \cdots + (-1)^d f(\xi_{d+1})\tilde{R}(\{\hat{\xi}_{d+1}\}) \right]. \tag{10.69}$$

For example, in case $d = 3$, we have

$$\Pi[\xi_1, \ldots, \xi_4](f) = \frac{f(\xi_1)D_{234} + f(\xi_2)D_{143} + f(\xi_3)D_{124} + f(\xi_4)D_{132}}{D_{234} + D_{143} + D_{124} + D_{132}},$$

$$\gamma_0 = -\frac{f(\xi_1)R_{234} + f(\xi_2)R_{143} + f(\xi_3)R_{124} + f(\xi_4)R_{132}}{D_{234} + D_{143} + D_{124} + D_{132}},$$

where $D_{ijm} = D(\xi_i, \xi_j, \xi_m)$ and

$$R_{ijm} = R(\xi_i, \xi_j) + R(\xi_j, \xi_m) + R(\xi_m, \xi_i).$$

As in case $d = 2$, we arrive at the following.

Proposition 10.12. *Let a family $\{\xi_1, \ldots, \xi_{d+1}\}$ in $\mathbf{R}^d$ satisfy condition (10.60), and let $f(\xi_1)$, $\ldots$, $f(\xi_{d+1})$ be arbitrary numbers. Then*

$$\Pi[\xi_1, \ldots, \xi_{d+1}](f) = \frac{1}{D} \sum_{i=1}^{d+1} [f(\xi_i)(-1)^{i+1} D(\{\hat{\xi}_i\})], \qquad (10.70)$$

and the minimum in (10.61) is attained on the single γ_0 given by (10.69).

Corollary 10.2. *Under the assumptions of Proposition 10.12,*

$$\Pi[\xi_1, \ldots, \xi_{d+1}](f) = \mathbf{E}f(\xi), \qquad (10.71)$$

$$\gamma_0 = \mathbf{E}\left[f(\xi) \frac{\tilde{R}(\{\hat{\xi}\})}{D(\{\hat{\xi}\})} \right], \qquad (10.72)$$

where the expectation is with respect to the probability law (10.65). This law is the unique risk-neutral probability law on $\{\xi_1, \ldots, \xi_{d+1}\}$, i.e. the one satisfying

$$\mathbf{E}(\xi) = \sum_{i=1}^{d+1} p_i \xi_i = 0. \qquad (10.73)$$

Proof. The only thing left to prove is that the law (10.65) satisfies (10.73). But as the r.h.s. of (10.73) can be written as the vector-valued determinant

$$\mathbf{E}f(\xi) = \det \begin{pmatrix} \xi_1 & \xi_2 & \cdots & \xi_{d+1} \\ \xi_1^1 & \xi_2^1 & \cdots & \xi_{d+1}^1 \\ & & \cdots & \\ \xi_1^d & \xi_2^d & \cdots & \xi_{d+1}^d \end{pmatrix},$$

that is a vector with co-ordinates

$$\det \begin{pmatrix} \xi_1^j & \xi_2^j & \cdots & \xi_{d+1}^j \\ \xi_1^1 & \xi_2^1 & \cdots & \xi_{d+1}^1 \\ & & \cdots & \\ \xi_1^d & \xi_2^d & \cdots & \xi_{d+1}^d \end{pmatrix}, \quad j = 1, \ldots, d.$$

it clearly vanishes.

To better visualize the above formulas, it is handy to delve a bit into their geometric meaning. Each term $(-1)^{i-1}D(\{\hat{\xi}_i\})$ in (10.64) equals $d!$ times the volume of the pyramid (polyhedron) with vertices $\{0 \cup \{\hat{\xi}_i\}\}$. The determinant D, being the volume of the parallelepiped built on $\xi_2 - \xi_1, \ldots, \xi_{d+1} - \xi_1$, equals $d!$ times the volume of the pyramid $\Pi[\xi_1, \ldots, \xi_{d+1}]$ in the affine space $\mathbf{R}^d$ with vertices being the end points of the vectors ξ_i, $i = 1, \ldots, d+1$. Consequently, formula (10.64) expresses the decomposition of the volume of the pyramid $\Pi[\xi_1, \ldots, \xi_{d+1}]$ into $d + 1$ parts, the volumes of the pyramids $\Pi[\{0 \cup \{\hat{\xi}_i\}\}]$ obtained by sectioning from the origin, and the weights of the distribution (10.65) are the ratios of these parts to the whole volume. Furthermore, the magnitude of the rotor $R(u_1, \ldots, u_{d-1})$ is known to equal the volume of the parallelepiped built on $u_1, \ldots, u_{d-1}$. Hence $\|\tilde{R}(\{\xi_i\})\|$ equals $(d-1)!$ times the volume (in the affine space $\mathbf{R}^d$) of the $(d-1)$-dimensional face of the pyramid $\Pi[\xi_1, \ldots, \xi_{d+1}]$ with vertices $\{\hat{\xi}_i\}$. Hence the magnitude of the ratios $\tilde{R}(\{\hat{\xi}_i\})/D(\{\hat{\xi}_i\})$, playing the roles of weights in (10.72), are the ratios of the $(d-1)!$ times $(d-1)$-dimensional volumes of the bases of the pyramids $\Pi[\{0 \cup \{\hat{\xi}_i\}\}]$ to the $d!$ times their full d-dimensional volumes. Consequently,

$$\frac{\|\tilde{R}(\{\hat{\xi}_i\})\|}{D(\{\hat{\xi}_i\})} = \frac{1}{h_i}, \tag{10.74}$$

where h_i is the length of the perpendicular from the origin to the affine hyperspace generated by the end points of the vectors $\{\hat{\xi}_i\}$. These geometric considerations lead directly to the following estimates for expressions (10.71) and (10.72).

Corollary 10.3.

$$|\Pi[\xi_1, \ldots, \xi_{d+1}](f)| \le \|f\|, \tag{10.75}$$

$$|\gamma_0| \le \|f\| \max_{i=1,\ldots,d+1} h_i^{-1}, \tag{10.76}$$

with h_i from (10.74).

These estimates are of importance for numerical calculations of γ_0 (yielding some kind of stability estimates with respect to the natural parameters). On the other hand, they are needed for nonlinear extensions of Proposition 10.12 (not discussed here).

Let us say that a finite family of non-vanishing vectors $\xi_1, \ldots, \xi_k$ in $\mathbf{R}^d$ are in *general position*, if the following conditions hold (extending naturally the corresponding conditions used in case $d = 2$):

(i) No d vectors out of this family are linearly dependent,

(ii) The collection $\{\xi_1, \ldots, \xi_k\}$ does not belong to any half-space, i.e. there is no $\omega \in \mathbf{R}^d$ such that $(\omega, \xi_i) > 0$ for all i.

It is worth noting that in case $k = d + 1$, assuming (i) and (ii) is equivalent to (10.60).

We are interested in evaluating the expression

$$\Pi[\xi_1, \ldots, \xi_k](f) = \min_{\gamma \in \mathbf{R}^d} \max_i [f(\xi_i) - (\xi_i, \gamma)]. \tag{10.77}$$

Theorem 10.3. *Let a family of non-vanishing vectors $\xi_1, \ldots, \xi_k$ in $\mathbf{R}^d$ satisfy (i) and (ii). Then*

$$\Pi[\xi_1, \ldots, \xi_k](f) = \max_{\{I\}} \mathbf{E}_I f(\xi), \tag{10.78}$$

where $\max$ *is taken over all families $\{\xi_i\}_{i \in I}$, $I \subset \{1, \ldots, k\}$ of size $|I| = d + 1$ that satisfy (ii) (i.e. such that the origin is contained in the interior of $\Pi[\xi_i, i \in I]$), and $\mathbf{E}_I$ denotes the expectation with respect to the unique risk-neutral probability on $\{\xi_i\}_{i \in I}$ (given by Proposition 10.12).*

Proof. This is the same as the proof of Theorem 10.2. The key geometrical observation, that any subset of the family $\{\xi_1, \ldots, \xi_k\}$ satisfying (i) and (ii) contains necessarily a subset with precisely $d + 1$ elements still satisfying (ii), is a direct consequence of the Carathéodory theorem.

Remark 10.2. As is easy to see, the max in (10.78) is attained on a family $\{\xi_i\}_{i \in I}$ if and only if

$$f(\xi_i) - (\gamma_I, \xi_i) \geq f(\xi_r) - (\gamma_I, \xi_r) \tag{10.79}$$

for any $i \in I$ and any r, where γ_I is the corresponding optimal value. Consequently, on the convex set of functions f satisfying inequalities (10.79) for all r, the mapping $\Pi[\xi_1, \ldots, \xi_k](f)$ is linear:

$$\Pi[\xi_1, \ldots, \xi_k](f) = \mathbf{E}_I f(\xi).$$

Let us expand a bit on the geometrical interpretation of the above results.

Let us call a probability law $\{p_1, \ldots, p_k\}$ on a finite set $\{\xi_1, \ldots, \xi_k\}$ of vectors in $\mathbf{R}^d$ *risk-neutral* (with respect to the origin) if the origin is its barycenter, that is

$$\mathbf{E}(\xi) = \sum_{i=1}^{k} p_i \xi_i = 0. \tag{10.80}$$

The geometrical interpretation we have in mind follows from the following simple observation.

Proposition 10.13. *For a family $\{\xi_1, \ldots, \xi_k\}$ satisfying (i) and (ii), the extreme points of the convex set of risk-neutral probabilities are risk-neutral probabilities with supports on subsets of size precisely $d + 1$, themselves satisfying conditions (i) and (ii).*

Proof. It is clear that risk-neutral probabilities with supports on subsets of size precisely $d+1$, themselves satisfying conditions (i) and (ii), are extreme points. In fact, if this were not the case for such a probability law, then it could be presented as a convex combination of other risk-neutral laws. But these risk-neutral laws would necessarily have the same support as the initial law, which would contradict the uniqueness of the risk-neutral law supported on $d + 1$ points in general position.

Assume $p = (p^1, \ldots, p^m)$ is a risk-neutral probability law on $m > d + 1$ points $\xi_1, \ldots, \xi_m$. Linear dependence of the vectors $\xi_2 - \xi_1, \ldots, \xi_m - \xi_1$ implies the existence of a non-vanishing vector $b = (b^1, \ldots, b^m)$ in $\mathbf{R}^m$ such that

$$\sum_{i=1}^m b^i = 0, \quad \sum_{i=1}^m b^i \xi_i = 0.$$

Hence for small enough ϵ, the vectors $p - \epsilon b$ and $p + \epsilon b$ are risk-neutral probability laws on $\xi_1, \ldots, \xi_m$. But

$$p = \frac{1}{2}(p - \epsilon b) + \frac{1}{2}(p + \epsilon b),$$

showing that p is not an extreme point.

Proposition 10.13 allows one to reformulate Theorem 10.3 as follows.

Theorem 10.4. *Let a family of non-vanishing vectors $\xi_1, \ldots, \xi_k$ in $\mathbf{R}^d$ satisfy (i) and (ii). Then the r.h.s. of formula (10.78), i.e.*

$$\Pi[\xi_1, \ldots, \xi_k](f) = \min_{\gamma \in \mathbf{R}^d} \max_i [f(\xi_i) - (\xi_i, \gamma)] = \max \mathbf{E}_I f(\xi) \qquad (10.81)$$

can be interpreted as the maximum of the averages of f with respect to all extreme points of the risk-neutral probabilities on $\xi_1, \ldots, \xi_k$. All these extreme probabilities are expressed in a closed form, given by (10.65).

It is natural to ask what happens if conditions (i) or (ii) do not hold. If (ii) does not hold, then $\Pi[\xi_1, \ldots, \xi_k](f)$ is not defined (equals to $-\infty$). If only (i) does not hold, one just has to take into account possible additional extreme risk-neutral probabilities coming from projections to subspaces. This leads to the following result obtained as a straightforward extension of Theorem 10.4.

Theorem 10.5. *Let a family of non-vanishing vectors $\xi_1, \ldots, \xi_k$ in $\mathbf{R}^d$ satisfy condition (ii). Then equation (10.81) still holds, where the maximum is taken over the averages of f with respect to all extreme points of the risk-neutral probabilities on $\xi_1, \ldots, \xi_k$. However, unlike the situation with condition (i) satisfied, these extreme risk-neutral measures may have support not only on families of size $d + 1$ in general positions, but also on families of any size $m + 1$, $m < d$, such that they belong to a subspace of dimension m and form a set of general position in this subspace.*

Remark 10.3. Notice that min and max in (10.81) are not interchangeable, as clearly

$$\max_{i} \min_{\gamma \in \mathbf{R}^d} [f(\xi_i) - (\xi_i, \gamma)] = -\infty.$$

Let us now formulate a mirror image of Theorem 10.4, where min and max are reversed. Its proof is almost literally the same as the proof of Theorem 10.4.

Theorem 10.6. *Under the assumptions of Theorem 10.4 the expression*

$$\tilde{\Pi}[\xi_1, \ldots, \xi_k](f) = \max_{\gamma \in \mathbf{R}^d} \min_{\xi_1, \ldots, \xi_k} [f(\xi_i) + (\xi_i, \gamma)] \tag{10.82}$$

can be evaluated by the formula

$$\Pi[\xi_1, \ldots, \xi_k](f) = \min_{I} \mathbf{E}_I f(\xi), \tag{10.83}$$

where min is taken over all families $\{\xi_i\}_{i \in I}$, $I \subset \{1, \ldots, k\}$ of size $|I| = d + 1$ that satisfy (ii), and $\mathbf{E}_I$ denotes the expectation with respect to the unique risk-neutral probability on $\{\xi_i\}_{i \in I}$. The min in (10.83) can be also interpreted as taken over all extreme points of the risk-neutral probabilities on $\xi_1, \ldots, \xi_k$.

Notice that the r.h.s. of (10.83) is similar to the formula for a coherent acceptability measure, see Artzner et al. [199] and Roorda, Schumacher and Engwerda [292]. However, in the theory of acceptability measures, the collection of measures with respect to which the minimization is performed is subjectively specified. In our model, this collection is the collection of all extreme points that arises objectively as an evaluation tool for our game-theoretic problem.

Remark 10.4. Coherent acceptability measures ϕ introduced in Artzner et al. [199] represent particular cases of nonlinear averages in the sense of Kolmogorov, see [266] and Maslov [281]. The distinguished feature that leads to the representation of ϕ as an infimum over probability measures is its super-additivity. Clearly, postulating sub-additivity, instead of super-additivity, would lead similarly to the representation as a supremum over probability measures, and hence to the analog of (10.81).

10.6   Interval model for option pricing

Here we develop a pure game-theoretic approach to option pricing in a multi-dimensional market (rainbow options), where risk-neutral probabilities emerge automatically from the robust control evaluation. The process of investment is considered as a zero-sum game of an investor with the Nature.

For basic examples of complete markets, like binomial model or geometric Brownian motion, our approach yields the same results as the classical (by now) risk-neutral evaluation developed by Cox-Ross-Rubinstein or Black-Scholes. However, for incomplete markets, like for rainbow options in multi-dimensional binomial or interval models, the coexistence of infinite number of risk-neutral measures precludes one from unified pricing of derivative securities by usual methods. Several competing methods were proposed for pricing options under these circumstances (see e.g. a review in Bingham and Kiesel [215]), most of them using certain subjective criteria, say a utility function for payoff or a certain risk measure. The difference in pricing arising from these methods is justified by referring vaguely to the intrinsic risk of incomplete markets. In our game-theoretic approach, no subjectivity enters the game. We define and calculate a hedge price, which is the minimal capital needed to meet the obligation for all performances of the markets, within the rules specified by the model (dominated hedging).

Though our price satisfies the so called 'no strictly acceptable opportunities' (NSAO) condition suggested in Carr, Geman and Madan [219], one still may argue of course that this is not a completely fair price, as the absence of an exogenously specified initial probability distribution does not allow us to speak about a.s. performance and implies perforce a possibility of an additional surplus. To address this issue, we observe that together with the hedging price for buying a security, that may be called an upper price, one can equally reasonable define a lower price, which can be looked as a hedge for selling the security. The difference of these two values can be considered as a precise measure of the intrinsic risk that is incident to incomplete markets. An alternative way to deal with possible unpredictable surplus, as suggested e.g. in Lyons [275] for models with unknown volatility, consists in specifying a possible cash-back, which should be due to the holder of an option when the moves of the prices (unpredictable at the beginning) turn out to be favorable.

Our method is robust enough to be able to accommodate various markets rules and settings including path dependent payoffs, American options, real options and transaction costs. The continuous-time limit is described by nonlinear and/or fractional Black-Scholes type equations.

1. Colored options as a game against Nature.

Recall that a European *option* is a contract between two parties where one party has right to complete a transaction in the future (with previously agreed amount, date and price) if he/she chooses, but is not obliged to do so. More precisely, consider a financial market dealing with several securities: the risk-free bonds (or bank account) and J common stocks, $J = 1, 2....$ In case $J > 1$, the corresponding options are called *colored or rainbow options* (J-colors option for a given J). Suppose the prices of the units of these securities, B_m and S_m^i, $i \in \{1, 2, ..., J\}$, change in discrete moments of time $m = 1, 2, ...$ according to the recurrent equations $B_{m+1} = \rho B_m$, where the $\rho \geq 1$ is an interest rate which remains unchanged over time, and $S_{m+1}^i = \xi_{m+1}^i S_m^i$, where $\xi_m^i, i \in \{1, 2, ..., J\}$, are unknown sequences taking values in some fixed intervals $M_i = [d_i, u_i] \subset \mathbf{R}$. This model generalizes the colored version of the classical CRR model in a natural way. In the latter a sequence ξ_m^i is confined to take values only among two boundary points d_i, u_i, and it is supposed to be random with some given distribution. In our model any value in the interval $[d_i, u_i]$ is allowed and no probabilistic assumptions are made.

The type of an option is specified by a given premium function f of J variables. The following are the standard examples:

option delivering the best of J risky assets and cash

$$f(S^1, S^2, ..., S^J) = \max(S^1, S^2, ..., S^J, K), \qquad (10.84)$$

calls on the maximum of J risky assets

$$f(S^1, S^2, ..., S^J) = \max(0, \max(S^1, S^2, ..., S^J) - K), \qquad (10.85)$$

multiple-strike options

$$f(S^1, S^2, ..., S^J) = \max(0, S^1 - K_1, S^2 - K_2, ..., S^J - K_J), \qquad (10.86)$$

portfolio options

$$f(S^1, S^2, ..., S^J) = \max(0, n_1 S^1 + n_2 S^2 + ... + n_J S^J - K), \qquad (10.87)$$

and spread options

$$f(S^1, S^2) = \max(0, (S^2 - S^1) - K). \qquad (10.88)$$

Here, the $S^1, S^2, ..., S^J$ represent the (in principle unknown at the start) expiration date values of the underlying assets, and $K, K_1, ..., K_J$ represent the (agreed from the beginning) strike prices. The presence of max in all these formulae reflects the basic assumption that the buyer is not obliged to exercise his/her right and would do it only in case of a positive gain.

The investor is supposed to control the growth of his/her capital in the following way. Let X_m denote the capital of the investor at the time $m = 1, 2, ...$ At each time $m - 1$ the investor determines his portfolio by choosing the numbers γ_m^i of common stocks of each kind to be held so that the structure of the capital is represented by the formula

$$X_{m-1} = \sum_{j=1}^{J} \gamma_m^j S_{m-1}^j + (X_{m-1} - \sum_{j=1}^{J} \gamma_m^j S_{m-1}^j),$$

where the expression in brackets corresponds to the part of his capital held in the bank account. The control parameters γ_m^j can take all real values, i.e. short selling and borrowing are allowed. The value ξ_m becomes known at the moment m and thus the capital at the moment m becomes

$$X_m = \sum_{j=1}^{J} \gamma_m^j \xi_m^j S_{m-1}^j + \rho(X_{m-1} - \sum_{j=1}^{J} \gamma_m^j S_{m-1}^j), \qquad (10.89)$$

if transaction costs are not taken into account.

If n is the prescribed *maturity date*, then this procedures repeats n times starting from some initial capital $X = X_0$ (selling price of an option) and at the end the investor is obliged to pay the premium f to the buyer. Thus the (final) income of the investor equals

$$G(X_n, S_n^1, S_n^2, ..., S_n^J) = X_n - f(S_n^1, S_n^2, ..., S_n^J). \qquad (10.90)$$

The evolution of the capital can thus be described by the n-step game of the investor with Nature, the behavior of the latter being characterized by unknown parameters ξ_m^j. The strategy of the investor is by definition any sequences of vectors $(\gamma_1, ..., \gamma_n)$ such that each γ_m could be chosen using the whole previous information: the sequences $X_0, ..., X_{m-1}$ and $S_0^i, ..., S_{m-1}^j$ (for every stock $j = 1, 2, ..., J$). The control parameters γ_m^j can take all real values, i.e. short selling and borrowing are allowed. A position of the game at any time m is characterized by $J + 1$ non-negative numbers $X_m, S_m^1, ..., S_m^J$ with the final income specified by the function

$$G(X, S^1, ..., S^J) = X - f(S^1, ..., S^J). \qquad (10.91)$$

The main definition of the theory is as follows. A strategy $\gamma_1, ..., \gamma_n$, of the investor is called a *hedge*, if for any sequence $(\xi_1, ..., \xi_n)$ the investor is able to meet his/her obligations, i.e.

$$G(X_n, S_n^1, ..., S_n^J) \geq 0.$$

The minimal value of the capital X_0 for which the hedge exists is called the *hedging price* H of an option.

Looking for the guaranteed payoffs means looking for the worst-case scenario (so called *robust-control approach*), i.e. for the minimax strategies. Thus if the final income is specified by a function G, the guaranteed income of the investor in a one-step game with the initial conditions $X, S^1, ..., S^J$ is given by the *Bellman operator*

$$\mathbf{B}G(X, S^1, ..., S^J)$$

$$= \frac{1}{\rho} \max_{\gamma} \min_{\{\xi^j \in [d_j, u_j]\}} G(\rho X + \sum_{i=1}^{J} \gamma^i \xi^i S^i - \rho \sum_{i=1}^{J} \gamma^i S^i, \xi^1 S^1, ..., \xi^J S^J),$$

$$(10.92)$$

and (as it follows from the standard backward induction argument, see e.g. [19] or Section 9.7) the guaranteed income of the investor in the n-step game with the initial conditions $X_0, S_0^1, ..., S_0^J$ is given by the formula

$$\mathbf{B}^n G(X_0, S_0^1, ..., S_0^J).$$

In our model G is given by (10.91). Clearly for G of the form

$$G(X, S^1, ..., S^J) = X - f(S^1, ..., S^J),$$

$$\mathbf{B}G(X, S^1, ..., S^J) = X - \frac{1}{\rho} \min_{\gamma} \max_{\xi} [f(\xi^1 S^1, \xi^2 S^2, ..., \xi^J S^J)$$

$$- \sum_{j=1}^{J} \gamma^j S^j (\xi^j - \rho)],$$

and hence

$$\mathbf{B}^n G(X, S^1, ..., S^J) = X - \frac{1}{\rho^n} (\mathcal{B}^n f)(S^1, ..., S^J),$$

where the *reduced Bellman operator* is defined as:

$$(\mathcal{B}f)(z^1,...,z^J) = \min_{\gamma} \max_{\{\xi^j \in [d_j, u_j]\}} [f(\xi^1 z^1, \xi^2 z^2, \ldots, \xi^J z^J) - \sum_{j=1}^{J} \gamma^j z^j (\xi^j - \rho)],$$

$$(10.93)$$

or, more concisely,

$$(\mathcal{B}f)(z) = \min_{\gamma} \max_{\{\xi^j \in [d_j, u_j]\}} [f(\xi \circ z) - (\gamma, \xi \circ z - \rho z)].$$

$$(10.94)$$

This leads to the following result from [87].

Theorem 10.7. *The minimal value of X_0 for which the income of the investor is non-negative (and which by definition is the hedge price H^n in the n-step game) is given by*

$$H^n = \frac{1}{\rho^n}(\mathcal{B}^n f)(S_0^1, \ldots, S_0^J).$$

$$(10.95)$$

2. Calculations of the Bellman operator via risk-neutral measures.

Let us now calculate the reduced Bellman operator of European colored options given by (10.93). Changing variables $\xi = (\xi^1, \ldots, \xi^J)$ to $\eta = \xi \circ z$ yields

$$(\mathcal{B}f)(z^1,...,z^J) = \min_{\gamma} \max_{\{\eta \in [z^i d_i, z^i u_i]\}} [f(\eta) - \sum_{i=1}^{J} \gamma^i (\eta^i - \rho z^i)],$$

$$(10.96)$$

or, by shifting,

$$(\mathcal{B}f)(z^1,...,z^J) = \min_{\gamma} \max_{\{\eta \in [z^i (d_i - \rho), z^i (u_i - \rho)]\}} [\tilde{f}(\eta) - (\gamma, \eta)]$$

$$(10.97)$$

with $\tilde{f}(\eta) = f(\eta + \rho z)$. Assuming f is convex (possibly not strictly), we find ourselves in the setting of Section 10.5 with Π being the rectangular parallelepiped

$$\Pi_{z,\rho} = \times_{i=1}^{J} [z^i(d_i - \rho), z^i(u_i - \rho)],$$

with vertices

$$\eta_I = \xi_I \circ z - \rho z,$$

where

$$\xi_I = \{d_i|_{i \in I}, u_j|_{j \notin I}\},$$

are the vertices of the normalized parallelepiped

$$\Pi = \times_{i=1}^{J} [d_i, u_i],$$

$$(10.98)$$

parametrized by all subsets (including the empty one) $I \subset \{1, \ldots, J\}$.

Since the origin is an internal point of Π (because $d_i < \rho < u_i$), condition (ii) of Theorem 10.4 is satisfied. Condition (i) is rough in the sense that it is fulfilled for an open dense subset of pairs (d_i, u_i). Applying Theorem 10.4 (and Remark 10.2) to (10.97) and returning back to ξ yields the following.

Theorem 10.8. *If the vertices ξ_I of the parallelepiped Π are in general position in the sense that for any J subsets $I_1, \ldots, I_J$, the vectors $\{\xi_{I_k} - \rho\mathbf{1}\}_{k=1}^{J}$ are independent in $\mathbf{R}^J$, then*

$$(\mathcal{B}f)(z) = \max_{\{\Omega\}} \mathbf{E}_\Omega f(\xi \circ z), \quad z = (z^1, \ldots, z^J), \tag{10.99}$$

where $\{\Omega\}$ is the collection of all subsets $\Omega = \xi_{I_1}, \ldots, \xi_{I_{J+1}}$ of the set of vertices of Π, of size $J + 1$, such that their convex hull contains $\rho\mathbf{1}$ as an interior point ($\mathbf{1}$ is the vector with all coordinates 1), and where $\mathbf{E}_\Omega$ denotes the expectation with respect to the unique probability law $\{p_I\}$, $\xi_I \in \Omega$, on the set of vertices of Π, which is supported on Ω and is risk-neutral with respect to $\rho\mathbf{1}$, that is

$$\sum_{I \subset \{1, \ldots, J\}} p_I \xi_I = \rho\mathbf{1}. \tag{10.100}$$

Moreover, if

$$f(\xi \circ z) - (\gamma_{I_1, \ldots, I_{J+1}}, (\xi - \rho\mathbf{1}) \circ z) \geq f(\zeta \circ z) - (\gamma_{I_1, \ldots, I_{J+1}}, (\zeta - \rho\mathbf{1}) \circ z)$$

for all vertices ξ, ζ such that $\xi \in \Omega$ and $\zeta \notin \Omega$, where $\gamma_{I_1, \ldots, I_{J+1}}$ is the corresponding optimal value for the polyhedron $\Pi[\xi_{I_1}, \ldots, \xi_{I_{J+1}}]$, then

$$(\mathcal{B}f)(z^1, \ldots, z^J) = \mathbf{E}_\Omega f(\xi \circ z). \tag{10.101}$$

Risk-neutrality now corresponds to its usual meaning in finance, i.e. (10.100) means that all discounted stock prices are martingales.

Notice that the max in (10.99) is over a finite number of explicit expressions, which is of course a great achievement as compared with initial minimax over an infinite set. In particular, it reduces the calculation of the iterations $\mathcal{B}^n f$ to the calculation for a control Markov chain. Let us also stress that the number of eligible Ω in (10.99) is the number of different pyramids (convex polyhedrons with $J + 1$ vertices) with vertices taken from the vertices of Π and containing $\rho\mathbf{1}$ as an interior point. Hence this number can be effectively calculated.

Let us point our some properties of the operator $\mathcal{B}$ given by (10.99) that are obvious, but important for practical calculations: it is non-expansive:

$$\|\mathcal{B}(f_1) - \mathcal{B}(f_2)\| \leq \|f_1 - f_2\|,$$

and homogeneous (both with respect to addition and multiplication):

$$\mathcal{B}(\lambda + f) = \lambda + \mathcal{B}(f), \quad \mathcal{B}(\lambda f) = \lambda \mathcal{B}(f)$$

for a function f and $\lambda \in \mathbf{R}$ (resp. $\lambda > 0$) for the first (resp. second) equation. Finally, if f_p is a power function, that is

$$f_p(z) = (z^1)^{i_1} \cdots (z^J)^{i_J},$$

then $f_p(\xi \circ z) = f_p(\xi) f_p(z)$, implying

$$(\mathcal{B}^n f_p)(z) = ((\mathcal{B}f_p)(1))^n f_p(z). \tag{10.102}$$

Therefore, power functions are invariant under $\mathcal{B}$ (up to a multiplication by a constant). Consequently, if for a payoff f one can find a reasonable approximation by a power function, that is there exists a power function f_p such that $\|f - f_p\| \le \epsilon$, then

$$\|\mathcal{B}^n f - \lambda^n f_p\| = \|f - f_p\| \le \epsilon, \quad \lambda = (\mathcal{B}f_p)(1), \tag{10.103}$$

so that an approximate calculation of $\mathcal{B}^n f$ is reduced to the calculation of one number λ. This implies the following scheme for an approximate evaluation of $\mathcal{B}$: first find the best fit to f in terms of functions $\alpha + f_p$ (where f_p is a power function and α a constant), and then use (10.103).

3. Upper and lower values, intrinsic risk.

The celebrated no-arbitrage property of the hedge price of an option in the CRR or Black-Scholes models means that, a.s., with respect to the initial probability distribution on paths, the investor cannot get an additional surplus when adhering to the hedge strategy that guarantees that there could be no loss. In our setting, even though our formula in case $J = 1$ coincides with the CRR formula, we do not assume any initial law on paths, so that the notion of 'no arbitrage' is not specified either.

It is a characteristic feature of our models that picking up an a priori probability law with support on all paths leads to an incomplete market, that is to the existence of infinitely many equivalent martingale measures, which fail to identify a fair price in a unique consistent way. Notice that our extreme points are absolutely continuous, but usually not equivalent to a measure with full support.

For the analysis of incomplete markets it is of course natural to look for some subjective criteria to specify a price. Lots of work of different authors has been devoted to this endeavor. Our approach is to search for objective bounds (price intervals), which are given by our hedging strategies.

Remark 10.5. Apart from supplying the lower price (as below), one can also argue about the reasonability of our main hedging price, noting that a chance for a possible surplus can be (and actually is indeed) compensated by inevitable inaccuracy of a model, as well as by transaction costs (if they are not properly taken into account).

For completeness, let us recall the general definitions of lower and upper prices, in the game-theoretic approach to probability, given in Shafer and Vovk [174]. Assume a process (a sequence of real numbers of a fixed length, say n, specifying the evolution of the capital of an investor) is specified by alternating moves of two players, an investor and the Nature, with complete information (all eligible moves of each player and their results are known to each player at any time, and the moves become publicly known at the moment when decision is made). Let us denote by $X_\gamma^\alpha(\xi)$ the last number of the resulting sequence, starting with an initial value α and obtained by applying the strategy γ (of the investor) and ξ (of Nature). By a random variable we mean just a function on the set of all possible paths. The upper value (or the upper expectation) $\overline{\mathbf{E}}f$ of a random variable f is defined as the minimal capital of the investor such that he/she has a strategy that guarantees that at the final moment of time, his capital is enough to buy f, i.e.

$$\overline{\mathbf{E}}f = \inf\{\alpha : \exists\gamma : \forall\xi,\, X_\gamma^\alpha(\xi) - f(\xi) \geq 0\}.$$

Dually, the lower value (or the lower expectation) $\underline{\mathbf{E}}f$ of a random variable f is defined as the maximum capital of the investor such that he/she has a strategy that guarantees that at the final moment of time, his capital is enough to sell f, i.e.

$$\underline{\mathbf{E}}f = \sup\{\alpha : \exists\gamma : \forall\xi,\, X_\gamma^\alpha(\xi) + f(\xi) \geq 0\}.$$

One says that the prices are *consistent* if $\overline{\mathbf{E}}f \geq \underline{\mathbf{E}}f$. If these prices coincide, we are in a kind of abstract analog of a complete market. In the general case, upper and lower prices are also referred to as *seller and buyer* prices respectively.

It is seen now that in this terminology our hedging price for a derivative security is the upper (or seller) price. The lower price can be defined similarly. Namely, in the setting of Section 10.6, lower price is given by

$$\frac{1}{\rho^n}(\mathcal{B}_{low}^n f)(S_0^1, \ldots, S_0^J),$$

where

$$(\mathcal{B}_{low}f)(z) = \max_\gamma \min_{\{\xi^j \in [d_j, u_j]\}} [f(\xi \circ z) - (\gamma, \xi \circ z - \rho z)]. \qquad (10.104)$$

In this simple interval model and for convex f this expression is trivial, it equals $f(\rho z)$. On the other hand, if our f is concave, or, more generally, if we allow only finitely many jumps, which leads, instead of (10.104), to the operator

$$(\mathcal{B}_{low}f)(z) = \max_{\gamma} \min_{\{\xi^j \in \{d_j, u_j\}\}} [f(\xi \circ z) - (\gamma, \xi \circ z - \rho z)], \qquad (10.105)$$

then Theorem 10.6 applies giving for the lower price the dual expression to the upper price (10.99), where maximum is turned to minimum (over the same set of extreme risk-neutral measures):

$$(\mathcal{B}_{low}f)(z) = \min_{\{\Omega\}} \mathbf{E}_\Omega f(\xi \circ z), \quad z = (z^1, \dots, z^J). \qquad (10.106)$$

The difference between lower and upper prices can be considered as a measure of intrinsic risk of an incomplete market.

4. Path dependence and other modifications.

The theory above is rough, in the sense that it can be easily modified to accommodate various additional mechanisms of price generations including transaction costs and American options. Here we shall discuss other three modifications: path-dependent payoffs, time-dependent jumps (including variable volatility) and nonlinear jump formations. For simplicity, we shall discuss these extensions separately, but any of their combinations can be easily dealt with.

Let us start with path dependent payoffs. That is, we generalize the setting above by making the payoff f at time m depend on the whole history of the price evolutions, i.e. being defined by a function $f(S_0, S_1, \dots, S_m)$, $S_i = (S_i^1, \dots, S_i^J)$, on $\mathbf{R}^{J(m+1)}$. The state of the game at time m must be now specified by $(m+1)J + 1$ numbers

$$X_m, \ S_i = (S_i^1, \dots, S_i^J), \quad i = 0, \dots, m.$$

The final payoff in the n-step game is now $G = X - f(S_0, \dots, S_n)$ and at the penultimate period $n-1$ (when $S_0, \dots, S_{n-1}$ are known) payoff equals

$$\mathbf{B}G(X, S_0, \dots, S_{n-1})$$

$$= X - \frac{1}{\rho} \min_{\gamma} \max_{\xi} [f(S_0, \dots, S_{n-1}, \xi \circ S_{n-1}) - (\gamma, S_{n-1} \circ (\xi - \rho \mathbf{1}))]$$

$$= X - \frac{1}{\rho} (\mathcal{B}_{n-1} f)(S_0, \dots, S_{n-1}),$$

where the modified *reduced Bellman operators* are now defined as

$$(\mathcal{B}_{m-1}f)(z_0, \ldots, z_{m-1})$$

$$= \min_{\gamma} \max_{\{\xi^j \in [d_j, u_j]\}} [f(z_0, \ldots, z_{m-1}, \xi \circ z_{m-1}) - (\gamma, \xi \circ z - \rho z)]. \quad (10.107)$$

Consequently, by dynamic programming, the guaranteed payoff at the initial moment of time equals

$$X - \frac{1}{\rho^n} \mathcal{B}_0(\mathcal{B}_1 \cdots (\mathcal{B}_{n-1}f) \cdots),$$

and hence the hedging price becomes

$$H^n = \frac{1}{\rho^n} \mathcal{B}_0(\mathcal{B}_1 \cdots (\mathcal{B}_{n-1}f) \cdots). \quad (10.108)$$

No essential changes are required if possible jump sizes are time-dependent. Only the operators $\mathcal{B}_{m-1}$ from (10.107) have to be generalized to

$$(\mathcal{B}_{m-1}f)(z_0, \ldots, z_{m-1})$$

$$= \min_{\gamma} \max_{\{\xi^j \in [d_j^m, u_j^m]\}} [f(z_0, \ldots, z_{m-1}, \xi \circ z_{m-1}) - (\gamma, \xi \circ z - \rho z)], \quad (10.109)$$

where the pairs (d_j^m, u_j^m), $j = 1, \ldots, J$, $m = 1, \ldots, n$ specify the model.

Let us turn to nonlinear jump patterns. Generalizing the setting of Section 10.6 let us assume, instead of the stock price evolution model $S_{m+1} = \xi \circ S_m$, that we are given k transformations $g_i : \mathbf{R}^J \to \mathbf{R}^J$, $i = 1, \ldots, k$, which give rise naturally to two models of price dynamics: either

(i) at time $m + 1$ the price S_{m+1} belongs to the closure of the convex hull of the set $\{g_i(S_m)\}$, $i = 1, \ldots, k$ (interval model), or

(ii) S_{m+1} is one of the points $\{g_i(S_m)\}$, $i = 1, \ldots, k$.

Since the first model can be approximated by the second one (by possibly increasing the number of transformations g_i), we shall work with the second model.

Remark 10.6. Notice that maximizing a function over a convex polyhedron is equivalent to its maximization over the edges of this polyhedron. Hence, for convex payoffs the two models above are fully equivalent. However, on the one hand, not all reasonable payoffs are convex, and on the other hand, when it comes to minimization (which one needs, say, for lower prices, the situation becomes rather different).

Assuming for simplicity that possible jump sizes are time independent and the payoffs depend only on the end-value of a path, the reduced Bellman operator (10.93) becomes

$$(\mathcal{B}f)(z) = \min_{\gamma} \max_{i \in \{1,\ldots,k\}} [f(g_i(z)) - (\gamma, g_i(z) - \rho z)], \quad z = (z^1, \ldots, z^J),$$
(10.110)

or equivalently

$$(\mathcal{B}f)(z) = \min_{\gamma} \max_{\eta_i \in \{g_i(z) - \rho z\}, i=1,\ldots,k} [f(\eta_i + \rho z) - (\gamma, \eta_i)].$$
(10.111)

The hedge price is still given by (10.95) and operator (10.110) is calculated by Theorem 10.3.

It is worth noting that if $k = d + 1$ and $\{g_i(z)\}$ form a collection of vectors in a general position, the corresponding risk-neutral probability is unique. Consequently our hedge price becomes fair in the sense of 'no arbitrage' in the strongest sense: no positive surplus is possible for all paths of the stock price evolutions (if the hedge strategy is followed). In particular, the evaluation of hedge strategies can be carried out in the framework of the standard approach to option pricing. Namely, choosing as an initial (real world) probability on jumps an arbitrary measure with a full support, one concludes that there exists a unique risk-neutral equivalent martingale measure, explicitly defined via formula (10.65), and the hedge price calculated by the iterations of operator (10.111) coincides with the standard risk-neutral evaluation of derivative prices in complete markets.

10.7   Options with sub-modular payoffs

One can get an essential reduction in the combinatorics of Theorem 10.8 (i.e. in the number of eligible Ω) under additional assumptions on the payoff f. The most natural one in the context of options turns out to be the notion of sub-modularity. A function $f : \mathbf{R}_+^2 \to \mathbf{R}_+$ is called *sub-modular*, if the inequality

$$f(x_1, y_2) + f(x_2, y_1) \geq f(x_1, y_1) + f(x_2, y_2)$$

holds whenever $x_1 \leq x_2$ and $y_1 \leq y_2$. Similarly, a function $f : \mathbf{R}_+^d \to \mathbf{R}_+$ is called *sub-modular* sub-modular if

$$f(x \bigvee y) + f(x \bigwedge y) \leq f(x) + f(y),$$

where $\bigvee$ (respectively $\bigwedge$) denotes the Pareto (coordinate-wise) maximum (respectively minimum). In other words, it is submodular, if it is submodular as a function of any two variables (for fixed values of other variables).

Remark 10.7. If f is twice continuously differentiable, then it is submodular if and only if $\frac{\partial^2 f}{\partial z_i \partial z_j} \leq 0$ for all $i \neq j$.

The payoffs of the first three examples of rainbow options, given in Section 10.6, that is those defined by (10.84), (10.85), (10.86), are submodular. Let us explain, on the examples of two and three colors J, how the assumptions of sub-modularity can simplify Theorem 10.8.

Let first $J = 2$. The polyhedron (10.98) is then a rectangle. The key observation is as follows. From sub-modularity of f it follows that if Ω in Theorem 10.8 is either

$$\Omega_{12} = \{(d_1, d_2), (d_1, u_2), (u_1, u_2)\}$$

or

$$\Omega_{21} = \{(d_1, d_2), (u_1, d_2), (u_1, u_2)\},$$

then $(f, \eta + \rho z) - (\gamma_0, \eta)$ coincide for all vertices ξ of Π, where γ_0 is the point of minimum in (10.97). In fact, in these cases

$$f(\eta_{12} + \rho z) - (\gamma_0, \eta_{12}) = f(\eta_\emptyset + \rho z) - (\gamma_0, \eta_\emptyset)$$

and moreover

$$f(\eta_{12} + \rho z) - (\gamma_0, \eta_{12}) \geq f(\eta_1 + \rho z) - (\gamma_0, \eta_1), \qquad (10.112)$$

$$f(\eta_\emptyset + \rho z) - (\gamma_0, \eta_\emptyset) \geq f(\eta_2 + \rho z) - (\gamma_0, \eta_2). \qquad (10.113)$$

By adding the last two inequalities and noting that $\eta_{12} + \eta_\emptyset = \eta_1 + \eta_2$ it follows that

$$f(\eta_{12} + \rho z) + f(\eta_\emptyset + \rho z) \geq f(\eta_1 + \rho z) + f(\eta_2 + \rho z).$$

But submodularity implies the opposite inequality. Hence

$$f(\eta_{12} + \rho z) + f(\eta_\emptyset + \rho z) = f(\eta_1 + \rho z) + f(\eta_2 + \rho z),$$

and both inequalities (10.112) and (10.113) are in fact equalities.

Therefore, Ω_{12} and Ω_{21} can be discarded in Theorem 10.8, i.e. the maximum is always achieved either on

$$\Omega_d = \{(d_1, d_2), (d_1, u_2), (u_1, d_2)\},$$

or on

$$\Omega_u = \{(d_1, u_2), (u_1, d_2), (u_1, u_2)\}.$$

But the interiors of the triangle formed by Ω_u and Ω_d do not intersect, so that each point of Π (in general position) lies only in one of them (and this position does not depend any more on f). Hence, depending on the position of $\rho 1$ in Π, the expression (10.99) reduces either to $\mathbf{E}_{\Omega_u}$ or to $\mathbf{E}_{\Omega_d}$. This yields the following result from [87].

Theorem 10.9. *Let $J = 2$ and f be convex sub-modular. Denote*

$$\kappa = \frac{(u_1 u_2 - d_1 d_2) - \rho(u_1 - d_1 + u_2 - d_2)}{(u_1 - d_1)(u_2 - d_2)} = 1 - \frac{\rho - d_1}{u_1 - d_1} - \frac{\rho - d_2}{u_2 - d_2}.$$
(10.114)

If $\kappa \geq 0$, then $(\mathcal{B}f)(z_1, z_2)$ equals

$$\frac{\rho - d_1}{u_1 - d_1} f(u_1 z_1, d_2 z_2) + \frac{\rho - d_2}{u_2 - d_2} f(d_1 z_1, u_2 z_2) + \kappa f(d_1 z_1, d_2 z_2),$$
(10.115)

and the corresponding optimal strategies are

$$\gamma^1 = \frac{f(u_1 z_1, d_2 z_2) - f(d_1 z_1, d_2 z_2)}{z_1(u_1 - d_1)}, \quad \gamma^2 = \frac{f(d_1 z_1, u_2 z_2) - f(d_1 z_1, d_2 z_2)}{z_2(u_2 - d_2)}.$$

If $\kappa \leq 0$, the $(\mathcal{B}f)(z_1, z_2)$ equals

$$\frac{u_1 - \rho}{u_1 - d_1} f(d_1 z_1, u_2 z_2) + \frac{u_2 - \rho}{u_2 - d_2} f(u_1 z_1, d_2 z_2) + |\kappa| f(u_1 z_1, u_2 z_2),$$
(10.116)

and

$$\gamma^1 = \frac{f(u_1 z_1, u_2 z_2) - f(d_1 z_1, u_2 z_2)}{z_1(u_1 - d_1)}, \quad \gamma^2 = \frac{f(u_1 z_1, u_2 z_2) - f(u_1 z_1, d_2 z_2)}{z_2(u_2 - d_2)}.$$

Clearly the linear operator $\mathcal{B}$ preserves the set of convex sub-modular functions. Hence one can use this formula recursively to obtain all powers of $\mathcal{B}$ in a closed form. For instance in case $\kappa = 0$ one obtains for the hedge price the following two-color extension of the classical Cox-Ross-Rubinstein formula:

$$\mathcal{B}^n f(S_0^1, S_0^2)$$

$$= \rho^{-n} \sum_{k=0}^{n} C_n^k \left(\frac{\rho - d_1}{u_1 - d_1} \right)^k \left(\frac{\rho - d_2}{u_2 - d_2} \right)^{n-k} f(u_1^k d_1^{n-k} S_0^1, d_2^k u_2^{n-k} S_0^2).$$
(10.117)

Now let $J = 3$. Then polyhedron (10.98) is a parallelepiped in $\mathbf{R}^3$. As above, from sub-modularity projected on the first two co-ordinates we conclude that whenever the vertices (d_1, d_2, d_3) and (u_1, u_2, d_3) are in Ω, then

$$f(\eta_I + \rho z) - (\gamma_0, \eta)$$

should coincide for I being $\{1,2,3\}$, $\{3\}$, $\{2,3\}$, $\{1,3\}$. In other word, the pair (d_1, d_2, d_3), (u_1, u_2, d_3) can be always substituted by the pair (u_1, d_2, d_3), (d_1, u_2, d_3). Consequently, those Ω containing the pair (d_1, d_2, d_3), (u_1, u_2, d_3) are superfluous and can be discarded from the possible Ω competing in formula (10.99). Similarly, we can discard all those Ω containing the following six pairs: three pairs containing (d_1, d_2, d_3) and one among (d_1, d_2, u_3), (d_1, u_2, d_3), (u_1, d_2, d_3), and three pairs containing (u_1, u_2, u_3) and one among (d_1, u_2, u_3), (u_1, u_2, d_3), (u_1, d_2, u_3).

These considerations reduce dramatically the number of eligible Ω. In particular, if $\rho 1$ lies in the tetrahedron Ω_d formed by the vertices (d_1, d_2, d_3), (d_1, d_2, u_3), (d_1, u_2, d_3), (u_1, d_2, d_3), then possible Ω containing such ρ should necessarily contain the vertex (d_1, d_2, d_3), and hence, excluding superfluous combinations, it remains only one eligible Ω, which is Ω_d. If $\rho 1$ lies in the tetrahedron Ω_u formed by the vertices (u_1, u_2, u_3), (d_1, u_2, u_3), (u_1, u_2, d_3), (u_1, d_2, u_3), then the only eligible Ω is Ω_u. Formally these cases are easily seen to be distinguished by the inequalities $\alpha_{123} > 0$ and $\alpha_{123} < -1$ respectively, where

$$\alpha_{123} = \left(1 - \frac{u_1 - \rho}{u_1 - d_1} - \frac{u_2 - \rho}{u_2 - d_2} - \frac{u_3 - \rho}{u_3 - d_3} \right). \tag{10.118}$$

This yields the following result (by continuity we are able to write $\alpha_{123} \geq 0$ instead of a strict inequality), where we use the following notation: for a set $I \subset \{1, 2, ..., J\}$, $f_I(z)$ is $f(\xi^1 z_1, ..., \xi^J z_J)$ with $\xi^i = d_i$ for $i \in I$ and $\xi_i = u_i$ for $i \notin I$.

Theorem 10.10. *Let $J = 3$ and f be continuous convex and sub-modular.*
(i) If $\alpha_{123} \geq 0$, then $(\mathcal{B}f)(z)$ equals

$$\frac{1}{\rho} \left[\alpha_{123} f_\emptyset(z) + \frac{u_1 - \rho}{u_1 - d_1} f_{\{1\}}(z) + \frac{u_2 - \rho}{u_2 - d_2} f_{\{2\}}(z) + \frac{u_3 - r}{u_3 - d_3} f_{\{3\}}(z) \right].$$

(ii) If $\alpha_{123} \leq -1$, then

$$(\mathcal{B}f)(z) = \frac{1}{\rho} \left[-(\alpha_{123} + 1) f_{\{1,2,3\}}(z) \right.$$

$$\left. + \frac{\rho - d_1}{u_1 - d_1} f_{\{2,3\}}(z) + \frac{\rho - d_2}{u_2 - d_2} f_{\{1,3\}}(z) + \frac{\rho - d_3}{u_3 - d_3} f_{\{1,2\}}(z) \right].$$

Hence in these cases, our $\mathcal{B}$ again reduces to a linear form, allowing for a straightforward calculation of its iterations, as in case $J = 2$ above.

Suppose now that $\rho 1$ lies neither in the tetrahedron Ω_d, nor in Ω_u (i.e. neither of the conditions of Theorem 10.10 are satisfied). From the above reductions of possible Ω, it follows that in that case one can discard all Ω containing either (d_1, d_2, d_3) or (u_1, u_2, u_3). Hence only six vertices are left for eligible Ω. From the consideration of general position we further deduce that altogether only six Ω are possible, namely the three tetrahedra containing the vertices (d_1, d_2, u_3), (d_1, u_2, d_3), (u_1, d_2, d_3) and one vertex from (d_1, u_2, u_3), (u_1, u_2, d_3), (u_1, d_2, u_3), and symmetrically the three tetrahedra containing the vertices (d_1, u_2, u_3), (u_1, u_2, d_3), (u_1, d_2, u_3) and one vertex from (d_1, d_2, u_3), (d_1, u_2, d_3), (u_1, d_2, d_3). However, any particular point in general position belongs to only three out of these six leaving in formula (10.99) the max over three possibilities only. The particular choice of these three tetrahedrons depends on the coefficients

$$
\begin{aligned}
\alpha_{12} &= \left(1 - \tfrac{u_1 - r}{u_1 - d_1} - \tfrac{u_2 - r}{u_2 - d_2}\right) \\
\alpha_{13} &= \left(1 - \tfrac{u_1 - r}{u_1 - d_1} - \tfrac{u_3 - r}{u_3 - d_3}\right) \\
\alpha_{23} &= \left(1 - \tfrac{u_2 - r}{u_2 - d_2} - \tfrac{u_3 - r}{u_3 - d_3}\right),
\end{aligned}
\tag{10.119}
$$

and leads to the following result obtained in Hucki and Kolokoltsov [74] (though with much more elaborate proof than here).

Theorem 10.11. *Let again f be convex and sub-modular, but now $0 > \alpha_{123} > -1$.*

(i) If $\alpha_{12} \geq 0$, $\alpha_{13} \geq 0$ and $\alpha_{23} \geq 0$, then

$$
(\mathcal{B}f)(\mathbf{z}) = \tfrac{1}{r} \max \left\{ \begin{array}{l} (-\alpha_{123}) f_{\{1,2\}}(\mathbf{z}) + \alpha_{13} f_{\{2\}}(\mathbf{z}) + \alpha_{23} f_{\{1\}}(\mathbf{z}) + \tfrac{u_3 - r}{u_3 - d_3} f_{\{3\}}(\mathbf{z}) \\ (-\alpha_{123}) f_{\{1,3\}}(\mathbf{z}) + \alpha_{12} f_{\{3\}}(\mathbf{z}) + \alpha_{23} f_{\{1\}}(\mathbf{z}) + \tfrac{u_2 - r}{u_2 - d_2} f_{\{2\}}(\mathbf{z}) \\ (-\alpha_{123}) f_{\{2,3\}}(\mathbf{z}) + \alpha_{12} f_{\{3\}}(\mathbf{z}) + \alpha_{13} f_{\{2\}}(\mathbf{z}) + \tfrac{u_1 - r}{u_1 - d_1} f_{\{1\}}(\mathbf{z}) \end{array} \right\}.
$$

(ii) If $\alpha_{ij} \leq 0$, $\alpha_{jk} \geq 0$ and $\alpha_{ik} \geq 0$, where $\{i, j, k\}$ is an arbitrary permutation of the set $\{1, 2, 3\}$, then

$$
(\mathcal{B}f)(\mathbf{z}) = \tfrac{1}{r} \max \left\{ \begin{array}{l} (-\alpha_{ijk}) f_{\{i,j\}}(\mathbf{z}) + \alpha_{ik} f_{\{j\}}(\mathbf{z}) + \alpha_{jk} f_{\{i\}}(\mathbf{z}) + \tfrac{u_k - r}{u_k - d_k} f_{\{k\}}(\mathbf{z}) \\ \alpha_{jk} f_{\{i\}}(\mathbf{z}) + (-\alpha_{ij}) f_{\{i,j\}}(\mathbf{z}) + \tfrac{u_k - r}{u_k - d_k} f_{\{i,k\}}(\mathbf{z}) - \tfrac{d_i - r}{u_i - d_i} f_{\{j\}}(\mathbf{z}) \\ \alpha_{ik} f_{\{j\}}(\mathbf{z}) + (-\alpha_{ij}) f_{\{i,j\}}(\mathbf{z}) + \tfrac{u_k - r}{u_k - d_k} f_{\{j,k\}}(\mathbf{z}) - \tfrac{d_j - r}{u_j - d_j} f_{\{i\}}(\mathbf{z}) \end{array} \right\}.
$$

(iii) If $\alpha_{ij} \geq 0$, $\alpha_{jk} \leq 0$ and $\alpha_{ik} \leq 0$, where $\{i, j, k\}$ is an arbitrary permutation of the set $\{1, 2, 3\}$, then

$$
(\mathcal{B}f)(\mathbf{z}) = \tfrac{1}{r} \max \left\{ \begin{array}{l} \alpha_{ij} f_{\{k\}}(\mathbf{z}) + (-\alpha_{jk}) f_{\{j,k\}}(\mathbf{z}) + \tfrac{u_i - r}{u_i - d_i} f_{\{i,k\}}(\mathbf{z}) - \tfrac{d_k - r}{u_k - d_k} f_{\{j\}}(\mathbf{z}) \\ \alpha_{ij} f_{\{k\}}(\mathbf{z}) + (-\alpha_{ik}) f_{\{i,k\}}(\mathbf{z}) + \tfrac{u_j - r}{u_j - d_j} f_{\{j,k\}}(\mathbf{z}) - \tfrac{d_k - r}{u_k - d_k} f_{\{i\}}(\mathbf{z}) \\ (\alpha_{123} + 1) f_{\{k\}}(\mathbf{z}) - \alpha_{jk} f_{\{j,k\}}(\mathbf{z}) - \alpha_{ik} f_{\{i,k\}}(\mathbf{z}) - \tfrac{d_k - r}{u_k - d_k} f_{\{i,j\}}(\mathbf{z}) \end{array} \right\}.
$$

One has to stress here that the application of Theorem 10.11 is rather limited: as $\mathcal{B}$ is not reduced to a linear form, it is not clear how to use it for the iterations of $\mathcal{B}$, because the sub-modularity does not seem to be preserved under such $\mathcal{B}$.

10.8 Nonlinear and fractional Black-Scholes equations

Our models and results are most naturally adapted to a discrete-time setting, which is not a disadvantage from the practical point of view, as all concrete calculations are anyway carried out on discrete data. However, for qualitative analysis, it is desirable to be able to see what is going on in the continuous-time limit. This limit can also be simpler sometimes and hence be used as an approximation to a less tractable discrete model. Having this in mind, let us analyze possible limits as the time between jumps and their sizes tend to zero.

Let us work with the general model of nonlinear jumps, with the reduced Bellman operator of form (10.110). Suppose the maturity time is T. Let us decompose the planning time $[0, T]$ into n small intervals of length $\tau = T/n$, and assume

$$g_i(z) = z + \tau^\alpha \phi_i(z), \quad i = 1, \ldots, k, \tag{10.120}$$

with some functions ϕ_i and a constant $\alpha \in [1/2, 1]$. Thus the jumps during time τ are of the order of magnitude τ^α. As usual, we assume that the risk-free interest rate per time τ equals

$$\rho = 1 + r\tau,$$

with $r > 0$.

From (10.110) we deduce for the one-period Bellman operator the expression

$$\mathcal{B}_\tau f(z) = \frac{1}{1 + r\tau} \max_I \sum_{i \in I} p_i^I(z, \tau) f(z + \tau^\alpha \phi_i(z)), \tag{10.121}$$

where I are subsets of $\{1, \ldots, n\}$ of size $|I| = J + 1$ such that the family of vectors $z + \tau^\alpha \phi_i(z)$, $i \in I$, are in general position and $\{p_i^I(z, \tau)\}$ is the risk-neutral probability law on such a family, with respect to ρz, i.e.

$$\sum_{i \in I} p_i^I(z, \tau)(z + \tau^\alpha \phi_i(z)) = (1 + r\tau)z. \tag{10.122}$$

Let us deduce the HJB equation for the limit, as $\tau \to 0$, of the approximate cost-function $\mathcal{B}_\tau^{T-t}$, $t \in [0, T]$, with a given final cost f_T, using the standard (heuristic) dynamic programming approach. Namely, from

(10.121) and assuming appropriate smoothness of f we obtain the approximate equation

$$f_{t-\tau}(z) = \frac{1}{1+r\tau} \max_I \sum_{i \in I} p_i^I(z, \tau)$$

$$\times \left[f_t(z) + \tau^\alpha \frac{\partial f_t}{\partial z} \phi_i(z) + \frac{1}{2} \tau^{2\alpha} \left(\frac{\partial^2 f_t}{\partial z^2} \phi_i(z), \phi_i(z) \right) + O(\tau^{3\alpha}) \right].$$

Since $\{p_i^I\}$ are probabilities and using (10.122), this rewrites as

$$f_t - \tau \frac{\partial f_t}{\partial t} + O(\tau^2) = \frac{1}{1+r\tau} [f_t(z) + r\tau(z, \frac{\partial f_t}{\partial z})$$

$$+ \frac{1}{2} \tau^{2\alpha} \max_I \sum_{i \in I} p_i^I(z) \left(\frac{\partial^2 f_t}{\partial z^2} \phi_i(z), \phi_i(z) \right)] + O(\tau^{3\alpha}),$$

where

$$p_i^I(z) = \lim_{\tau \to 0} p_i^I(z, \tau)$$

(clearly well-defined non-negative numbers). This leads to the following equations:

$$rf = \frac{\partial f}{\partial t} + r(z, \frac{\partial f}{\partial z}) + \frac{1}{2} \max_I \sum_{i \in I} p_i^I(z) \left(\frac{\partial^2 f}{\partial z^2} \phi_i(z), \phi_i(z) \right) \qquad (10.123)$$

in case $\alpha = 1/2$, and to the trivial first order equation

$$rf = \frac{\partial f}{\partial t} + r(z, \frac{\partial f}{\partial z}), \qquad (10.124)$$

with the obvious solution

$$f(t, z) = e^{-r(T-t)} f_T(e^{-r(T-t)} z) \qquad (10.125)$$

in case $\alpha > 1/2$.

Equation (10.123) is a nonlinear extension of the classical Black-Scholes equation. Well posedness of the Cauchy problem for such a nonlinear parabolic equation in the class of viscosity solutions is well known in the theory of controlled diffusions, as well as the fact that the solutions solve the corresponding optimal control problem, see e.g. Fleming and Soner [232].

In case $J = 1$ and the classical CCR (binomial) setting with

$$\sqrt{\tau} \phi_1 = (u-1)z = \sigma\sqrt{\tau} z, \quad \sqrt{\tau} \phi_2 = (d-1)z = -\sigma\sqrt{\tau} z,$$

(10.123) becomes the usual Black-Scholes equation.

More generally, if $k = J + 1$, the corresponding market in discrete time becomes complete and equation (10.123) reduces to

$$rf = \frac{\partial f}{\partial t} + r(z, \frac{\partial f}{\partial z}) + \frac{1}{2} \sum_{i=1}^{J+1} p_i(z) \left(\frac{\partial^2 f}{\partial z^2} \phi_i(z), \phi_i(z) \right). \tag{10.126}$$

This is a generalized Black-Scholes equation describing a complete market (with randomness coming from J correlated Brownian motions), whenever the diffusion matrix

$$(\sigma^2)_{jk} = \sum_{i=1}^{J+1} p_i(z) \phi_i^j(z), \phi_i^k(z)$$

is non-degenerate.

As a more nontrivial example, let us consider the case of $J = 2$ and a sub-modular final payoff f_T, so that Theorem 10.9 applies to the approximations $\mathcal{B}_\tau$. Assume the simplest (and usual) symmetric form for upper and lower jumps (further terms in Taylor expansion are irrelevant for the limiting equation):

$$u_i = 1 + \sigma_i \sqrt{\tau}, \quad d_i = 1 - \sigma_i \sqrt{\tau}, \quad i = 1, 2. \tag{10.127}$$

Hence

$$\frac{u_i - \rho}{u_i - d_i} = \frac{1}{2} - \frac{r}{2\sigma_i} \sqrt{\tau}, \quad i = 1, 2,$$

and

$$\kappa = -\frac{1}{2} r \sqrt{\tau} (\frac{1}{\sigma_1} + \frac{1}{\sigma_2}).$$

As $\kappa < 0$, we find ourselves in the second case of Theorem 10.9. Hence the only eligible collection of three vectors ϕ is $(d_1, u_2), (u_1, d_2), (u_1, u_2)$, and the probability law p_i^I is $(1/2, 1/2, 0)$. Therefore, equation (10.123) takes the form

$$rf = \frac{\partial f}{\partial t} + r(z, \frac{\partial f}{\partial z}) + \frac{1}{2} \left[\sigma_1^2 z_1^2 \frac{\partial^2 f}{\partial z_1^2} - 2\sigma_1 \sigma_2 z_1 z_2 \frac{\partial^2 f}{\partial z_1 \partial z_2} + \sigma_2^2 z_2^2 \frac{\partial^2 f}{\partial z_2^2} \right]. \tag{10.128}$$

The limiting Black-Scholes type equation is again linear in this example, but with degenerate second-order part. In the analogous stochastic setting, this degeneracy would mean that only one Brownian motion governs the behavior of both underlying stocks. This is not surprising in our approach, where Nature was assumed to be a single player. One could expect uncoupled second derivatives (non-degenerate diffusion) in the limit, if one would choose two independent players for Nature, each playing for each stock.

Thus we are still in the setting of an incomplete market. The hedge price calculated from equation (10.128) is actually the upper price, in the terminology of Section 10.6. To get a lower price, we shall use approximations of type (10.106), leading, instead of (10.123), to the equation

$$rf = \frac{\partial f}{\partial t} + r(z, \frac{\partial f}{\partial z}) + \frac{1}{2} \min_I \sum_{i \in I} p_i^I(z) \left(\frac{\partial^2 f}{\partial z^2} \phi_i(z), \phi_i(z) \right). \qquad (10.129)$$

If $J = 2$ and the payoff is submodular, the maximum can be taken over the triples $(d_1, d_2), (d_1, u_2), (u_1, d_2)$ or $(d_1, u_2), (u_1, d_2), (u_1, u_2)$ (under (10.127) only the second triple works). Similarly the minimum can be taken only over the triples $(d_1, d_2), (d_1, u_2), (u_1, u_2)$ or $(d_1, d_2), (u_1, d_2), (u_1, u_2)$. Under (10.127) both these cases give the same limit as $\tau \to 0$, yielding for the lower price the equation

$$rf = \frac{\partial f}{\partial t} + r(z, \frac{\partial f}{\partial z}) + \frac{1}{2} \left[\sigma_1^2 z_1^2 \frac{\partial^2 f}{\partial z_1^2} + 2\sigma_1 \sigma_2 z_1 z_2 \frac{\partial^2 f}{\partial z_1 \partial z_2} + \sigma_2^2 z_2^2 \frac{\partial^2 f}{\partial z_2^2} \right],$$
$$(10.130)$$

that differs only by sign at the mixed derivative from the equation for the upper price.

As f was assumed sub-modular, so that its mixed second derivative is negative, we have

$$\sigma_1 \sigma_2 \frac{\partial^2 f}{\partial z_1 \partial z_2} \leq 0 \leq -\sigma_1 \sigma_2 \frac{\partial^2 f}{\partial z_1 \partial z_2}.$$

Hence, for the solution f_u of the upper value equation (10.128), the solution f_l of the lower value equation (10.130), and the solution f_c of the classical Black-Scholes equation of a complete market based on two independent Brownian motions, i.e. equation (10.128) or (10.130) without the term with the mixed derivative (with the same sub-modular initial condition f_T) we have the inequality

$$f_l \leq f_c \leq f_u,$$

as expected.

Equations (10.128) and (10.130) can be solved explicitly via the Fourier transform, as with the standard Black-Scholes equation. Namely, changing the unknown function f to g by

$$f(z_1, z_2) = e^{-r(T-t)} g(\frac{1}{\sigma_1} \log z_1, \frac{1}{\sigma_2} \log z_2),$$

so that

$$\frac{\partial f}{\partial z_i} = e^{-r(T-t)} \frac{1}{\sigma_i z_i} \frac{\partial g}{\partial y_i} (\frac{1}{\sigma_1} \log z_1, \frac{1}{\sigma_2} \log z_2),$$

transforms these equations to the equations

$$\frac{\partial g}{\partial t} + \frac{1}{2}(2r - \sigma_1)\frac{\partial g}{\partial y_1} + \frac{1}{2}(2r - \sigma_2)\frac{\partial g}{\partial y_2} + \frac{1}{2}\left[\frac{\partial^2 g}{\partial y_1^2} \mp 2\frac{\partial^2 g}{\partial y_1 \partial y_2} + \frac{\partial^2 g}{\partial y_2^2}\right] = 0$$

(10.131)

(with $\mp$ respectively). Equation (10.131) has constant coefficients and the equation for the Fourier transform $\tilde{g}(p)$ of g is obviously

$$\frac{\partial \tilde{g}}{\partial t} = \frac{1}{2}[(p_1 \mp p_2)^2 - i(2r - \sigma_1)p_1 - i(2r - \sigma_2)p_2]\tilde{g}.$$

(10.132)

Hence the inverse Cauchy problem for equation (10.131) with a given final function g_T equals the convolution of g_T with the inverse Fourier transform of the functions

$$\exp\{-\frac{1}{2}(T - t)[(p_1 \mp p_2)^2 - i(2r - \sigma_1)p_1 - i(2r - \sigma_2)p_2]\},$$

which equals (after changing the integration variables p_1 and p_2 to $q_1 = p_1 - p_2$, $q_2 = p_1 + p_2$)

$$\frac{1}{2(2\pi)^2}\int_{\mathbf{R}^2} dq_1 dq_2 \exp\{-\frac{1}{2}(T - t)q_{1,2}^2$$

$$+\frac{iq_1}{2}\left(y_1 - y_2 - \frac{(\sigma_1 - \sigma_2)(T - t)}{2}\right) + \frac{iq_2}{2}\left(y_1 + y_2 + (2r - \frac{\sigma_1 + \sigma_2}{2})(T - t)\right)\}$$

(with $q_{1,2}$ corresponding to $\mp$), or explicitly

$$\frac{1}{2}\frac{1}{\sqrt{2\pi(T - t)}}\delta\left(\frac{y_1 + y_2}{2} + (r - \frac{\sigma_1 + \sigma_2}{4})(T - t)\right)$$

$$\times \exp\{-\frac{1}{8(T - t)}\left(y_1 - y_2 - \frac{(\sigma_1 - \sigma_2)(T - t)}{2}\right)^2\}$$

and

$$\frac{1}{2}\frac{1}{\sqrt{2\pi(T - t)}}\delta\left(\frac{y_1 - y_2}{2} - \frac{(\sigma_1 - \sigma_2)(T - t)}{4}\right)$$

$$\times \exp\{-\frac{1}{8(T - t)}\left(y_1 + y_2 + (2r - \frac{\sigma_1 + \sigma_2}{2})(T - t)\right)^2\}$$

respectively, where δ denotes the Dirac δ-function. Returning to equations (10.128) and (10.130) we conclude that the solutions f_u and f_l respectively of the inverse time Cauchy problem for these equations are given by the formula

$$f_{u,l}(t, z_1, z_2) = \int_0^\infty \int_0^\infty G_{T-t}^\mp(z_1, z_2; w_1, w_2)f_T(w_1, w_2)\, dw_1 dw_2,$$

(10.133)

with the Green functions or transition probabilities being

$$G_{T-t}^{-}(z_1, z_2; w_1, w_2)$$

$$= \frac{e^{-r(T-t)}}{2\sqrt{2\pi(T-t)}\sigma_1\sigma_2 w_1 w_2} \delta\left(\frac{1}{2\sigma_1}\log\frac{z_1}{w_1} + \frac{1}{2\sigma_2}\log\frac{z_2}{w_2} + (r - \frac{\sigma_1+\sigma_2}{4})(T-t)\right)$$

$$\exp\{-\frac{1}{8(T-t)}\left(\frac{1}{\sigma_1}\log\frac{z_1}{w_1} - \frac{1}{\sigma_2}\log\frac{z_2}{w_2} - \frac{(\sigma_1-\sigma_2)(T-t)}{2}\right)^2\} \quad (10.134)$$

and

$$G_{T-t}^{+}(z_1, z_2; w_1, w_2)$$

$$= \frac{e^{-r(T-t)}}{2\sqrt{2\pi(T-t)}\sigma_1\sigma_2 w_1 w_2} \delta\left(\frac{1}{2\sigma_1}\log\frac{z_1}{w_1} - \frac{1}{2\sigma_2}\log\frac{z_2}{w_2} - \frac{(\sigma_1-\sigma_2)(T-t)}{4}\right)$$

$$\exp\{-\frac{1}{8(T-t)}\left(\frac{1}{\sigma_1}\log\frac{z_1}{w_1} + \frac{1}{\sigma_2}\log\frac{z_2}{w_2} + (2r - \frac{\sigma_1+\sigma_2}{2})(T-t)\right)^2\}$$
$$(10.135)$$

respectively. Of course, formulas (10.133) can be further simplified by integrating over the δ-function. Singularity, presented by this δ-function, is due to the degeneracy of the second-order part of the corresponding equations.

So far we have analyzed models where the jumps (from a given set) occur with regular frequency. However, it is natural to allow the periods between jumps to be more flexible. One can also have in mind an alternative picture of the model: instead of instantaneous jumps at fixed periods one can think about waiting times for the distance from a previous price to reach certain levels. It is clear then that these periods need not be constant. In the absence of a detailed model, it is natural to take these waiting times as random variables. In the simplest model, they can be i.i.d. Their intensity represents a kind of stochastic volatility. Slowing down the waiting periods is, in some sense, equivalent to decreasing the average jump size per period.

Assume now for simplicity that we are dealing with 2 colored options and sub-modular payoffs, so that Theorem 10.9 applies, yielding a unique eligible risk-neutral measure. Hence the changes in prices (for investor choosing the optimal γ) follow the Markov chain $X_n^\tau(z)$ described by the recursive equation

$$X_{n+1}^\tau(z) = X_n^\tau(z) + \sqrt{\tau}\phi(X_n^\tau(z)), \quad X_0^\tau(z) = z,$$

where $\phi(z)$ is one of three points $(z^1 d_1, z^2 u_2), (z^1 u_1, z^2 d_2), (z^1 u_1, z^2 u_2)$ that are chosen with the corresponding risk-neutral probabilities. As was shown above, this Markov chain converges, as $\tau \to 0$ and $n = [t/\tau]$ (where $[s]$ denotes the integer part of a real number s), to the diffusion process X_t solving the Black-Scholes type (degenerate) equation (10.128), i.e. a sub-Markov process with the generator

$$Lf(x) = -rf + r(z, \frac{\partial f}{\partial z}) + \frac{1}{2}\left[\sigma_1^2 z_1^2 \frac{\partial^2 f}{\partial z_1^2} - 2\sigma_1\sigma_2 z_1 z_2 \frac{\partial^2 f}{\partial z_1 \partial z_2} + \sigma_2^2 z_2^2 \frac{\partial^2 f}{\partial z_2^2}\right].$$
(10.136)

Assume now that the times between jumps $T_1, T_2, \cdots$ are i.i.d. random variables with a power-law decay, that is

$$\mathbf{P}(T_i \geq t) \sim \frac{1}{\beta t^\beta}$$

with $\beta \in (0,1)$ (where $\mathbf{P}$ denotes probability and $\sim$ means, as usual, that the ratio between the l.h.s. and the r.h.s. tends to one as $t \to \infty$). It is well known that such T_i belong to the domain of attraction of the β-stable law (see e.g. [302]), meaning that the normalized sums

$$\Theta_t^\tau = \tau^\beta(T_1 + \cdots + T_{[t/\tau]})$$

(where $[s]$ denotes the integer part of a real number s) converge, as $\tau \to 0$, to a β-stable Lévy motion Θ_t, which is a Lévy process on $\mathbf{R}_+$ with the fractional derivative of order β as the generator:

$$Af(t) = -\frac{d^\beta}{d(-t)^\beta}f(t) = -\frac{1}{\Gamma(-\beta)}\int_0^\infty (f(t+r) - f(t))\frac{1}{y^{1+\beta}}dr.$$

We are now interested in the process $Y^\tau(z)$ obtained from $X_n^\tau(z)$ by changing the constant times between jumps by scaled random times T_i, so that

$$Y_t^\tau(z) = X_{N_t^\tau}^\tau(z),$$

where

$$N_t^\tau = \max\{u : \Theta_u^\tau \leq t\}.$$

The limiting process

$$N_t = \max\{u : \Theta_u \leq t\}$$

is therefore the inverse (or hitting time) process of the β-stable Lévy motion Θ_t.

By Theorems 4.2 and 5.1 of [253], (see also Chapter 8 in [263]), we obtain the following result.

Theorem 10.12. *The process Y_t^τ converges (in the sense of distribution on paths) to the process $Y_t = X_{N_t}$, whose averages $f(T - t, x) = \mathbf{E}f(Y_{T-t}(x))$, for continuous bounded f, have the explicit integral representation*

$$f(T-t,x) = \int_0^\infty \int_0^\infty \int_0^\infty G_u^-(z_1, z_2; w_1, w_2) Q(T-t, u) f_T(w_1, w_2)\, du\, dw_1\, dw_2,$$

where G^-, the transition probabilities of X_t, are defined by (10.134), and where $Q(t, u)$ denotes the probability density of the process N_t.

Moreover, for $f \in C_\infty^2(\mathbf{R}^d)$, $f(t, x)$ satisfy the (generalized) fractional evolution equation (of Black-Scholes type)

$$D_{T-*}^\beta f(t,x) = Lf(t,x), \quad f(T,x) = f_T(x), \tag{10.137}$$

where on the l.h.s. sits the Caputo-Dzherbashyan fractional derivative.

Remark 10.8. An alternative way of writing equation (10.137) is

$$\frac{d^\beta}{dt^\beta} f(t,x) = Lf(t,x) + \frac{t^{-\beta}}{\Gamma(1-\beta)} f(x),$$

where on the l.h.s. sits the fractional derivative in the so-called generator form.

Similar procedure with a general nonlinear Black-Scholes type equation (10.123) will lead of course to its similar fractional extension. However, a rigorous analysis of the corresponding limiting procedure is beyond the scope of the present paper.

'Fractional everything' becomes a popular topic in modern literature, see e.g. the recent monograph Tarasov [300]. For the study of financial markets, this is of course a natural step to move from the discussion of power laws in economics (see e.g. various perspectives in Uchaikin and Zolotarev [302], Newman [291], Maslov [281] and references therein) to the applicability of fractional dynamics in financial markets, see e.g. Meerschaert and Scala [284], Meerschaert, Nane and Xiao [285], Jumarie [248], Wang [303] and references therein. Our game-theoretic analysis leads to degenerate and/or nonlinear versions of fractional Black-Scholes type equations. A survey on probabilistic approach to the analysis of fractional equations can be found in [258].

Chapter 11

Elements of more advanced analysis

11.1 Short overview

In this chapter we discuss certain characteristic game theoretic results (mostly original) and indicate further possible directions of the mathematical analysis of game.

Starting in the next section with putting forward various ways of proving basic classical result on the existence of Nash equilibria in mixed strategies extensions of finite game, we move on to the stability theory discussing two ways of thinking: (i) continuous stability based on general topological tools, and (ii) smooth stability based on the methods of differential and algebraic geometry. Section 11.4 is devoted to an abstract (or axiomatic) approach to differential games, based on the notion of a generalized dynamic system. This approach allows for rather general as well technically simpler development of the theory that includes classical differential games as a particular case.

In the next two sections we present two curious results, on a non-cooperative game representation of the classical (Neumann-Morgenstern) solution to a cooperative game and on the turnpikes (stable dynamic equilibria) for general stochastic games.

In Section 11.7 an exotic algebra is discussed, which was designed to transform a nonlinear optimization to a (sort of) linear theory, in order to be able to use linear tolls in some non-linear settings. This section is concluded with an elementary example illustrating the relevance to game theory and in the following Section 11.8 we use this algebra to deduce a remarkable differential equation of Belmann's type that describes the evolution of Pareto sets (non-smooth objects of a peculiar geometry) in dynamic multi-criteria optimization problems or differential games.

The last section is devoted to the sketch of the exciting theory of Markov models of interacting particles and their dynamic laws of large numbers that include both the famous statistical mechanics models of Boltmann, Landau, Vlasov and Smolichovski and the far reaching generalizations (deterministic as well as stochastic) of replicator dynamics of the evolutionary game theory.

Each section begins with a short introduction and can be read almost independently of the others.

11.2 Two proofs of the Nash-Gliksberg theorem on the existence of equilibria

In this section two proofs are given of the Nash-Fantszi-Gliksberg theorem on the existence of equilibria for games with compact spaces of strategies, one being based on an appropriate extension of the Nash's arguments used for the case of finite state space, and another based on the reduction to the finite case via discretization. It is worth noting however that this general result is already by no means the strongest available (though the basic idea of the proof based on a version of fixed point theorem remains the same). For more general facts (including noncompact and even non locally compact state spaces) we refer to [190], [194], [192].

For a metric compact space Y (for instance, a bounded subset of a Euclidean space $\mathbf{R}^d$) we denote the space of continuous real functions on Y by $C(Y)$ and the set of probability measures on Y by Y^*. The following well known analytic fact is crucial: the set Y^* is compact in the weak topology, where $\mu_n \to \mu$ as $n \to \infty$ means that $\int f(y)\mu_n(dy) \to \int f(y)\mu(dy)$ for all $f \in C(Y)$.

Remark. Let us explain how this fact can be deduced from the basic theorems of Functional Analysis. For a compact space Y the Banach dual to the Banach space $C(Y)$ (equipped with the sup-norm $\|f\| = \sup_y |f(y)|$) is precisely the space of finite (signed) measures $\mathcal{M}(Y)$ on Y. By the classical Banach-Alaoglu theorem the unit ball in any dual is compact if considered in its weak (sometimes called $\star$-weak) topology. The set Y^* of probability measures is clearly a convex closed subset of $\mathcal{M}(Y)$ in the weak topology. Hence Y^* is a compact convex set.

Consider a non-cooperative n-person game

$$\Gamma_H = \langle I = \{1, ..., n\}, \{X_i\}_1^n, \{H_i\}_1^n \rangle, \qquad (11.1)$$

where I is the set of players, X_i denotes a compact metric space of the strategies of the player i, $i \in I$, $X = X_1 \times \ldots \times X_n$ is the set of all profiles and $H_i \in C(X)$ is a payoff function of i. Let

$$\Gamma_H^* = \langle\{I\}, \{X_i^*\}_1^n, \{H_i^*\}_1^n\rangle$$

denotes the mixed strategy extension of Γ_H. Here the set X_i^* is interpreted as the set of mixed strategies of the player i, whose payoff in a profile $P = (p_1, ..., p_n) \in \mathbf{X} = X_1^* \times ... \times X_n^*$ equals

$$H_i^*(P) = \int_X H_i(x_1, ..., x_n) dp_1 ... dp_n.$$

Let

$$H_i^*(P\|x_i) = \int_{X_1 \times ... X_{i-1} \times X_{i+1} \times ... \times X_n} H_i(x_1, ..., x_n) \, dp_1 ... dp_{i-1} dp_{i+1} ... dp_n.$$

(11.2)

Recall that a situation P is called an *equilibrium*, if

$$H_i^*(P) \geq H_i^*(P\|x_i) \tag{11.3}$$

for all i and $x_i \in X_i$.

Theorem 11.1. *(Fant Szi-Gliksberg) The game Γ_H^* has an equilibrium.*

First proof. Let us choose some probability measures μ_i on X_i, $i = 1, ..., n$, such that any open set in X_i has a (strictly) positive measure (the existence of such a measure is known and is rather easy to prove by considering measures supported on a countable dense subsets of X_i). Let

$$C_i(P, y) = \max\{0, H_i^*(P\|y) - H_i^*(P)\}, \qquad y \in X_i.$$

The key idea of the proof is to introduce the following mapping $F_{\Gamma_H} = \mathbf{X} \mapsto \mathbf{X}$. If $P = (p_1, ..., p_n) \in \mathbf{X}$, then $F_{\Gamma_H}(P) = \bar{P} = (\bar{p}_1, ..., \bar{p}_n)$ is such that for each Borel set $A_i \subset X_i$

$$\bar{p}_i(A_i) = \frac{p_i(A_i) + \int_{A_i} C_i(p, y)\mu_i(dy)}{1 + \int_{X_i} C_i(p, y)\mu_i(dy)},$$

or equivalently for any $g \in C(X_i)$

$$\int g(y)\bar{p}_i(dy) = \frac{\int g(y)\bar{p}_i(dy) + \int g(y)C_i(p, y)\mu_i(dy)}{1 + \int C_i(p, y)\mu_i(dy)}. \tag{11.4}$$

From the definition of the weak convergence and expression (11.4) it is clear that the mapping F_{Γ_H} is continuous in the weak topology. Applying the famous Schauder-Tikhonov fixed point theorem stating that any continuous mapping of a convex metric compact set into itself has a fixed point results in the existence of a fixed point of the mapping F_{Γ_H}.

The final observation is that a profile P is a fixed point of this mapping if and only if it is an equilibrium. In fact, if P is an equilibrium, then all functions $C_i(P, y)$ vanish implying that $F_{\Gamma_H}(P) = P$. Conversely, assume $F_\Gamma(p) = p$. Let us show that all functions $c_i(P, y)$ vanish. Suppose this does not hold for at least one $i \in I$. Since c_i is continuous, the set $B_i = \{y \in X_i : C_i(P, y) > 0\}$ is open and hence $\mu_i(B_i) > 0$. In particular, $\int C_i(p, y)\mu_i(dy) > 0$, so that for $D_i = X_i \setminus B_i$

$$\bar{p}_i(D_i) = \frac{p_i(D_i)}{1 + \int_{X_i} c_i(p, y)\mu_i(dy)} < p_i(D_i)$$

whenever $p_i(D_i) > 0$. But this is true, because assuming $p_i(D_i) = 0$ leads to

$$H_i^\star(P) = \int H_i^\star(P\|y)p_i(dy) = \int_{B_i} H_i^\star(P\|y)p_i(dy) > H_i^\star(P)p_i(B_i) = H_i^\star(P),$$

which is a contradiction.

Second proof. Choose an arbitrary number $\varepsilon > 0$ and a finite ε-net

$$X_i^\varepsilon = \{x_{i1}^\varepsilon, ..., x_{iN_i}^\varepsilon\},$$

in X_i, i.e. the subset $X_i^\varepsilon \subset X_i$ enjoys the property that for any point $x \in X_i$ there exists x_{ik}^ε such that $\rho_i(x, x_{ik}^\varepsilon) < \varepsilon$, where ρ_i is the metric of X_i (existence of such a net follows from compactness of X_i). Consider a finite game

$$\Gamma_\varepsilon = \langle I, \{X_i^\varepsilon\}_1^n, \{H_i\}_1^n \rangle,$$

where $H_i : X^\varepsilon = \prod_1^n X_i^\varepsilon \to R_1$ is the restriction of H_i on X^ε, and its mixed strategy extension

$$\Gamma_\varepsilon^\star = \langle I, \{(X_i^\varepsilon)^\star\}_1^n, \{H_i^\star\}_1^n \rangle.$$

Choose an equilibrium $P^\varepsilon = (p_1^\varepsilon, ..., p_n^\varepsilon)$ for $\Gamma_\varepsilon^\star$ (which exists according to the Nash theorem on the existence of equilibria for finite games). Here $p_i^\varepsilon = (\xi_i^1, ..., \xi_i^{N_i})$ is the probability distribution on the set X_i^ε that can be considered also as a probability measure on X_i with a finite support. The equilibrium condition means that for all $i \in I$, $k_i = 1, ..., N_i$

$$H_i^\star(P^\varepsilon) \geq H_i^\star(P^\varepsilon\|x_{ik_i}^\varepsilon), \tag{11.5}$$

where the r.h.s. is defined in (11.2). In the discrete case under consideration it can be written also as

$$H_i^\star(P^\varepsilon \| x_{ik_i}^\varepsilon) = \sum_{j_l=1, l \neq i}^{N_l} H_i(x_{1j_1}^\varepsilon, ..., x_{(i-1)j_{i-1}}^\varepsilon, x_{(i+1)j_{i+1}}^\varepsilon, ..., x_{nj_n}^\varepsilon) \prod_{k \neq i} \xi_k^{j_k}.$$

Since X_i^ε is a ε-net and due to the continuity (and hence uniform continuity) of H, for arbitrary $y \in X_i$ there exists $x_{ik_i}^\varepsilon \in X_i^\varepsilon$ such that

$$|H_i(y_1, ..., y_{i-1}, y, y_{i+1}, ..., y_n) - H_i(y_1, ..., y_{i-1}, x_{ik_i}^\varepsilon, y_{i+1}, ..., y_n)| < \varepsilon$$

and in particular

$$H_i(y_1, ..., y_{i-1}, x_{ik_i}^\varepsilon, y_{i+1}, ..., y_n) > H_i(y_1, ..., y_{i-1}, y, y_{i+1}, ..., y_n) - \varepsilon$$

for all $y_l \in X_l$, $l \neq i$. Integrating this inequality with respect to the probability measures p_l^ε, $l \neq i$, yields

$$H_i^\star(P^\varepsilon \| x_{ik_i}^\varepsilon) \geq H_i^\star(P^\varepsilon \| y) - \varepsilon,$$

which together with (11.5) implies

$$H_i^\star(P^\varepsilon) \geq H_i^\star(P^\varepsilon \| y) - \varepsilon \tag{11.6}$$

for all $y \in X_i$.

Situations satisfying (11.6) are called *ε-equilibria*.

Let us take a sequence $\varepsilon_k = 1/k$ and choose a corresponding sequence $\{P^k\}$ of $1/k$-equilibria. Since $\mathbf{X}$ is weakly compact, there exists a subsequence $\{P^{k_m}\}_{m=1}^\infty$ weakly converging to a certain $P^0 \in \mathbf{X}$. As

$$H_i^\star(P^{k_m}) \geq H_i^\star(P^{k_m} \| y) - 1/k_m$$

for any $y \in X_i$ it follows by passing to the limit $m \to \infty$ that

$$H_i^\star(P^0) \geq H_i(P^0 \| y),$$

which means that P^0 is an equilibrium in $\Gamma_H^\star$.

Let us get a stronger result in the case of symmetric games, which is of particular importance in the light of evolutionary games.

A game Γ_H of type (11.1) is called *symmetric* if all strategy spaces X_i are the same, i.e. $X_i = Y$ with a certain compact space Y for all i and the payoffs are symmetric, i.e. there exists a function $H(y; x_1, ..., x_{n-1})$ on Y^n, which is symmetric with respect to $x_1, ..., x_{n-1}$ (i.e. is invariant under permutations of all but the first variables) such that

$$H_i(x_1, ..., x_n) = H(x_i; x_1, ..., x_{i-1}, x_{i+1}, ..., x_n).$$

Theorem 11.2. *The mixed strategy extension $\Gamma_H^\star$ of a symmetric game Γ_H (with compact Y and continuous H) has a symmetric equilibrium, that is an equilibrium of the form $P = (p, ..., p)$ with $p \in Y^\star$.*

Proof. Looking at the first proof of the previous theorem one observes that under the condition of symmetry the set of symmetric profiles $(p, ..., p)$ is invariant under the mapping F_{Γ_H} so that this mapping has a symmetric fixed point.

In many practical situations one can not expect to find an equilibrium exactly, and hence one is looking only for its approximation, i.e. for ε-*equilibria*. In order these approximation to exist weaker assumptions are needed. For instance the following result holds.

Theorem 11.3. *Let X_i be metric compact spaces, and H_i are bounded on $X = X_1 \times ... \times X_n$ and upper uniformly semi-continuous with respect to x_i, i.e. for each $x_i \in X_i$ there exists its neighborhood $V(x_i)$ such that $x'_i \in V(x_i)$ implies*

$$H_i(y_1, ..., y_{i-1}, y, y_{i+1}, ..., y_n) - H_i(y_1, ..., y_{i-1}, x^{\varepsilon}_{ik_i}, y_{i+1}, ..., y_n) < \varepsilon$$

for all $y_1, ..., y_n$. Then the game Γ_H has (mixed) ε-equilibria for any $\varepsilon > 0$.

Proof. Looking at the second proof of Theorem 11.1 observe that the above condition is actually all one needs to establish the existence of ε-equilibrium.

11.3 Introduction to structural stability

This section is an adapted introduction to the books [120], [122], to which we refer for details and bibliographical comments.

It is shown in particular (under rather general assumptions) that any game with a non empty set of solutions can be approximated by stable games in the sense of this solution. The notion of strategic equivalence is introduced and it is shown that the sets of solutions of equivalent games coincide (as well as the sets of their stable solutions). At the end we shortly discuss smooth stability showing that the property of having finitely many equilibria in mixed strategy extensions of finite game is a generic property.

1. Preliminaries.

Let Y be a compact metric space with the metric (distance) d. Denote $K(Y)$ the set of all closed (and hence compact) subset of Y. For $A \in K(Y)$ define

$$d(x, A) = \inf\{d(x, y) : y \in A\} = \min\{d(x, y) : y \in A\}.$$

Then for $A, B \in K(Y)$ one can define

$$D(A, B) = \max\left(\sup\{d(x, B) : x \in A\}, \sup\{d(x, A) : x \in B\}\right). \qquad (11.7)$$

Exercise. The function D is a metric on $K(Y)$, called *Hausdorf metric or distance.*

An important (not very deep, but not at all obvious) topological theorem states that the set $K(Y)$ with the Hausdorf metric D is a compact space (see e.g. [107]).

Suppose now X is another metric space. A mapping $F : X \mapsto K(Y)$ is called *lower semi-continuous (lsc)* in $x_0 \in X$ if for any sequence $\{x_n\}_1^\infty$, $x_n \in X$, $x_n \to x_0$ and a point $y \in F(x_0)$ there exists a sequence $\{y_n\}_1^\infty$, $y_n \in F(x_n)$ such that $y_n \to y$. A mapping $F : X \mapsto K(Y)$ is called *upper semi-continuous (usc)* in $x_0 \in X$ if for any sequence $\{x_n\}_1^\infty$, $x_n \in X$, $x_n \to x_0$ and any sequence $\{y_n\}_1^\infty$, $y_n \in F(x_n)$ there exists a subsequence $\{y_{n_k}\}$ such that $y_{n_k} \to y \in F(x_0)$. A mapping $F : X \mapsto K(Y)$ is called *continuous* in $x_0 \in X$ if it is simultaneously usc and lsc in x_0.

Exercise. The following are equivalent: (i) A mapping $F : X \mapsto K(Y)$ is continuous in $x_0 \in X$ in the sense of the above definition; (ii) it is continuous as a mapping between metric spaces with $K(Y)$ equipped with the metric (11.7); (iii) for any $x \in X$ and any $\epsilon > 0$ there exists a $\delta > 0$ such that if the distance between y and x does not exceed δ, then $K(x)$ belongs to the ϵ-neighborhood of $K(y)$ and $K(y)$ belongs to the ϵ-neighborhood of $K(x)$.

A mapping $F : X \mapsto K(Y)$ is called *closed* in $x_0 \in X$ if for any sequences $\{x_n\}_1^\infty$, $x_n \in X$, $x_n \to x_0$ and $\{y_n\}_1^\infty$, $y_n \in F(x_n)$ $y_n \to y$ it follows that $y \in F(x_0)$.

Exercise. A mapping $F : X \mapsto K(Y)$ is usc in $x_0 \in X$ if and only if it is closed.

A subset M in X is called *non-dense (or nowhere dense)* if its closure does not contain a non-void open set. A subset M in X is said to be of *the first category (in the sense of Baire)* if it is the union of a countable number of non-dense subsets of X. A subset M in X is said to be of *the second category (in the sense of Baire)* if it is not of the first category. The complements in X of the sets of the first category are called *residual sets*. A crucial for the functional analysis *Baire's theorem* states that residual sets in complete metric spaces X are dense in X. A key topological result for the stability theory of games is the following (see e.g. [107]): if $F : X \mapsto K(Y)$ is closed, then the set of its discontinuity points is of the first category. In particular, in the light of the Baire theorem it implies the following

Proposition 11.1. *If X is a complete metric space, Y is a compact metric space, and $F : X \mapsto K(Y)$ is a closed mapping, then the set of those points*

where $F : X \mapsto K(Y)$ is continuous is dense in X.

2. Stability of Nash equilibria.

Consider the metric space (γ, ρ) of non cooperative n person games

$$\Gamma_H = \langle I = \{1, 2, \ldots, n\}, \{(X_i, d_i)\}_1^n, \{H_i\}_1^n \rangle.$$

Here I denotes the set of players, (X_i, d_i) is a compact metric space of the strategies of the player i, $X = X_1 \times \ldots \times X_n$ is the set of all profiles, $H_i \in C(X)$ is a payoff function of i and ρ is the metric in γ:

$$\rho(\Gamma_H, \Gamma_{H'}) = \max_{x \in X} \| H(x) - H'(x) \|$$

($\| \cdot \|$ denotes the usual norm of the n-dimensional Euclidean space). For $x, x' \in X$ put

$$d(x, x') = \sum_{i=1}^{n} d_i(x_i, x_i')$$

and $x \| x_i' = (x_1, x_2, \ldots, x_{i-1}, x_i', x_{i+1}, \ldots, x_n)$.

Recall that a situation (or a profile) $x = (x_1, \ldots, x_n) \in X$ is called an equilibrium in Γ, if for arbitrary $x_i' \in X_i$, $i = 1, 2, \ldots, n$

$$H_i(x) \geq H_i(x \| x_i').$$

Let us denote by $\mathcal{E}_H$ the set (clearly a compact one) of equilibrium profiles in Γ_H.

Let γ' denote a closed subset of γ such that all games from γ' have a non-empty set of equilibrium situations.

We have in mind two basic examples of γ': 1) γ' is the set of all games from γ with a non-empty set of equilibrium situations (Exercise: check that this set is in fact closed in γ); 2) γ' is the class of the mixed strategy extensions of the n-person games with fixed compact spaces of (pure) strategies (it is shown in the previous section that these games do have equilibria).

Recall that an equilibrium situation $x \in \mathcal{E}$ is called *(structurally) stable*, if for any $\varepsilon > 0$ there exists a $\delta > 0$ such that for any $\Gamma_{H'} \in \gamma'$ with $\rho(\Gamma_H, \Gamma_{H'}) < \delta$ there exists a situation $x' \in \mathcal{E}_{H'}$ such that $d(x, x') < \varepsilon$. Let $\mathcal{E}_H^*$ denote the set of all stable profiles of Γ_H. We shall say that Γ_H is *stable* whenever $\mathcal{E}_H = \mathcal{E}_H^*$.

Proposition 11.2. *The mapping $E \colon \gamma' \to K(X)$ acting as $E(\Gamma_H) = \mathcal{E}_H$ is closed.*

Proof. Consider the sequence of games $\{\Gamma_{H^k}\}_{k=1}^{\infty}$, $\Gamma_{H^k} \in \gamma'$, $\Gamma_{H^k} \to \Gamma \in \gamma'$. Let $x^k = (x_1^k, \ldots, x_n^k) \in \mathcal{E}_{H^k}$, $x^k \to x \in X$. Let us show that $x \in \mathcal{E}_H$. For all $i \in I$, $k \in Z = \{1, 2, \ldots\}$, $x_i' \in X_i$ the inequalities

$$H_i^k(x^k) \geq H_i^k(x^k \| x_i') \tag{11.8}$$

hold true. Let us choose $x_i' \in X_i, i$ and consider numerical sequences $\{H_i^k(x^k)\}_{k=1}^{\infty}$, $\{H_i^k(x^k \| x_i')\}_{k=1}^{\infty}$. They converge to $H_i(x)$ and $H_i(x \| x_i')$ respectively. In fact, say, for the first sequence, one has

$$|H_i(x) - H_i^k(x^k)| \leq |H_i(x) - H_i(x^k)| + |H_i(x^k) - H_i^k(x^k)|.$$

As $k \to \infty$, the first term on the r.h.s. tends to zero by the continuity of H_i (which is the uniform limit of a sequence of continuous functions H_i^k), and the second term tends to zero by the convergence of the sequence H_i^k to H_i. Similarly one shows the convergence of the second sequence. As the inequality (11.8) holds for each $k \in Z$, one can pass to the limit as $k \to \infty$ in it implying that for all $i \in I$ and $x_i' \in X$

$$H_i(x) \geq H_i(x \| x_i').$$

Consequently $x \in \mathcal{E}_H$.

Proposition 11.3. *A game $\Gamma_H \in \gamma'$ is stable if and only if Γ_H is a continuity point of the mapping E.*

Proof. Necessity. By Proposition 11.2, it is enough to check only the semi-continuity from below of the mapping E, i.e. that if for any sequence $\{H_n\}_1^{\infty}$, $H_n \to H$ and a point $y \in \mathcal{E}_H$ there exists a sequence $\{y_n\}_1^{\infty}$, $y_n \in \mathcal{E}_{H_n}$ such that $y_n \to y$. But this holds, because by stability for arbitrary $\varepsilon > 0$ and any H_n close enough to H there exists $y_n \in \mathcal{E}_{H_n}$ such that $d(y, y_n) < \varepsilon$.

Sufficiency. Let Γ_H be a continuity point of E. Then, for any $\varepsilon > 0$ there exists a $\delta > 0$ such that for any game $\Gamma_{H'} \in \gamma'$ with $\rho(\Gamma_H, \Gamma_{H'}) < \delta$, one has $D(\mathcal{E}_H, \mathcal{E}_{H'}) < \varepsilon)$. This proves the stability of the game Γ_H.

Theorem 11.4. *The set of stable games from γ' is dense in γ'.*

Proof. Straightforward from Propositions 11.1-11.3.

Theorem 11.5. *If the set $\mathcal{E}_H$ contains only one point x, then x is a stable equilibrium and Γ_H is a stable game.*

Proof. If it would not be stable, there would exist $\varepsilon > 0$ and a sequence $H_n \to H$ and $y_n \in \mathcal{E}_{H_n}$ such that $d(y_n, x) > \varepsilon$. Passing (if necessary) to a subsequence one can choose y_n to be converging (by compactness) to a point $y \neq x$. By usc $y \in \mathcal{E}_H$ - contradiction.

3. Strategic equivalence.

Let us say that a game Γ_H is *strategically equivalent to the* $\Gamma_{H'}$ ($\Gamma_H \sim \Gamma_{H'}$), if there exist functions $\lambda_i = \lambda_i(x) > 0$, $\mu_i = \mu_i(x)$ $i = 1, 2, \ldots, n$, not depending on x_i such that $H_i = \lambda_i H_i' + \mu_i$.

Proposition 11.4. *The binary relation $\sim$ is an equivalence relation in the space γ.*

Proof. The reflectivity and symmetry are obvious. To prove transitivity observe that if $\Gamma_H \sim \Gamma_{H'}$, $\Gamma_{H'} \sim \Gamma_{H''}$, then $\Gamma_H \sim \Gamma_{H''}$, because from $H_i = \lambda_i H_i' + \mu_i$, $H_i' = \lambda_i' H_i'' + \mu_i'$ it follows that

$$H_i = \lambda_i(\lambda_i' H_i'' + \mu_i') + \mu_i = \lambda_i \lambda_i' H_i'' + (\lambda_i \mu_i' + \mu_i).$$

Proposition 11.5. *If $\Gamma_H, \Gamma_{H'} \in \gamma$, $\Gamma_H \sim \Gamma_{H'}$, then $\mathcal{E}_H = \mathcal{E}_{H'}$.*

Proof. Let $H_i = \lambda_i H_i' + \mu_i$, $i \in I$, $x \in \mathcal{E}_H$. Then for all $i \in I$, $x_i' \in X_i$

$$H_i(x) \geq H_i(x \| x_i'), \qquad (11.9)$$

and hence

$$\lambda_i(x) H_i(x) + \mu_i(x) \geq \lambda_i(x \| x_i') H_i(x \| x_i') + \mu_i(x \| x_i'),$$

so that (because λ_i, μ_i do not depend on x_i) $x \in \mathcal{E}_{H'}$, and consequently $\mathcal{E}_H \subset \mathcal{E}_{H'}$. Similarly one shows that $\mathcal{E}_{H'} \subset \mathcal{E}_H$, and thus $\mathcal{E}_H = \mathcal{E}_{H'}$.

Proposition 11.6. *If $\Gamma_H \sim \Gamma_{\widetilde{H}}$, then $\mathcal{E}_H^* = \mathcal{E}_{\widetilde{H}}^*$.*

Proof. Follows directly from Proposition 11.5.

4. Stability of Pareto sets and compromise sets.

Let us now discuss the stability of the Pareto optimal sets. A situation $x \in X$ is called *Pareto optimal* in the game $\Gamma_H \in \gamma$ if there are no points $x' \in X$ such that $H_i(x') > H_i(x)$ for all $i \in I$.

Let us denote the set of Pareto optimal situation in Γ_H by $\mathcal{P}_H$.

Proposition 11.7. *The set $\mathcal{P}_H$ is compact.*

Proof. It is sufficient to show that $\mathcal{P}_H$ is closed. Choose a sequence $\{x^k\}_{k=1}^{\infty}$, $x^k \in \mathcal{P}_H$, $x^k \to x$. To show that $x \in \mathcal{P}_H$ assume that this is not the case.

Let γ denote the Banach space of continuous mapping $H\colon X \to R_n$ equipped with the sup-norm.

Proposition 11.14. *The mapping* $C\colon \gamma \to K(X)$, $C(H) = C_H$, *is semi-continuous from above.*

Proof. Assume $H^k \xrightarrow[k\to\infty]{} H_0$, $x \in X$, $x_k^* \in C_{H^k}$, $x_k^* \xrightarrow[k\to\infty]{} x_0^*$. Let us check that $x_0^* \in C_{H_0}$. From the assumption one deduces that $M_i^k \xrightarrow[k\to\infty]{} M_i^0$. Let us estimate the difference

$$\|H_i^k(x_k^*) - H_i^0(x_0^*)\| \le \|H_i^k(x_k^*) - H_i^0(x_k^*)\| + \|H_i^0(x_k^*) - H_i^0(x_0^*)\|.$$

The continuity of H_i^0 and the uniform convergence of H^k to H^0 imply that the r.h.s. tends to zero. Consequently $H_i^k(x_k^*) \xrightarrow[k\to\infty]{} H_i^0(x_0^*)$. Hence for $i \in I$

$$\{M_i^k - H_i^k(x_k^*)\} \xrightarrow[k\to\infty]{} \{M_i^0 - H_i^0(x_0^*)\}.$$

Hence the sequence of vectors $\{\cdot\}_{k=1}^n$ converges in R_n, and consequently the sequence of their norms converges as well:

$$\max_i \{M_i^k - H_i^k(x_k^*)\}.$$

Similarly

$$\max_i \{M_i^k - H_i^k(x)\} \xrightarrow[k\to\infty]{} \max_i \{M_i^0 - H_i^0(x)\}.$$

For each $k = 1, 2, \ldots$ the inequality

$$\max_i \{M_i^k - H_i^k(x_k^*)\} \le \max_i \{M_i^k - H_i^k(x)\}$$

holds true. Passing to the limit as $k \to \infty$ one obtains for each $x \in X$

$$\max_i \{M_i^0 - H_i^0(x_0^*)\} \le \max_i \{M_i^0 - H_i^0(x)\},$$

i.e. $x_0^* \in C_{H^0}$.

By arguments similar to those given above, one can prove the following.

Proposition 11.15. *The set of multi-criteria optimization problems* $H \in \gamma$ *with a stable compromise set* C_H *is everywhere dense in* γ.

5. On the smooth stability.

In the case of smooth payoff functions (for instance, in the case of mixed strategies extensions of finite games and taking into account the application to differential games), it is natural to investigate the smooth dependence of the solutions on the parameters of the game, using the methods of differential topology and algebraic geometry.

It turns out that the space of the games can be decomposed into the union of the open set of regular (stable) games and the closed nowhere dense bifurcation set of unstable games. On each regular (connected) component there exists a finite (in fact odd) number of equilibria, which can be represented as smooth functions of the parameters of the games (see [122]).

As a convenient technical tool for these results one can use the parametric theorem of transversality, which we shall recall now. Some elementary knowledge of differential topology (basic definition of manifolds and related notions) would be advantageous when reading this section further, but let us note for those not acquainted with it that for our purposes it is essentially enough to think about manifolds used below as being just the open subsets of Euclidean spaces, in which case all tangent spaces can be identified with this Euclidean space and the tangent mapping is a linear operator specified by the matrix of partial derivatives.

Let X and Y be smooth manifolds (without a boundary) of dimensions n and m respectively and of the smoothness class C^r, $r \geq 1$, and $W \subset Y$ be a submanifold of Y of co-dimension q (which is the difference between the dimensions of Y and W). A smooth mapping $f : X \mapsto Y$ is called *transversal to* W whenever for each $x \in X$ such that $f(x) = y \in W$ the tangent space $T_y Y$ (to Y at the point y) is generated by the spaces $T_y W$ and $d_x f(T_x X)$ (the image of $T_x X$ under the differential or tangent mapping $d_x f : T_x X \mapsto T_y Y$). It is known (and easy to see by the implicit function theorem) that if this is the case, then $f^{-1}(W)$ is a submanifold of X. The following statement is called the *parametric theorem of transversality* (see e.g. [70] for a proof). Let $r > \max(n - q, 0)$ and let N be another manifold of class C^r, and assume we are given a smooth (also of class C^r) mapping $f : N \times X \mapsto Y$ that is transversal to W. Then the set N_f of all points $n \in N$ such that the mapping $x \mapsto f(n, x)$ from X to Y is transversal to W is a residual (in particular dense) subset of N.

Let us consider the set $\gamma = \gamma_{m_1,\ldots,m_p}$ of the standard mixed strategies extensions of finite non-cooperative p person games with the number of pure strategies being $m_1, \ldots, m_p$ respectively.

The set $\gamma = \gamma_{m_1,\ldots,m_n}$ can be identified in a natural way with the Euclidean space of dimension $N = n \cdot m_1 \ldots m_n$, which allows to consider it as a metric space and a smooth manifold.

By the Nash theorem, the set of (Nash) equilibria in each $\Gamma_H \in \gamma$ is not empty.

Proposition 11.16. *The set γ' of the games from γ having a finite* *number of equilibria is a residual set (in particularly it is dense) in γ.*

Proof. **Step 1.** We shall establish first that the set of games with an infinite number of fully mixed equilibria (with all probabilities being not zeros) is of the first category in γ.

To shorten the formula let us give the proof only for the case of three players. The space $\gamma_{m_1 m_2 m_3}$ is then parametrized by the three three-dimensional arrays $A = (a_{i_1 i_2 i_3})$, $B = (b_{i_1 i_2 i_3})$ and $C = (c_{i_1 i_2 i_3})$ with $i_j = 1, \ldots, m_j$, $j = 1, 2, 3$, defining the payoffs for the players as

$$\bar{A} = \sum_{i_j=1}^{m_j-1} a_{i_1 i_2 i_3} x_{i_1} y_{i_2} z_{i_3}, \quad \bar{B} = \sum_{i_j=1}^{m_j-1} b_{i_1 i_2 i_3} x_{i_1} y_{i_2} z_{i_3}, \quad \bar{C} = \sum_{i_j=1}^{m_j-1} c_{i_1 i_2 i_3} x_{i_1} y_{i_2} z_{i_3},$$

with $\{x_1, \ldots, x_{m_1}\}$, $\{y_1, \ldots, y_{m_2}\}$ and $\{z_1, \ldots, z_{m_3}\}$ being the probability distributions. Taking into account that $x_1 + \ldots + x_{m_1} = 1$ (and similar for y, z), the sets of strategies could be identified with

$$X = \{x = (x_1, \ldots, x_{m_1-1}) : x_i \geq 0, \quad \sum_{i=1}^{m_1-1} x_i \leq 1\} \subset \mathbf{R}^{m_1-1},$$

$$Y = \{y = (y_1, \ldots, y_{m_2-1}) : y_i \geq 0, \quad \sum_{i=1}^{m_2-1} y_i \leq 1\} \subset \mathbf{R}^{m_2-1},$$

$$Z = \{z = (z_1, \ldots, z_{m_3-1}) : z_i \geq 0, \quad \sum_{i=1}^{m_3-1} z_i \leq 1\} \subset \mathbf{R}^{m_3-1},$$

so that one can write

$$\bar{A}(x, y, z) = \sum_{i_1=1}^{m_1-1} \sum_{i_2=1}^{m_2-1} \sum_{i_3=1}^{m_3-1} a_{i_1 i_2 i_3} x_{i_1} y_{i_2} z_{i_3} + \ldots,$$

where by $\ldots$ are denoted the terms depending on those $a_{i_1 i_2 i_3}$, where at least one of i_j equals m_j. Similar formulae hold for $\bar{B}$ and $\bar{C}$. In particular, for $i < m_1$, $j < m_2$

$$\frac{\partial \bar{A}}{\partial x_i} = \sum_{i_2=1}^{m_2-1} \sum_{i_3=1}^{m_3-1} a_{i i_2 i_3} y_{i_2} z_{i_3} + \ldots, \tag{11.13}$$

$$\frac{\partial \bar{B}}{\partial y_j} = \sum_{i_1=1}^{m_1-1} \sum_{i_3=1}^{m_3-1} b_{i_1 j i_3} x_{i_1} z_{i_3} + \ldots, \tag{11.14}$$

and similarly for the derivatives of $\bar{C}$, where by ... are again denoted the terms depending on those $a_{i_1 i_2 i_3}$ (respectively $b_{i_1 i_2 i_3}$), where at least one of i_j equals m_j.

Consider now a mapping

$$(F, G, H) : \mathbf{R}^{m_1 + m_2 + m_3 - 3} \times \mathbf{R}^{3m_1 m_2 m_3} \mapsto \mathbf{R}^{m_1 + m_2 + m_3 - 3},$$

given by

$$(x, y, z, A, B, C) \mapsto \left(F = \frac{\partial \bar{A}}{\partial x}, G = \frac{\partial \bar{B}}{\partial y}, H = \frac{\partial \bar{C}}{\partial z} \right). \tag{11.15}$$

The point in defining this mapping lies in the observation that by the Lemma on the equality of payoffs, for a completely mixed Nash equilibrium x, y, z of the game A, B, C the values of F, G, H should be zeros. Our aim is to show that this mapping is regular whenever x, y, z are non vanishing vectors.

Consider $(m_1 + m_2 + m_3 - 3) \times (m_1 + m_2 + m_3 - 3 + 3m_1 m_2 m_3)$-matrix of Jacobi of this mapping. In order to show that it has the full rank $(m_1 + m_2 + m_3 - 3)$ it is enough to show that its $(m_1 + m_2 + m_3 - 3) \times 3(m_1 - 1)(m_2 - 1)(m_3 - 1)$ sub-matrix

$$\begin{pmatrix} \frac{\partial F_1}{\partial a_{i_1 i_2 i_3}} & \frac{\partial F_1}{\partial b_{i_1 i_2 i_3}} = 0 & \frac{\partial F_1}{\partial c_{i_1 i_2 i_3}} = 0 \\ \cdots & \cdots & \cdots \\ \frac{\partial F_{m_1 - 1}}{\partial a_{i_1 i_2 i_3}} & \frac{\partial F_{m_1 - 1}}{\partial b_{i_1 i_2 i_3}} = 0 & \frac{\partial F_{m_1 - 1}}{\partial c_{i_1 i_2 i_3}} = 0 \\ \frac{\partial G_1}{\partial a_{i_1 i_2 i_3}} = 0 & \frac{\partial G_1}{\partial b_{i_1 i_2 i_3}} & \frac{\partial G_1}{\partial c_{i_1 i_2 i_3}} = 0 \\ \cdots & \cdots & \cdots \\ \frac{\partial F_1}{\partial a_{i_1 i_2 i_3}} = 0 & \frac{\partial G_{m_2 - 1}}{\partial b_{i_1 i_2 i_3}} & \frac{\partial G_{m_2 - 1}}{\partial c_{i_1 i_2 i_3}} = 0 \\ \frac{\partial H_1}{\partial a_{i_1 i_2 i_3}} = 0 & \frac{\partial H_1}{\partial b_{i_1 i_2 i_3}} = 0 & \frac{\partial H_1}{\partial c_{i_1 i_2 i_3}} \\ \cdots & \cdots & \cdots \\ \frac{\partial H_{m_3 - 1}}{\partial a_{i_1 i_2 i_3}} = 0 & \frac{\partial H_{m_3 - 1}}{\partial b_{i_1 i_2 i_3}} = 0 & \frac{\partial H_{m_3 - 1}}{\partial c_{i_1 i_2 i_3}} \end{pmatrix} \tag{11.16}$$

has the full rank $(m_1 + m_2 + m_3 - 3)$, where $i_j = 1, \ldots, m_j - 1$, $j = 1, 2, 3$. And for this it is clearly enough to show that the matrices

$$\frac{\partial F}{\partial a_{i_1 i_2 i_3}}, \quad \frac{\partial G}{\partial b_{i_1 i_2 i_3}}, \quad \frac{\partial H}{\partial c_{i_1 i_2 i_3}}$$

(where again $i_j = 1, \ldots, m_j - 1$, $j = 1, 2, 3$) have their full ranks $m_1 - 1$, $m_2 - 1$, $m_3 - 1$ respectively. Consider for example the first of these matrices. Observing that

$$\frac{\partial F_i}{\partial a_{i_1 i_2 i_3}} = \delta_i^{i_1} y_{i_2} z_{i_3}, \quad i_j < m_j - 1, j = 1, 2, 3,$$

one can write

$$\frac{\partial F}{\partial a_{i_1 i_2 i_3}} = \begin{pmatrix} \frac{\partial F_1}{\partial a_{1 i_2 i_3}} = y_{i_2} z_{i_3} & \frac{\partial F_1}{\partial a_{2 i_2 i_3}} = 0 & \cdots & \frac{\partial F_1}{\partial a_{(m_1-1) i_2 i_3}} = 0 \\ \cdots & \cdots & \cdots & \cdots \\ \frac{\partial F_{m_1-1}}{\partial a_{1 i_2 i_3}} = 0 & \frac{\partial F_{m_1-1}}{\partial a_{2 i_2 i_3}} = 0 & \cdots & \frac{\partial F_{m_1-1}}{\partial a_{(m_1-1) i_2 i_3}} = y_{i_2} z_{i_3} \end{pmatrix}$$

and it has full rank $m_1 - 1$ whenever at least one of the numbers $y_{i_2} z_{i_3}$ does not vanish, i.e. whenever vectors y, z are nonzero vectors.

Applying the parametric transversality theorem with $N = \{(A, B, C)\}$, $W = \{0\}$ to the mapping (11.15) one concludes that the set γ' of those A, B, C such that the mapping

$$E_{A,B,C} : (x, y, z) \mapsto \left(\frac{\partial \bar{A}}{\partial x}, \frac{\partial \bar{B}}{\partial y}, \frac{\partial \bar{C}}{\partial z} \right) \tag{11.17}$$

from $(\mathbf{R}^{m_1-1} \setminus \{0\}) \times (\mathbf{R}^{m_2-1} \setminus \{0\}) \times (\mathbf{R}^{m_3-1} \setminus \{0\})$ to $\mathbf{R}^{m_1+m_2+m_3-3}$ is transversal to W is residual and hence dense in N. For these (A, B, C) the pre-image $E_{(A,B,C)}^{-1}(0)$ (consisting of Nash equilibria) is a zero dimensional manifold, which therefore consists of a discrete collection of points, whose number in any compact set is finite. Hence they can accumulate only to either $x = 0$ or $y = 0$, or $z = 0$, i.e. either to $x_{m_1} = 1$ or to $y_{m_2} = 1$, or to $z_{m_3} = 1$. But repeating this procedure excluding x_1, y_1, z_1 instead of x_{m_1}, y_{m_2}, z_{m_3} we get another residual set for which only the accumulation to $x_1 = 1$, $y_1 = 1$, $z_1 = 1$ is allowed. On the intersection only a finite number of completely mixed equilibria is possible.

Remark. In a simpler case of only two players the corresponding mapping

$$E_{A,B} : (x, y) \mapsto \left(\frac{\partial \bar{A}}{\partial x}, \frac{\partial \bar{B}}{\partial y} \right)$$

(replacing (11.17) above) turns out to be linear, which would allow to avoid trasversality theory by using linear algebra. In this case it follows directly that the pre-image $E_{(A,B)}^{-1}(0)$ consists of either a single point (non-degenerate case) or a linear subspace. Only in case of more than two players the methods of differential geometry become relevant.

Step 2. If an equilibrium is not fully mixed in Γ, then it is necessarily fully mixed in a certain subgame Γ', formed by an index $m'_1 m'_2 \ldots m'_p$ such that $m'_i \leq m_i$. The game Γ' is the projection of the game Γ on $\gamma' = \gamma_{m'_1 \ldots m'_n}$, which is a proper subspace of the space $\gamma = \gamma_{m_1 \ldots m_n}$. Consequently the set of games $\Gamma \in \gamma_{m_1 \ldots m_n}$ having an infinite set of equilibria in the interior of the boundary $m'_1 \ldots m'_n$, (and hence forming the set $\widehat{\gamma}_{m'_1 \ldots m'_n}$), are of the first category in $\gamma_{m_1 \ldots m_n}$, for otherwise $\widehat{\gamma}_{m'_1 \ldots m'_n}$

would contain a non empty open set of the space γ with the projection on $\gamma_{m'_1...m'_n}$ being also open in $\gamma_{m'_1...m'_n}$, which would contradict the statement proved on the Step 1 (applied to $\gamma = \gamma_{m'_1...m'_n}$). Hence the set of games $\Gamma \in \gamma_{m_1...m_n}$ with an infinite set of equilibria is of the first category in $\gamma_{m_1...m_n}$. The proof is complete.

Proposition 11.17. *The set of games Γ with a finite number of equilibria is open in the space $\gamma_{m_1...m_n} = \gamma$. For these games the completely mixed equilibria can be expressed as smooth functions of the payoff coefficients.*

Proof. Consider the set $\{x_1,...,x_k\}$ of completely mixed (all probabilities are positive) equilibria of the game Γ. They are regular points of the mapping $E_{(A,B)}$ constructed above (or its natural extensions in case of more than three players) for the value $0 \in R^*$. Let us choose neighborhoods $V_1,...,V_k$ of the points $x_1,...,x_k$ such that $V_i \cap V_j = \emptyset$ for $i \neq j$. Then $F_\Gamma^{-1}(0)$ does not contain completely mixed equilibria, different from $x_1,...,x_k$, and for any game Γ' that is close enough to Γ, $F_{\Gamma'}^{-1}(0)$ does not contain equilibria outside the neighborhoods $V_1,...,V_k$. Since the Jacobian of the mapping $F_{\Gamma'}$ does not vanish, for any game Γ' that is close enough to Γ the Jacobian of $F_{\Gamma'}$ does not vanish as well. Applying the implicit function theorem yields that for each $j = 1,...,k$ there exists a unique point $x'_j \in V_j$ such that $F_{\Gamma'}(x'_j) = 0$, which represents an equilibrium for Γ'. The case of equilibria lying on the boundary of X, is considered analogously. As the number of pure strategies is obviously finite, Proposition is proved.

Corollary. If a game Γ_H has a finite number of mixed equilibria, and for each pure equilibrium all inequalities entering the definition are strict, the function $|\varepsilon_H|$ (the number of equilibria) is locally constant.

Remarks. Using algebraic geometry it is possible to strengthen the above results in the following direction. One can show that the space $\gamma = \gamma_{m_1...m_n}$ of the games of a given size (or the space $\gamma/\sim$ of the classes of strategic equivalence) can be decomposed into a finite union of nonintersecting semi-algebraic sets $\gamma = \gamma_S \cup \gamma_{\alpha_1} \cup ... \cup \gamma_{\alpha_\gamma}$, where γ_S is a set of lower dimension containing all games with an infinite set of equilibria and each γ_{α_j} is a connected open set such that each game in it has one and the same number of equilibria that depend smoothly on the parameters of games. Moreover, this number is necessarily odd.

11.4 Introduction to abstract differential games

In this section we introduce an approach to the analysis of differential games and their generalizations based on the notion of a generalized dynamic system. As an illustration we shall prove a result on the existence of the value and ε-equilibria in the simplest situation of zero-sum games with separated dynamics. For the full story (including the bibliography) we refer to the book [120], see also [117]-[119].

We shall consider the games evolving in a complete locally compact metric space X.

Generalized (time homogeneous) dynamic system $\mathcal{D}$ in X is defined by means of a family of multi-valued mappings of the space X into itself, denoted by $\mathcal{D}(x,t)$, (i.e. $\mathcal{D}(x,t)$ is a subset of X for each $x \in X, t \geq 0$) and called the *accessibility function* that satisfies the following conditions:

1. $\mathcal{D}(x,t)$ *is a non empty compact subset of* X *for all* $x \in X$, $t \geq 0$.
2. *Initial condition:* $\mathcal{D}(x,0) = x$ *for all* $x \in X$.
3. *Semigroup property: for arbitrary numbers* $t_1 \leq t_2$, $x \in X$,

$$\mathcal{D}(x,t_2) = \bigcup_{y \in \mathcal{D}(x,t_1)} \mathcal{D}(y,t_2 - t_1).$$

4. *The function* $\mathcal{D}(x,t)$ *is continuous in the Huasdorf metric (defined by* (11.7)) *in both variables.*

Intuitively $\mathcal{D}(x,t)$ denotes the set of points of X, which a controlled object (whose dynamics can be defined, for instance, by a controlled ordinary differential equation of the type $\dot{x} = f(x,u)$) can reach from the point x in time $t \geq 0$.

The function $\widehat{x} : [t_0, t_1] \to X$ is called a *trajectory of a generalized dynamic system* $\mathcal{D}$, if $\widehat{x}(\tau_1) \in \mathcal{D}(\widehat{x}(\tau_0), \tau_1 - \tau_0)$ whenever $t_0 \leq \tau_0 \leq \tau_1 \leq t_1$. Let us denote by $\widehat{\mathcal{D}}(B,t)$ the set of all trajectories of the generalized dynamic system $\mathcal{D}$ on the interval $[0,t]$, coming out of the set $B \subset X$.

One can show (see [13]) that the following main properties hold:

1. *A trajectory is always continuous.*
2. *If the sets* $A \subset X$, $B \subset [0,\infty)$ *are compact, then the set*

$$\mathcal{D}(A,B) = \bigcup_{x \in A,\, t \in B} \mathcal{D}(x,t)$$

is compact.

3. *For an arbitrary point* $y \in \mathcal{D}(x,t)$ *and all* $x \in X$, $t \in [0,\infty)$ *there exists a trajectory of the generalized dynamic system* $\mathcal{D}$ *on* $[0,t]$ *that starts from* x *and ends at* y.

4. *If B is compact, then the set $\widehat{\mathcal{D}}_l(B,t)$ is compact in the uniform metric $\widehat{\rho}_t$:*

$$\widehat{\rho}_t(\widehat{x}^l, \widehat{y}^l) = \max_{\tau \in [0,t]} \{\rho(\widehat{x}^l(\tau), \widehat{y}^l(\tau))\}.$$

5. *The compact subset $\widehat{\mathcal{D}}(x,t)$ of the metric space of continuous mappings from the interval $[0,t]$ to X with the uniform metric depends continuously (in the Hausdorf metric) on x.*

Consider now a game between two players, whose dynamics is given by generalized dynamic system $\mathcal{D}_I$, $\mathcal{D}_{II}$ in X. We shall denote by $\widehat{x}^l[x,t]$ a trajectory of the player l starting at x and of the duration t.

Denote by Σ_T the set of finite partitions σ of the interval $[0,T]$, $T < \infty$:

$$\sigma = \{0 < t_1 < t_2 < \ldots < t_{N_\sigma} = T\}.$$

The games $\Gamma(x_0^I, x_0^{II}, T)$ considered below are games with complete information. Namely, at each moment $t \in [0,T]$ every player knows precisely the positions of both players $\widehat{x}^l(t)$, $l = I, II$, as well as the dynamic possibilities of both players specified by the functions $\mathcal{D}_l$, $l = I, II$, and the total duration of the game $T < \infty$. Let us define now the strategies in the games $\Gamma(x_0^I, x_0^{II}, T)$.

A *strategy φ_l of the player l in the game* $\Gamma(x_0^I, x_0^{II}, T)$ is a pair $(\sigma_{\varphi_l}, K_\sigma^l)$, where $\sigma_{\varphi_l} \in \Sigma_T$, and K_σ^l is a mapping that to each possible pair of positions

$$\widehat{x}^I[x_0^I, t_k] \in \widehat{\mathcal{D}}_I(x_0^I, t_k), \qquad \widehat{x}^{II}[x_0^{II}, t_k] \in \widehat{\mathcal{D}}_{II}(x_0^{II}, t_k), \qquad t_k \in \sigma_{\varphi_l} = \sigma_l,$$

of the players at time t_k set into correspondence a trajectory

$$\widehat{x}^l[\widehat{x}^l[x_0^l, t_k](t_k), t_{k+1} - t_k] \in \widehat{\mathcal{D}}_l(\widehat{x}^l[x_0^l, t_k](t_k), t_{k+1} - t_k).$$

The set of all strategies of the player l in the game $\Gamma(x_0^I, x_0^{II}, T)$ will be denoted by Φ_l, $l = I, II$. The pair

$$(\varphi_I, \varphi_{II}) = \left((\sigma_{\varphi_I}, K_{\sigma_I}^I), (\sigma_{\varphi_{II}}, K_{\sigma_{II}}^{II})\right) \in \Phi_I \times \Phi_{II}$$

is called a *situation (or a profile) in the game* $\Gamma(x_0^I, x_0^{II}, T)$.

Let us choose now a situation $(\varphi_I, \varphi_{II}) \in \Phi_I \times \Phi_{II}$, and let

$$\sigma_I = \left\{0 < t_1^I < \ldots < t_{N_{\sigma_I}}^I = T\right\},$$

$$\sigma_{II} = \left\{0 < t_1^{II} < \ldots < t_{N_{\sigma_{II}}}^{II} = T\right\}.$$

Assume that $t_1^I \leq t_1^{II}$, say. Then in accordance with the definition of strategies the partial trajectories of the players I, II on the intervals $[0, t_1^I]$ and $[0, t_1^{II}]$ respectively are the images of the mappings $K_{\sigma_I}^I$, $K_{\sigma_{II}}^{II}$, i.e.

$$\hat{x}^l[x_0^l, t_1^l] = K_{\sigma_l}^l(x_0^I, x_0^{II}), \quad l = I, II.$$

Then the trajectory of player I on the interval $[t_1^I, t_2^I]$ is defined by applying $K_{\sigma_I}^I$ to the already known pair of the trajectories that are realized to time t_1^I. In this way we can build the trajectory recursively on the intervals $[t_k^l, t_{k+1}^l]$, $k = 0, 1, \ldots, N_{\sigma_l} - 1$, $l = I, II$, so that on each such interval the initial points are chosen as the end points of the pieces constructed on the previous step, and the whole trajectory $\hat{x}^l[x_0^l, T]$ is "glued" from these pieces. Thus one obtains the well defined mapping

$$\chi: \ \Phi_I \times \Phi_{II} \to \hat{\mathcal{D}}_I(x_0^I, T) \times \hat{\mathcal{D}}_{II}(x_0^{II}, T).$$

To completely specify the games $\Gamma(x_0^I, x_0^{II}, T)$ in the normal form it remain to define the payoffs on the set $\Phi_I \times \Phi_{II}$. For simplicity we shall reduce our attention to the case of games with terminal payoff only. Namely, let a continuous real function H is defined on the product $X \times X$. Then in the game $\Gamma(x_0^I, x_0^{II}, T)$ the payoff of the player II in a situation $(\varphi_I, \varphi_{II}) = \varphi$ is defined as

$$H(\chi(\varphi)(T)) = \overline{H}(\varphi).$$

The player II, choosing φ_{II}, tries to maximise this payoff the aim of I is precisely opposite. Recall that if G is a real function on $\Phi_I \times \Phi_{II}$, then a situation $(\varphi_I, \ \varphi_{II})$ is called a *saddle point* of the game Γ_G (*ε-saddle point* Γ_G), if for all $\varphi_I' \in \Phi_I$, $\varphi_{II}' \in \Phi_{II}$ the inequalities

$$G(\varphi_I, \varphi_{II}') \leq G(\varphi_I, \varphi_{II}) \leq G(\varphi_I', \varphi_{II}),$$

$$(G(\varphi_I, \varphi_{II}') - \varepsilon \leq G(\varphi_I, \varphi_{II}) \leq G(\varphi_I', \varphi_{II}) + \varepsilon)$$

hold true.

Remark. In the strategies considered a partition σ of the interval $[0, T]$ was chosen before the start of the game. In some cases it is convenient to allow to a player to choose the next point t_{k+1} of the partition σ at the moment t_k (assuming that the resulting partition σ belongs to the set Σ_T of finite partitions of $[0, T]$. Nothing changes in the following expositions if one would adhere to this new kind of strategies.

Let us introduce the auxiliary games $\underline{\Gamma}^\sigma(x_0^I, x_0^{II}, T)$ and $\overline{\Gamma}^\sigma(x_0^I, x_0^{II}, T)$, which we shall call respectively *lower and upper for the game* $\Gamma^\sigma(\cdot)$. Here $\sigma \in \Sigma_T$.

The dynamics of the players I, II in auxiliary games $\overline{\Gamma}^\sigma(\cdot)$, $\underline{\Gamma}^\sigma(\cdot)$ is considered to be the same as in $\Gamma(\cdot)$ and is defined by the generalized dynamic systems $\mathcal{D}_I$, $\mathcal{D}_{II}$.

The game $\underline{\Gamma}^\sigma(x_0^I, x_0^{II}, T)$ develops in the following way. At the moment $t_0 = 0$ the player II observes the initial positions of both players x_0^I, x_0^{II} and chooses the trajectory $\hat{x}^{II}[x_0^{II}, t_1]$, t_1. Then the player I chooses the trajectory $\hat{x}^I[x_0^I, t_1]$ taking into account not only the initial positions, but also the trajectory $\hat{x}^{II}[x_0^{II}, t_1]$, chosen by the opponent. On the second step at time t_1 the player II is supposed to know the trajectories $\hat{x}^l[x_0^l, t_1]$, $l = I, II$, and chooses a trajectory on the next interval $[t_1, t_2]$, taking this information into account. Then the player I chooses a trajectory on $[t_1, t_2]$ taking into account the trajectories $\hat{x}^{II}[x_0^{II}, t_2]$, $\hat{x}^I[x_0^I, t_1]$. Analogously the process develops till the time T, when the game stops and the player II receives from the player I the payoff $\overline{H}(\hat{x}_T^I, \hat{x}_T^{II})$. Here $(\hat{x}_T^I, \hat{x}_T^{II})$ denotes the resulting trajectory in the game $\underline{\Gamma}^{\sigma_n}(\cdot)$.

The game $\overline{\Gamma}^{\sigma_n}(\cdot)$ develops in the dual (symmetric) way with the player I taking the lead. Namely, at the moment $t_0 = 0$ the player I observes the initial positions of both players x_0^I, x_0^{II} and chooses the trajectory $\hat{x}^I[x_0^I, t_1]$. Then the player II chooses the trajectory $\hat{x}^{II}[x_0^{II}, t_1]$ taking into account not only the initial positions, but also the trajectory $\hat{x}^I[x_0^I, t_1]$, chosen by the opponent. Similarly the process develops on the next steps $2, 3, \ldots, N_{\sigma_n}$. On the step N_{σ_n} the game stops and the payer II receives from the player I the payoff $\overline{H}(\hat{x}_T^I, \hat{x}_T^{II})$. Here $(\hat{x}_T^I, \hat{x}_T^{II})$ denotes the resulting trajectory in the game $\overline{\Gamma}^{\sigma_n}(\cdot)$. Let us stress for clarity that the players know the conditions of the games, i.e. the dynamics of the players, the duration of the game and the partition σ_n.

A *strategy* $\overline{\varphi}^{I\sigma}(\varphi^{II\sigma})$ *of the player* $I(II)$ *in the game* $\overline{\Gamma}^\sigma(\cdot)(\underline{\Gamma}^\sigma(\cdot))$ is by definition an arbitrary mapping that for any moment $t_k \in \sigma_n$, $k = 0, \ldots, N_\sigma - 1$ and a given couple of trajectories

$$\left(\hat{x}^I[x_0^I, t_k], \hat{x}^{II}[x_0^{II}, t_k]\right) \in \mathcal{D}_I(\cdot) \times \mathcal{D}_{II}(\cdot)$$

specifies uniquely a trajectory

$$\hat{x}^I[x_k^I, \delta_n] \qquad (\hat{x}^{II}[x_k^{II}, \delta_n]),$$

where $x_k^l = \hat{x}^l[x_0^l, t_k](t_k)$ denotes the end position (position at time t_k) of the trajectory $\hat{x}^l[x_0^l, t_k]$.

A strategy $\underline{\varphi}^{I_\sigma}$ ($\overline{\varphi}^{II_\sigma}$) of the player $I, (II)$ in the game $\underline{\Gamma}^\sigma(\cdot)(\overline{\Gamma}^\sigma(\cdot))$ is a mapping that for any moment $t_k \in \sigma_n$, $k = 0, \ldots, N_\sigma - 1$ and a given couple of trajectories

$$\left(\widehat{x}^I[x_0^I, t_k],\ \widehat{x}^{II}[x_0^{II}, t_{k+1}]\right)\ \left(\widehat{x}^I[x_0^I, t_{k+1}], \widehat{x}^{II}[x_0^{II}, t_k]\right)$$

specifies uniquely a trajectory

$$\widehat{x}^I[x_k^I, \delta_n] \qquad (\widehat{x}^{II}[x_k^{II}, \delta_n]).$$

The set of the strategies of the player l in the game $\underline{\Gamma}^\sigma(\cdot)$ ($\overline{\Gamma}^\sigma(\cdot)$) we shall denote by $\underline{\Phi}_l^\sigma$ ($\overline{\Phi}_l^\sigma$).

As in the case of the game $\Gamma(\cdot)$ it follows from the definition of the game $\overline{\Gamma}^\sigma(\cdot), \underline{\Gamma}^\sigma(\cdot)$ that an arbitrary situation $\overline{\varphi}^\sigma$, $\underline{\varphi}^\sigma$ uniquely specifies a trajectory of the game $\overline{\Gamma}^\sigma(\cdot)$ ($\underline{\Gamma}^\sigma(\cdot)$).

Proposition 11.18. *In the games* $\overline{\Gamma}^\sigma(x_0^I, x_0^{II}, T)$, $\underline{\Gamma}^\sigma(\cdot)$ *there exist saddle points in pure strategies, the value function* $val(\overline{\Gamma}^\sigma(\cdot))$, $val(\underline{\Gamma}^\sigma(\cdot))$ *is continuous with respect to* x_0^I, x_0^{II}, *and for any partition* $\sigma \in \Sigma_T^2$

$$val\left(\overline{\Gamma}^\sigma(x_0^I, x_0^{II}, T)\right) \geq val\left(\underline{\Gamma}^\sigma(x_0^I, x_0^{II}, T)\right). \tag{11.18}$$

Proof. Note that since the payoffs depend only on the terminal positions, from the point of view of the optimization problem considered, choosing pieces of trajectories on each step is equivalent to choosing the end point of those trajectories. And then we are falling precisely under the condition of the general dynamic programming Theorem 9.15, which implies all the statements, but for the equation (11.18). But this equation is a consequence of trivial induction and the elementary fact that for any continuous function of two variables

$$\max_y \min_x h(x, y) \leq \min_x \max_y h(x, y) \tag{11.19}$$

(as long as these min and max exist of course).

Proposition 11.19. *For any pair of partitions* $\sigma, \sigma' \in \Sigma_T$ *such that* σ' *is a subpartition of* σ,

$$val\left(\overline{\Gamma}^\sigma(x_0^I, x_0^{II}, T)\right) \geq val\left(\overline{\Gamma}^{\sigma'}(x_0^I, x_0^{II}, T)\right).$$

$$val\left(\underline{\Gamma}^\sigma(x_0^I, x_0^{II}, T)\right) \leq val\left(\underline{\Gamma}^{\sigma'}(x_0^I, x_0^{II}, T)\right).$$

Proof. Let us prove the first inequality (others are proved analogously). It is sufficient to show it in the case when σ' is obtained from σ by adding an additional point s: $\sigma' = \sigma \cup s$, as the general case is easily reduced to this particular one.

Suppose t_k, t_{k+1} are the neighboring points of σ and $s \in (t_k, t_{k+1})$. Let $v(x^I, x^{II})$ denote the value of the upper game starting at (x^I, x^{II}), of the duration $[t_{k+1}, T]$ and specified by the partition σ restricted to $[t_{k+1}, T]$. One has to show that for any x^I, x^{II}

$$\min_{x^I_{k+1} \in \mathcal{D}^I(x^I, t_{k+1}-t_k)} \max_{x^{II}_{k+1} \in \mathcal{D}^{II}(x^{II}, t_{k+1}-t_k)} v(x^I_{k+1}, x^{II}_{k+1})$$

$$\geq \min_{x^I_s \in \mathcal{D}^I(x^I, s-t_k)} \max_{x^{II}_s \in \mathcal{D}^{II}(x^{II}, s-t_k)} \min_{x^I_{k+1} \in \mathcal{D}^I(x^I_s, t_{k+1}-s)} \max_{x^{II}_{k+1} \in \mathcal{D}^{II}(x^I, t_{k+1}-s)}$$
$$v(x^I_{k+1}, x^{II}_{k+1}). \tag{11.20}$$

But the l.h.s. of this inequality can be written as

$$\min_{x^I_s \in \mathcal{D}^I(x^I, s-t_k)} \min_{x^I_{k+1} \in \mathcal{D}^I(x^I_s, t_{k+1}-s)} \max_{x^{II}_s \in \mathcal{D}^{II}(x^{II}, s-t_k)} \max_{x^{II}_{k+1} \in \mathcal{D}^{II}(x^I, t_{k+1}-s)}$$
$$v(x^I_{k+1}, x^{II}_{k+1})$$

and consequently (11.20) follows from (11.19).

To get better continuity properties of payoffs we shall introduce a stronger continuity assumption on our generalized dynamics. We shall assume that for any compact set K there exists a constant $L = L(K)$ such that

$$\rho(\mathcal{D}^0_l(t, x), \mathcal{D}^0_l(t, y) \leq \|x - y\|(1 + Lt) \tag{11.21}$$

for all $x, y \in K$ and both $l = I, II$. For instance, this clearly holds whenever the dynamics is described by a usual controlled differential equation $\dot{x} = f(x, u)$ with a locally Lipschitz continuous in x function f.

For a $\sigma \in \Sigma_T$ let $|\sigma| = \max_{1 \leq i \leq N_\sigma}(t_i - t_{i-1})$.

Proposition 11.20. *Under assumption* (11.21) *for any sequence of partitions* $\{\sigma_n\}_{n=1}^{\infty}$ *of the interval* $[0, T]$, $\sigma \in \Sigma_T$, *with* $|\sigma_n| \to 0$ *as* $n \to \infty$ *one has*

$$\lim_{n \to \infty} val\left(\overline{\Gamma}^{\sigma_n}(x^I_0, x^{II}_0, T)\right) = \lim_{n \to \infty} val\left(\underline{\Gamma}^{\sigma_n}(x^I_0, x^{II}_0, T)\right) \tag{11.22}$$

whenever the limits on the left and right hand sides exist. In particular they exist if each σ_{n+1} *is obtained by a subdivision of* σ_n.

Proof. The last statement follows from the monotonicity obtained in Proposition 11.19 so one needs to show (11.22) assuming the limits exist. But under (11.21) it follows from Theorem 9.15 that if the terminal payoff function H is Lipschitz continuous with a Lipschitz constant L then $val\left(\overline{\Gamma}^{\sigma_n}(x_0^I, x_0^{II}, T)\right)$ is Lipschitz continuous with the constant

$$\prod_j (1 + L(t_j - t_{j-1})) = \exp\{\sum_j \ln(1+L(t_j-t_{j-1}))\} \le \exp\{L\sum_j (t_j-t_{j-1})\}$$

$$= \exp\{Lt\}.$$

Similarly, if H is continuous, then it is uniformly continuous on each compact set and hence the whole family of functions $val\left(\overline{\Gamma}^{\sigma}(x_0^I, x_0^{II}, T)\right)$ (as well as $val\left(\underline{\Gamma}^{\sigma}(x_0^I, x_0^{II}, T)\right)$) is uniformly equi-continuous for all partitions σ of $[0, T]$ if reduced to an arbitrary compact set. Hence by passing if necessary to a subsequence one can consider the limits in (11.22) to be uniform. Hence the equation (11.22) follows from Theorem 9.16.

Theorem 11.7. *Under assumption (11.21) for any sequence of partitions $\{\sigma_n\}_1^\infty$ with $|\sigma_n| \to 0$ as $n \to \infty$ one has*

$$\lim_{n\to\infty} val\left(\overline{\Gamma}^{\sigma_n}(x_0^I, x_0^{II}, T)\right) = \lim_{n\to\infty} val\left(\underline{\Gamma}^{\sigma_n}(x_0^I, x_0^{II}, T)\right)$$

$$= \inf_{\sigma\in\Sigma_T} val\left(\overline{\Gamma}^{\sigma}(x_0^I, x_0^{II}, T)\right) = \sup_{\sigma\in\Sigma_T} val\left(\underline{\Gamma}^{\sigma}(x_0^I, x_0^{II}, T)\right) \quad (11.23)$$

and this expression defines a function of (x_0^I, x_0^{II}) that is continuous (respectively locally or globally Lipschitz continuous) if so is H.

Proof. Taking into account Propositions 11.19, 11.20 one only needs to show that for an arbitrary sequences $\{\sigma_n'\}_1^\infty$, $\{\sigma_n\}_1^\infty$, σ_n, $\sigma_n' \in \Sigma_T$ with $|\sigma_n| \to 0$, $|\sigma_n'| \to 0$ as $n \to \infty$ the equation

$$\lim_{n\to\infty} val\left(\underline{\Gamma}^{\sigma_n}(x_0^I, x_0^{II}, T)\right) = \lim_{n\to\infty} val\left(\underline{\Gamma}^{\sigma_n'}(x_0^I, x_0^{II}, T)\right)$$

holds true. Let us give a proof by contradiction. Assume that

$$\lim_{n\to\infty} val\left(\underline{\Gamma}^{\sigma_n}(x_0^I, x_0^{II}, T)\right) > \lim_{n\to\infty} val\left(\underline{\Gamma}_1^{\sigma_n'}(\cdot)\right). \quad (11.24)$$

Then by Proposition 11.20 one can find integers $n_1 > 0$, $m_1 > 0$ such that

$$val(\overline{\Gamma}^{\sigma_{m_1}}(\cdot)) \ge val(\underline{\Gamma}^{\sigma_{m_1}}(\cdot)) > val(\overline{\Gamma}^{\sigma_{n_1}'}(\cdot)) \ge val(\underline{\Gamma}^{\sigma_{n_1}'}(\cdot)).$$

Let us introduce a new partition $\overline{\sigma} = \sigma_{m_1} \cup \sigma_{n_1}^1$. Then Proposition 11.19 implies

$$val(\overline{\Gamma}^{\overline{\sigma}}(\cdot)) \leq val(\overline{\Gamma}^{\sigma'_n}(\cdot)) < val(\underline{\Gamma}^{\sigma_{m_1}}(\cdot)) \leq val(\underline{\Gamma}^{\overline{\sigma}}(\cdot)),$$

which contradicts (11.18).

Theorem 11.8. *For arbitrary* $x_0^I, x_0^{II} \in X$, $T < \infty$, *and* $\varepsilon > 0$, *there exist* ε-*equilibria in the game* $\Gamma_1(x_0^I, x_0^{II}, T)$. *Moreover*

$$val\left(\Gamma_1(x_0^I, x_0^{II}, T)\right) = \lim_{n\to\infty} val\left(\overline{\Gamma}_1^{\sigma_n}(x_0^I, x_0^{II}, T)\right)$$

where $\{\sigma_n\}_1^\infty$ *is an arbitrary decreasing sequence of partitions of* $[0, T]$ *with* $|\sigma_n| \to 0$ *as* $n \to \infty$.

Proof. Given a $\varepsilon > 0$ let us show that there exist strategies $\varphi_I^\varepsilon \in \Phi_I$, $\varphi_{II}^\varepsilon \in \Phi_{II}$ such that for arbitrary strategies $\varphi_I \in \Phi_I$, $\varphi_{II} \in \Phi_{II}$

$$\bar{H}(\varphi_I^\varepsilon, \varphi_{II}) - \varepsilon \leq \bar{H}(\varphi_I^\varepsilon, \varphi_{II}^\varepsilon) \leq \bar{H}(\varphi_I, \varphi_{II}^\varepsilon) + \varepsilon.$$

In fact, by Theorem 11.9 there exist partitions $\sigma_{I,\varepsilon}$, $\sigma_{II,\varepsilon} \in \Sigma_T$, such that

$$val(\overline{\Gamma}^{\sigma_{I,\varepsilon}}(\cdot)) - \lim_{n\to\infty} val(\overline{\Gamma}^{\sigma_n}(\cdot)) < \varepsilon,$$

$$\lim_{n\to\infty} val(\underline{\Gamma}^{\sigma_n}(\cdot)) - val(\underline{\Gamma}^{\sigma_{II,\varepsilon}}(\cdot)) < \varepsilon.$$

Put $\varphi_l^\varepsilon = (\sigma_{l,\varepsilon}, K_{\sigma_{l,\varepsilon}}^l)$, $l = I, II$, where $K_{\sigma_{l,\varepsilon}}^l$ are the optimal strategies of the players I,II in the games $\overline{\Gamma}^{\sigma_{I,\varepsilon}}(x_0^I, x_0^{II}, T)$, $\underline{\Gamma}^{\sigma_{II,\varepsilon}}(x_0^I, x_0^{II}, T)$. From the definition of the strategy φ_l^ε it follows that the pair $(\varphi_I^\varepsilon, \varphi_{II}^\varepsilon)$ specifies a unique trajectory of the game. Moreover, the choice of the mapping $K_{\sigma_{l,\varepsilon}}^l$, $l = I, II$, ensures that the payoff of player II using φ_{II}^ε could not be less than $(\lim_{n\to\infty} val(\underline{\Gamma}_1^{\sigma_n}(\cdot)) - \varepsilon)$, and the payoff of player I (that equals the negation of its loss) using φ_I^ε could not be less than $(\lim_{n\to\infty} val(\underline{\Gamma}_1^{\sigma_n}(\cdot)) + \varepsilon)$. Consequently φ_l^ε, $l = I, II$, are ε-optimal strategies.

As this holds for any ε, the function

$$val\left(\Gamma(x_0^I, x_0^{II}, T)\right) = \lim_{n\to\infty} val\left(\overline{\Gamma}^{\sigma_n}(x_0^I, x_0^{II}, T)\right)$$

represents the value of the game $\Gamma_1(x_0^I, x_0^{II}, T)$.

Remark. From the Lipschitz continuity of the value (see Theorem 11.7) one can deduce that it satisfies almost everywhere a first order equation called Isaacs-Bellman equation. An important achievement of the theory of game was to identify the obtained solutions with the so called *viscosity solutions* (constructed in [45]) that are playing a major role on modern theory of differential equations. For this development one can refer e.g. to [180].

Let us now consider shortly the time minimization $\Gamma(x_0^I, x_0^{II})$, which are defined on the half-line $[0, \infty)$, unlike the games with a fixed duration. The game $\Gamma(x_0^I, x_0^{II})$ also evolves in a complete locally compact metric space X, the dynamics of the players I, II is specified by generalized dynamic systems $\mathcal{D}_I$, $\mathcal{D}_{II}$. The information available to players I, II in $\Gamma(x_0^I, x_0^{II})$ is the same as in the games $\Gamma(x_0^I, x_0^{II}, T)$. A strategy φ_l, $l = I, II$, of the player l in the game $\Gamma(x_0^I, x_0^{II})$ is a pair (σ_l, K_{σ_l}), where $\sigma_l = \{t_0 = 0 < t_1 < \ldots < t_k < \ldots\}$ is a partition of $[0, \infty)$, not containing finite limit points, and K_{σ_l} is a mapping, that to each state of information of the player l at each time moment $t_k \in \sigma_l$ set into correspondence a trajectory $\widehat{x}^l \in \widehat{\mathcal{D}}_l(x_k^l, t_{k+1} - t_k)$. The set of all strategies of the player l in the game $\Gamma(x_0^I, x_0^{II})$ will be denoted by Φ_l.

As above each situation $(\varphi_I, \varphi_{II})$ specifies uniquely a pair of trajectories of the players I, II on $[0, \infty)$:

$$(\widehat{x}_\infty^I, \widehat{x}_\infty^{II}) = \mathcal{X}(\varphi_I, \varphi_{II}).$$

Let $\mathcal{M}$ be a nonempty closed subset of the space $X \times X$. Let us define the payoff in a situation $(\varphi_I, \varphi_{II})$ as

$$H_\alpha(\varphi_I, \varphi_{II}) = \min_{t \in [0, \infty)} \{t \,|\, \mathcal{X}(\varphi_I, \varphi_{II})(t) \in \mathcal{M}_\alpha\}, \qquad (11.25)$$

where

$$\mathcal{M}_\alpha = \{z \in X \times X \,|\, \rho(z, M) \leq \alpha\}.$$

If $H_\alpha(\varphi_I, \varphi_{II}) = \infty$, one says that in the situation $(\varphi_I, \varphi_{II})$ the game $\Gamma(x_0^I, x_0^{II})$ can not be completed in a finite time. By choosing a strategy φ_I, the player I (respectively II) tries to minimize (respectively maximize) the payoff.

A strategy $\varphi_I \in \Phi_I$ is called *successful* whenever for any strategy φ_{II} in the situation $(\varphi_I, \varphi_{II})$ the game $\Gamma(x_0^I, x_0^{II})$ is completed in a finite time.

Thus we defined the sets of the strategies of I, II and the payoffs on the product of these sets. Hence the game $\Gamma(x_0^I, x_0^{II})$ is defined in the normal form.

Theorem 11.9. *If the player I has a successful strategy for arbitrary $\alpha > 0$ there exists a ε-equilibrium for each $\varepsilon > 0$ and all $\alpha > 0$.*

Proof. It can be deduced from the previous theorems. Details can be found e.g. in [122] and are omitted here.

11.5   Cooperative games versus zero-sum games

The aim of this section is to present a curious connection between the solutions to cooperative games and lower values of auxiliary zero-sum games.
The following definitions are standard:

Cooperative game with non-transferable utility is a triple $G = (I, v, H)$,
where $I = \{1, 2, \ldots, n\}$ is the set of players, H is a non-empty compact
set from R^n, and v is a mapping from the set of all coalitions (non-empty
subset $S \subset I$) to the set of non-empty closed subsets $v(S) \subset H$.

For $x, y \in H$ one says that x *dominates* y $(x \succ y)$, if there exists a
coalition $S \subseteq I$ such that

1) $x, y \in v(S)$;
2) $x_i > y_i$ for any $i \in S$.

A *(von Neumann-Morgenstern) solution* to $G = (I, v, H)$ is called a
subset $V \subseteq H$ such that

1) *(internal stability)* there are no pairs of vectors from V such that one
of them dominates the other one;

2) *(external stability)* for any $y \in H \backslash V$ there exists $x \in V$ such that
$x \succ y$.

For $A \subset R^n$ let us denote by A_ε the ε-neighborhood of A, i.e. $A_\varepsilon = A + B_\varepsilon$ with

$$B_\varepsilon = \left\{ x \in R^n \left| \sum_{i=1}^{n} x_i^2 < \varepsilon \right. \right\}.$$

A closed subset V is called a ε-*solution* whenever V is internally stable
and for any $y \in H \backslash V_\varepsilon$ there exists $x \in V$ such that $x \succ y$.

It is clear that if a closed A is a ε-solution in the game $G = (I, v, H)$
for any $\varepsilon > 0$, then A is a solution.

Consider the function

$$L(x, y) = \max_{S : x, y \in v(S)} \min_{i \in S} (x_i - y_i).$$

Then $L(x, y) > 0 \Leftrightarrow x \succ y$. Let $\mathcal{A}(G) = \{B \in 2^H : \max_{x, y \in B} L(x, y) = 0\}$.

Proposition 11.21. *Let $\varepsilon > 0$. The game $G = (I, v, H)$ has a ε-solution
if and only if*

$$\sup_{A \in \mathcal{A}(G)} \left\{ \min_{y \in H \backslash A_\varepsilon} \max_{x \in A} L(x, y) \right\} > 0. \tag{11.26}$$

Proof. Let (11.26) holds. Then there exist $A \subseteq H$ such that

$$\min_{y \in H \backslash A_\varepsilon} \max_{x \in A} L(x, y) > 0 \tag{11.27}$$

and

$$\max_{x,y \in A} L(x, y) = 0. \tag{11.28}$$

It follows from (11.27) that $\max_{x \in A} L(x, \overline{y}) > 0$ for each $\overline{y} \in H \backslash A_\varepsilon$. Consequently for arbitrary $\overline{y} \in H \backslash A_\varepsilon$ one can find a $\overline{x} \in A$ such that $L(\overline{x}, \overline{y}) > 0$, i.e. $\overline{x} \succ \overline{y}$. It now follows from (11.28) that $L(x, y) \leq 0$ for all $x, y \in A$, i.e. neither x dominates y, nor vice versa. Consequently the set A is a ε-solution.

Now let A be a ε-solution. Then $L(x, y) \leq 0$ for all $x, y \in A$, but $L(x, x) = 0$ implying (11.28).

With A being a ε-solution, for any $\overline{y} \in H \backslash A_\varepsilon$ there exists a $\overline{x} \in A$ such that $\overline{x} \succ \overline{y}$, i.e. $L(\overline{x}, \overline{y}) > 0$. Consequently $\max_{x \in A} L(x, \overline{y}) > 0$ for any $\overline{y} \in H \backslash A_\varepsilon$ and therefore

$$\min_{y \in H \backslash A_\varepsilon} \max_{x \in A} L(x, y) > 0.$$

The proof is complete.

Let us choose now an $\varepsilon > 0$ and introduce a two-person zero-sum game $\Gamma_\varepsilon(I, v, H)$, in which the first player makes a first move by choosing $A \in \mathcal{A}$ and then the second player replies by choosing $y \in H \backslash A_\varepsilon$. Finally the first player makes the third (and the last) move by choosing $x \in A$. The income of the first player in this game equals $L(x, y)$ (the second one gets $-L(x, y)$).

Proposition 11.22. *The game $G = (I, v, H)$ has a ε-solution if and only if the maximal guaranteed gain of the first plater in $\Gamma_\varepsilon(I, v, H)$ is positive. Moreover any $A \in \mathcal{A}$ yielding such gain is a ε-solution to G.*

Proof. Follows directly from Proposition 11.21.

One can now define a two-player zero-sum game in the normal form $N_\varepsilon(I, v, H)$, where the strategies of the first players are the sets $A \in \mathcal{A}$ and the strategies of the second player are the mappings $f : A \in \mathcal{A} \mapsto H \backslash A_\varepsilon$. Let F denote the set of the strategies of the second player. Let the payoff to the first player in this game $N_\varepsilon(I, v, H)$ be

$$h(A, f) = \max_{x \in A} L(x, f(A)).$$

Proposition 11.23. *The game $N_\varepsilon(I, v, H)$ has a value, i.e.*

$$\sup_{A \in \mathcal{A}} \inf_{f \in F} h(A, f) = \inf_{f \in F} \sup_{A \in \mathcal{A}} h(A, f).$$

Proof. This follows from the dynamic programming Theorem 9.13. But let us give a direct proof. Consider the strategy f^* such that

$$\min_{y \in H \setminus A_\varepsilon} \max_{x \in A} L(x, y) = \max_{x \in A} L(x, f^*(A))$$

for any $A \in \mathcal{A}$.
Then

$$\sup_{A \in \mathcal{A}} \inf_{f \in F} h(A, f) \leq \inf_{f \in F} \sup_{A \in \mathcal{A}} h(A, f) \leq \sup_{A \in \mathcal{A}} h(A, f^*)$$

$$= \sup_{A \in \mathcal{A}} \max_{x \in A} L(x, f^*(A)) = \sup_{A \in \mathcal{A}} \min_{y \in H \setminus A_\varepsilon} \max_{x \in A} L(x, y) = \sup_{A \in \mathcal{A}} \inf_{f \in F} h(A, f).$$

Proposition is proved.

Theorem 11.10. *([122]) The game $G = (I, v, H)$ has a ε-solution if and only if*

$$\sup_{A \in \mathcal{A}} \inf_{f \in F} h(A, f) > 0$$

holds for the game $N_\varepsilon(I, v, H)$.

Proof. Follows from Propositions 11.21, 11.23.

11.6   Turnpikes for stochastic games

This section is based on [89], [90]. Full story with extensive bibliography is presented in [265]. A version of this theory for 'nonlinear Markov games' is developed in [98] and [263].

By the *turnpike* theory a class of results in mathematical economics is meant that states (under certain given rules) the existence of an optimal way of exchange and production (turnpike), which is worth being stuck to for the main part of a large enough period of development independently of both the initial conditions (e.g. at the beginning you have lots of bread and butter) and the final goal (e.g. at the end you would like to have lots of cars and trains). The word "turnpike" comes from the comparison with route planning, when going from a village near London to a village near Edinburgh you have to find the quickest path to M1 motor-way (turnpike), drive along it the rest of your journey and turn to a small road leading to our destination at the end of it.

Here we shall prove a turnpike theorem in the context of stochastic games.

A stochastic zero-sum game on a finite state space $X = \{1, \ldots, n\}$ is usually specified by the collection $p_{ij}(\alpha, \beta)$ of the probabilities of transitions from state i to state j if the two players choose strategies α and β respectively (α and β belong to some fixed metric spaces), and by the incomes $b_{ij}(\alpha, \beta)$ of the second player from this transition. The functions p_{ij} and b_{ij} are assumed to be bounded continuous functions of α, β. The dynamic n-step game under these rules, certain terminal income $y \in \mathbb{R}^n$ and a given initial position i is played in the following way. At the first step the players choose independently certain strategies α, β, and with the probability $p_{ij}(\alpha, \beta)$ the game moves to the state j and the first player pays $b_{ij}(\alpha, \beta)$ to the second player (if $b_{ij}(\alpha, \beta)$ is negative this means of course that the second player pays $-b_{ij}(\alpha, \beta)$ to the first one). Then the second step is played analogously starting from the position j. After the last n-th step the second player receives additional payment y^k depending on the position k, where the game terminates. This general scheme models a variety of the conflict control problems in a random environment.

The stochastic game is called a *game with a value* if

$$\min_{\alpha} \max_{\beta} \sum_{j=1}^{n} p_{ij}(\alpha, \beta)(y^j + b_{ij}(\alpha, \beta)) = \max_{\beta} \min_{\alpha} \sum_{j=1}^{n} p_{ij}(\alpha, \beta)(y^j + b_{ij}(\alpha, \beta))$$

(11.29)

for all $y \in \mathbb{R}^n$ and $i \in X$. In that case, the operator $B \colon \mathbb{R}^n \to \mathbb{R}^n$ such that $B_i(y)$ is equal to (11.29) is called the *Bellman operator* of the game. This condition means that for any i there exist *minimax strategies* (α_i, β_i) providing a saddle point of the average payoff function $\sum_{j=1}^{n} p_{ij}(\alpha, \beta)(y^j + b_{ij}(\alpha, \beta))$, i.e. such that

$$\sum_{j=1}^{n} p_{ij}(\alpha_i, \beta)(y^j + b_{ij}(\alpha_i, \beta))$$

$$\leq \sum_{j=1}^{n} p_{ij}(\alpha_i, \beta_i)(y^j + b_{ij}(\alpha_i, \beta_i)) \leq \sum_{j=1}^{n} p_{ij}(\alpha, \beta_i)(y^j + b_{ij}(\alpha, \beta_i)) \quad (11.30)$$

for all α, β.

By the method of backward induction or dynamic programming method, see Chapter 3 or [19] for the context of stochastic games, we can show that the value of the k-step game defined by the initial position i and the terminal income $h \in \mathbb{R}^n$ of the first player exists and is equal to $B_i^k(h) = (B^k(h))_i$.

It is clear that the operator B has the following two properties:

$$B(ae + h) = ae + B(h) \quad \forall a \in \mathbb{R}, \ h \in \mathbb{R}^n, \tag{11.31}$$

$$\|B(h) - B(g)\| \leq \|h - g\| \quad \forall h, g \in \mathbb{R}^n, \tag{11.32}$$

where $\|h\| = \max |h^i|$ and $e = (1, \dots, 1) \in \mathbb{R}^n$.

Remarkably enough, these two properties are characteristic of the game Bellman operator. As it was shown in [88], each mapping satisfying (11.31), (11.32) can be represented in form (11.29).

Of course the most important examples of games with value represent games with finitely many pure strategies, when the value is attained at the mixed strategies.

Now we define the quotient space Φ of the space $\mathbb{R}^n$ by the one-dimensional subspace generated by the vector $e = (1, \dots, 1)$. Let $\Pi \colon \mathbb{R}^n \mapsto \Phi$ be the natural projection. The quotient norm on Φ is obviously defined by the formula

$$\|\Pi(h)\| = \inf_{a \in \mathbb{R}} \|h + ae\| = \frac{1}{2} \Big(\max_j h^j - \min_j h^j \Big).$$

It is clear that Π has a unique isometric (but not linear) section $S \colon \Phi \mapsto \mathbb{R}^n$. The image $S(\Phi)$ consists of all $h \in \mathbb{R}^n$ such that $\max_j h^j = -\min_j h^j$. Thus one can identify $H\Phi$ with its image $h = S(H) \in \mathbb{R}^n$. By virtue of the properties (11.31) and (11.32) of B, the continuous quotient map $\widetilde{B} \colon \Phi \mapsto \Phi$ is well defined.

To state the main result of this section, we need some additional properties of the transition probabilities:

$$\exists \delta > 0 : \forall i, j \quad \exists \beta : \forall \alpha \quad p_{ij}(\alpha, \beta) \geq \delta, \tag{11.33}$$

$$\exists \delta > 0 : \forall i, j, \alpha, \beta \quad \exists m : p_{im}(\alpha, \beta) > \delta, \ p_{jm}(\alpha, \beta) > \delta. \tag{11.34}$$

Let all $|b_{ij}(\alpha, \beta)|$ be bounded by some constant C.

Lemma 11.1. *(i) If* (11.33) *holds and* $\delta < 1/n$, *then* $\widetilde{B}$ *maps each ball of radius* $R \geq C\delta^{-1}$ *centered at the origin into itself.*

(ii) If (11.34) *holds, then*

$$\|\widetilde{B}(H) - \widetilde{B}(G)\| \leq (1 - \delta)\|H - G\|, \qquad \forall H, G \in \Phi.$$

Proof. (i) It follows from the definition that for arbitrary $h \in \mathbb{R}^n$, β, i, j

$$B_i(h) - B_j(h) \leq \sum_{m=1}^{n} p_{im}(\alpha_1, \beta_1)(b_{im}(\alpha_1, \beta_1) + h^m)$$

$$- \sum_{m=1}^{n} p_{jm}(\alpha_0, \beta)(b_{jm}(\alpha_0, \beta) + h^m)$$

for certain α_1, β_1, and α_0. Hence,

$$B_i(h) - B_j(h) \leq 2C + \|h\| \sum_{m=1}^{n} |p_{im}(\alpha_1, \beta_1) - p_{jm}(\alpha_0, \beta)|.$$

Let us choose m_0 so that $p_{im_0}(\alpha_1, \beta_1) > \delta$. By (11.33), we can take β so that $p_{jm_0}(\alpha_0, \beta) > \delta$. Then

$$\sum_{m=1}^{n} |p_{im}(\alpha_1, \beta_1) - p_{jm}(\alpha_0, \beta)|$$

$$\leq |p_{im_0} - p_{jm_0}| + (1 - p_{im_0}) + (1 - p_{jm_0}) \leq 2(1 - \delta).$$

Consequently,

$$B_i(h) - B_j(h) \leq (2C + 2(1 - \delta))\|h\|.$$

Using the definition of the norm in Φ, we have

$$\|\Pi(B(h))\| \leq C + \|h\|(1 - \delta).$$

Thus, for $h = S(H)$ we obtain

$$\|\widetilde{B}(H)\| \leq C + \|H\|(1 - \delta).$$

It follows that the map $\widetilde{B}$ takes the ball of radius R into itself provided that $C + R(1 - \delta) \leq R$, i.e., $R \geq C\delta^{-1}$.

(ii) Clearly

$$\min_{\alpha, \beta} \sum_{m=1}^{n} p_{im}(\alpha, \beta)(h^m - g^m) \leq (Bh - Bg)_i \leq \min_{\alpha, \beta} \sum_{m=1}^{n} p_{im}(\alpha, \beta)(h^m - g^m).$$

Hence

$$(Bh - Bg)_i - (Bh - Bg)_j \leq \max_{\alpha_1, \beta_1, \alpha_2, \beta_2} \sum_{m=1}^{n} |p_{im}(\alpha_1, \beta_1) - p_{jm}(\alpha_2, \beta_2)||h^j - g^j|.$$

Choosing now m from (11.34) one completes the proof as above.

Theorem 11.11. (On the average income) *(i) If* (11.33) *holds, then there exists a unique $\lambda \in \mathbb{R}$ and a vector $h \in \mathbb{R}^n$ such that*

$$B(h) = \lambda + h \tag{11.35}$$

and for all $g \in \mathbb{R}^n$ we have

$$\|B^m g - m\lambda\| \leq \|h\| + \|h - g\|, \tag{11.36}$$

$$\lim_{m \to \infty} \frac{B^m g}{m} = \lambda. \tag{11.37}$$

(ii) If (11.34) *holds, then h is unique up to equivalence, i.e. $H = \Pi(h)$ is unique, and*

$$\|\widetilde{B}^m G - H\| \leq (1 - \delta)^m \|G - H\| \quad \forall G \in \Phi. \tag{11.38}$$

Proof. This follows readily from the Lemma above and the standard fixed point theorems.

Let $E(g) = E(G)$ be the set of equilibrium (minimax) strategies for the vector $g \in G \in \Phi$, i.e. $E(g) = (E^1(g), \ldots, E^n(g))$, where $E^i(g) = E^i(G) = \{(\alpha_i, \beta_i)\}$ such that (11.30) holds with $y = g$. In particular, if h solves (11.35), $E(h)$ contains the stationary strategies in the infinite-time game. Assume that $E(g)$ depends continuously on g. Then (11.38) implies directly the following.

Theorem 11.12. (**Turnpikes on the set of strategies**) *Let* (11.34) *hold. Then for arbitrary* $\Omega > 0$ *and a neighborhood* $U(E(h))$ *of the set* $E(h)$ *there exists an* $M \in \mathbb{N}$ *such that if* $E(B^{T-t}g)$ *is the set of equilibrium strategies on the step* t *of the* T*-step game,* $T > M$, *with terminal income of the second player defined by a vector* g *with* $\|\Pi(g)\| \leq \Omega$, *then*

$$E(B^{T-t}g) \subset U(E(h))$$

for all $t < T - M$.

Theorem 11.13. (**Turnpikes on the state space [88]**) *Assume additionally that for each* i, $E^i(h)$ *contains only one pair of strategies* α_i, β_i. *Let* $Q^* = (q_1^*, \ldots, q_n^*)$ *denote the stationary distribution for the stationary Markov chain defined on the state space* X *by these strategies. Assume also that the equilibrium transitions depend locally Lipschitz continuous on* g *around* h, *i.e. that there exists a constant* κ *such that*

$$|p_{ij}(\alpha, \beta) - p_{ij}(\alpha_i, \beta_i)| \leq \kappa \|G - H\| \tag{11.39}$$

for all $i, j \in X$, $(\alpha, \beta) \in E^i(G)$ *and* G *sufficiently close to* H. *Then for all* $\epsilon > 0$ *and* $\Omega > 0$ *there exists an* $M \in \mathbb{N}$ *such that for each* T*-step game,* $T > 2M$, *with terminal income* $g \in \mathbb{R}^n$, $\|\Pi(g)\| < \Omega$, *of the second player we have*

$$\|Q(t) - Q^*\| < \varepsilon$$

for $t \in [M, T - M]$, *where* $Q(t) = (q_1(t), \ldots, q_n(t))$ *and* $q_i(t)$ *is the probability that the process is in a state* $i \in X$ *at time* t *if the game is carried out with the equilibrium strategies.*

In other words, q_j^* is the mean amount of time that each sufficiently long game with equilibrium strategies spends in position j.

Proof. It follows from (11.38) and (11.39) that for each $\varepsilon_1 > 0$ there exists an $M_1 \in \mathbb{N}$ such that for any t-step equilibrium game, $t > M_1$, with the second player's terminal income g, $\|\Pi(g)\| \le \Omega$, the transition probabilities at the first $t - M_1$ steps are ε_1^t-close to the transition probabilities $p_{ij}(\alpha_i, \beta_i)$. Consequently,

$$Q(t) = Q^0(P + \delta_1) \cdots (P + \delta_t) = Q^0(P^t + \Delta(t)),$$

where $\delta_k = O(\varepsilon_1^{t-k})$ and hence, $\Delta(t) = O(\varepsilon_1)$. By the theorem on the convergence of probability distributions in homogeneous Markov chains to a stationary distribution, we have

$$\|Q^0 P^t - Q^*\| \le (1 - \delta)^{t-1}.$$

Thus, we can successively choose M_2 and ε_1 so that

$$\|Q(M_2) - Q^*\| < \varepsilon \quad \text{for all } Q^0.$$

Then

$$\|Q(t) - Q^*\| < \varepsilon \quad \text{for all } t \in [M_2, T - M_1].$$

There is a natural generalization of conditions (11.33) and (11.34) under which the cited results can still be proved. Namely, we require these conditions to be valid for some iteration of the operator B. This is the case of cyclic games. For generalizations to n-person games see [124].

If each coordinate of the operator B is convex, B is the Bellman operator of some controlled Markov process. Thus, Theorems 11.11 and 11.12, in particular, contain the turnpike theorems for Markov processes [92], [164].

Problem. Extend the above theory to the case of general (not necessarily finite) state space X (in case of Markov processes the case of continuous time and general state spaces was analyzed in [103]).

11.7 Games and tropical (or idempotent) mathematics

1. Idempotent mathematics. The origins of $(max, +)$-*algebra* go back to the 60s of the last century, one of the first systematic paper being [185], where it was noted that discrete equations of dynamic programming become linear in a exotic algebra, where the real numbers are considered to be equipped with two distributing binary operations $a \oplus b = \max(a, b)$ and $a \odot b = a + b$, some non-trivial results for this $(max, +)$-algebra were obtained and the program of its systematic study was put forward. As the *generalised addition* $\oplus$ is *idempotent* in this algebra meaning that $a \oplus a = a$, this algebra became the basic example of what was later called *idempotent algebra* (or *idempotent dioid*). A couple of monographs on this algebra were published (see [11] and references therein) to the time when V.P. Maslov observed (see [130], [132]) that this exotic linearity can be fruitfully pursued further from discrete dynamic programming to the differential Bellman equation of continuous time optimization theory, and the *Idempotent analysis* was initiated. It was baptized in [100] and its first 10 years of development were summarized in monographs [102], [103].

The idempotent algebra can be obtained by a certain scaling procedure from the usual one. This transformation can be looked at the transformation of probability theory to the optimization theory and is sometimes referred to as *de-quantization* (see [109]).

Apart from optimization theory a spectacular application of idempotency appeared in the theory of finite automata and formal languages. In honor of the contribution to this field made by the Brazilian mathematician Imre Simon, the idempotent algebras were started to be called *tropical* by some authors. The appearance and rapid development of tropical geometry (see e.g. reviews in [109]) gave a new twist to the development. Referring the interested readers to above cited sources for the full story, we confine ourselves in this section to basic definitions and a simple example showing the relevance of idempotent techniques to the analysis of certain games. A more involved application of idempotent analysis to multi-criteria optimization will be shortly described in Section 11.8. Some important recent application for numerical solutions of stochastic optimization are developed in [137].

2. Main definitions. An *idempotent semigroup* is a set M equipped with a commutative, associative operation $\oplus$ (generalized addition) that has a unit element $\mathbf{0}$ such that $\mathbf{0} \oplus \mathbf{a} = \mathbf{a}$ for each $a \in M$ and satisfies the idempotency condition $a \oplus a = a$ for any $a \in M$. There is a naturally defined partial order on any idempotent semigroup; namely, $a \leq b$ if and only if $a \oplus b = a$. Obviously, the reflexivity of $\leq$ is equivalent to the idempotency of $\oplus$, whereas the transitivity and the anti-symmetricity ($a \leq b, b \leq a \implies a = b$) follow, respectively, from the associativity and the commutativity of the semigroup operation. The unit element $\mathbf{0}$ is the greatest element; that is, $a \leq \mathbf{0}$ for all $a \in M$. The operation $\oplus$ is uniquely determined by the relation $\leq$ via the formula

$$a \oplus b = \inf\{a, b\} \tag{11.40}$$

(recall that the infimum of a subset X in a partially ordered set $(Y, \leq)$ is the element $c \in Y$ such that $c \leq x$ for all $x \in X$ and any element $d \in Y$ satisfying the same condition also satisfies $d \leq c$). Furthermore, if every subset of cardinality 2 in a partially ordered set M has an infimum, then equation (11.40) specifies the structure of an idempotent semigroup on M.

It is often more convenient to use the opposite order, which is related to the semigroup operation by the formula $a \oplus b = \sup\{a, b\}$.

An idempotent semigroup is called an *idempotent semiring* if it is equipped with yet another associative operation $\odot$ (generalized multiplication) that has a unit element $\mathbf{1}$, distributes over $\oplus$ on the left and on the right, i.e.,

$$a \odot (b \oplus c) = (a \odot b) \oplus (a \odot c), \qquad (b \oplus c) \odot a = (b \odot a) \oplus (c \odot a),$$

and satisfies the property $\mathbf{0} \odot \mathbf{a} = \mathbf{0}$ for all a. Distributivity obviously implies the following property of the partial order:

$$\forall a, b, c \quad a \leq b \implies a \odot c \leq b \odot c.$$

An idempotent semiring is said to be *commutative* (or *abelian*) if the operation $\odot$ is commutative.

Let X be a set, and let $M = (M, \oplus, \rho)$ be an idempotent semigroup. The set $B(X, M)$ of bounded mappings $X \to M$ (i.e., mappings with order-bounded range) is an idempotent semigroup with respect to the point-wise addition $(\varphi \oplus \psi)(x) = \varphi(x) \oplus \psi(x)$. If $A = (A, \oplus, \odot, \rho)$ is a semiring, then $B(X, A)$ bears the structure of an *A-semimodule*; namely, the multiplication by elements of A is defined on $B(X, A)$ by the formula $(a \odot \varphi)(x) = a \odot \varphi(x)$. This A-semimodule is also often referred to as the space of (bounded) A-valued functions on X. If X is finite, $X = \{x_1, \ldots, x_n\}$, $n \in \mathbb{N}$, then the semimodule $B(X, A)$ can be identified with the semimodule $A^n = \{(a_1, \ldots, a_n) : a_j \in A\}$ of A-valued vectors. Any vector $a \in A^n$ can be uniquely represented as a linear combination $a = \bigoplus_{j=1}^{n} a_j \odot e_j$, where $\{e_j, \ j = 1, \ldots, n\}$ is the standard basis of A^n (the jth coordinate of e_j is equal to $\mathbf{1}$, and the other coordinates are equal to $\mathbf{0}$). As in the conventional linear algebra, we can readily prove that any homomorphism $m \colon A^n \to A$ (such homomorphisms are also often called *linear functionals* on A^n) has the form

$$m(a) = \bigoplus_{i=1}^{n} m^i \odot a_i,$$

where $m^i \in A$. Therefore, the semimodule of linear functionals on A^n is isomorphic to A^n. Similarly, any endomorphism $H \colon A^n \to A^n$ (a linear operator on A^n) has the form

$$(Ha)_j = \bigoplus_{k=1}^{n} h^k_j \odot a_k,$$

i.e. is determined by an A-valued $n \times n$ matrix. By analogy with the case of Euclidean spaces, one can define an inner product on A^n by setting

$$\langle a, b \rangle_A = \bigoplus_{k=1}^{n} a_k \odot b_k. \tag{11.41}$$

The inner product (11.41) is bilinear with respect to the operations $\oplus$ and $\odot$, and the standard basis of A^n is orthonormal, that is,

$$\langle e_i, e_j \rangle_A = \delta_A^{ij} = \begin{cases} \mathbf{1}, & \mathbf{i} = \mathbf{j} \\ \mathbf{0}, & \mathbf{i} \neq \mathbf{j}. \end{cases}$$

Examples. 1. $A = \mathbb{R} \cup \{+\infty\}$ with the operations $\oplus = \max$ and $\odot = +$, the unit elements $\mathbf{0} = +\infty$ and $\mathbf{1} = 0$. This is the simplest and the most important example of idempotent semiring, involving practically all specific features of idempotent arithmetic. 2. $A = \mathbb{R} \cup \{-\infty\}$ with the operations $\oplus = \min$ and $\odot = +$. 3. $A = \mathbb{R}_+$ with the operations $\oplus = \min$ and $\odot = \times$ (the usual multiplication). 4. $A = \mathbb{R} \cup \{\pm\infty\}$ with the operations $\oplus = \min$ and $\odot = \max$, the unit elements $\mathbf{0} = +\infty$ and $\mathbf{1} = -\infty$. 5. Let $\mathbb{R}_+^n$ be the nonnegative octant in $\mathbb{R}^n$ with the inverse Pareto order ($a = (a_1, \ldots, a_n) \leq b = (b_1, \ldots, b_n)$ if and only if $a_i \geq b_i$ for all $i = 1, \ldots, n$). Then $\mathbb{R}_+^n$ is an idempotent semiring with respect to the idempotent addition $\oplus$ corresponding to this order by (11.40) and the generalized multiplication given by $(a \odot b)_i = a_i + b_i$. 6. The subsets of a given set form an idempotent semiring with respect to the operations $\oplus$ of set union and $\odot$ of set intersection.

3. Idempotent linear equations. Let X be a finite set of cardinality n and A be an idempotent semiring. Let H be a linear operator with matrix (h_j^i) on $B(X, A) = A^n$, and let $F \in B(X, A)$. The discrete-time equation

$$S_{t+1} = HS_t \oplus F, \qquad S_t \in B(X, A), \ t = 0, 1, \ldots, \tag{11.42}$$

in the space (vector semimodule) $B(X, A)$ is called the *generalized evolution Bellman equation*, and the equation

$$S = HS \oplus F \tag{11.43}$$

is called the *generalized stationary (or steady-state) Bellman equation*. These equations are said to be *homogeneous* if $F = \mathbf{0}$ and *nonhomogeneous* otherwise.

Both for the theory and the applications the following geometric representation of idempotent linear operators and equations is crucial. To each Bellman equation there corresponds a *(directed weighted) graph* (X, Γ, L, A), where X is the set of nodes, $\Gamma = \{(x_i, x_j) \in X \times X : h_j^i \neq \mathbf{0}$ is the set of (directed) arcs, and h_j^i are the arc weights valued in A. Set $\Gamma(x_i) = \{x' \in X : (x', x_i) \in \Gamma\}$. Then $\Gamma(x_i)$ is the set of points linked with x_i often called the *neighborhood* of x_i. The operator H can be represented as

$$(H\varphi)(x_i) = \bigoplus_{x_j \in \Gamma(x_i)} h_j^i \odot \varphi(x_j). \tag{11.44}$$

This is just the form in which H and the corresponding equations (11.42) and (11.43) arise when optimization problems on graphs are solved by dynamic programming.

Let us now derive the basic formulas for the solutions of the Bellman equation directly from linearity. First, the solution S_t^F of the evolution equation (11.42) with the zero initial function $S_0 = \mathbf{0}$ obviously has the form

$$S_t^F = H^{(t-1)}F \equiv \bigoplus_{k=0}^{t-1} H^k F \tag{11.45}$$

(here H^k is the kth power of H). This is the simplest *Duhamel type formula* expressing a particular solution of the nonhomogeneous equation as a generalized sum of solutions of the homogeneous equation.

Next, the solution $S_t = H^t S_0$ of the homogeneous equation (11.42) with $F \equiv \mathbf{0}$ and with an arbitrary initial function $S_0 = \bigoplus_{i=1}^n S_0^i \odot e_i$, where e_i is the standard basis in A^n, is obviously equal to

$$S_t = \bigoplus_{i=0}^n S_0^i \odot H^t e_i,$$

that is, is a linear combination of the source functions $H^t e_i$, which are the solutions of equation (11.42) with the localized initial perturbations $S_0 = e_i$. Similarly, if $S = G_i \in A^n$ is the solution of the stationary equation (11.43) with localized nonhomogeneous term $F = e_i$, i.e., $G_i = HG_i \oplus e_i$, then by multiplying these equations by arbitrary coefficients $F_i \in A$ and by taking the $\oplus$-sum of the resultant products, we find that the function

$$S = \bigoplus_{i=1}^n F_i \odot G_i$$

is a solution of equation (11.43) with arbitrary $F = \bigoplus F^i \odot e_i$. Thus, we have obtained a source representation of solutions of the stationary A-linear Bellman equation.

A majority of methods for constructing solutions of the stationary Bellman equation are based on the following simple assertion.

Proposition 11.24. *Set* $H^{(t)} = \bigoplus_{k=0}^{t} H^k = (H \oplus I)^t$, *where* I *is the identity operator. If the sequence* $H^{(t)}$ *converges to a limit* H^* *as* $t \to \infty$, $t \in \mathbb{N}$, *then* $S = H^*F$ *is a solution to equation* (11.43) *called the Duhamel solution.*

Proof. The proof is by straightforward substitution:

$$H(S_*^F) \oplus F = H \lim_{t \to \infty} (H^{(t)} F \oplus F)$$

$$= \lim_{t \to \infty} H((H^t F) \oplus F) = \lim_{t \to \infty} H^{(t+1)} F = S_*^F.$$

An important supplement to Proposition 11.24 is given by the following assertion.

Proposition 11.25. *Let the operator sequences* $H^{(t)}$ *and* H^t *tend to some limits* H^* *and* H^∞, *respectively, as* $t \to \infty$. *Then each solution of* (11.43) *has the form*

$$S = g \oplus S_*^F, \tag{11.46}$$

where g *is a solution of the homogeneous equation* $Hg = g$. *Moreover,* S_*^F *is a unit element with respect to the operation* $\oplus$ *on the set of all solutions of* (11.43), *i.e., for each solution* S *one has* $S = S \oplus S_*^F$.

Proof. Let S be a solution of (11.43). On substituting $S = H\varphi \oplus F$ into the right-hand side, we obtain

$$S = H^2 S \oplus HF \oplus F.$$

After k successive applications of this procedure, we have

$$S = H^k S \oplus H^{(k-1)} F.$$

This is valid for all $k \in \mathbb{N}$. By passing to the limit as $k \to \infty$, we obtain

$$S = H^\infty S \oplus S_*^F,$$

which proves (11.46), since $H^\infty S$ is obviously a solution of the homogeneous equation $H(H^\infty S) = H^\infty S$. The second part of the proposition readily follows from (11.46) and from the fact that the operation $\oplus$ is idempotent.

Thus, to solve the stationary equation (11.43) effectively, one should be capable of evaluating the limits of the operator sequences $H^{(t)} = (H \oplus I)^t$ and H^t. We refer to the sources cited above for the development of the theory. Below, we shall use Proposition 11.25 in a situation, where H^* can be easily calculated explicitly.

We shall give now an example showing how this technique can be applied to the analysis of queues in models including conflict (game theoretic) interactions of agents.

4. Example. Synchronization of services. Consider a queue at the entrance to a system of m servers S_i, $i = 1, \ldots, m$ with a buffer at the first one. Each customer k, $(k = 1, 2, \ldots, N)$ is to be successfully served by m in the given order: $S_1, S_2, \ldots, S_m$. A finite number o customers arrive at the buffer simultaneously and each should wait for its turn l_k. Let us denote by $k[l_k]$ the customer, taking the place $[l_k]$ in the queue. The times $\tau_i(k[l_k])$ necessary for S_i to serve the customer k are known.

Denote $x_i(k[l_k])$ the instant at which S_i begins to serve the customer k. Before S_i can start serving the customer $k[l_k + 1]$ it is necessary that S_i completes serving the customer $k[l_k]$ and S_{i-1} completes serving the customer $k[l_k + 1]$. Hence

$$x_1(k[l_k + 1]) = \tau_1(k[l_k]) + x_1(k[l_k]),$$

$$x_i(k[l_k+1]) = \max(\tau_i(k[l_k])+x_i(k[l_k]), \tau_{i-1}(k[l_k+1])+x_{i-1}(k[l_k+1]), \ i > 1.$$

In terms of idempotent $(max, +)$-algebra with generalized addition $a \oplus b = \max(a, b)$ these condition on the vector $x = (x_1, \ldots, x_m)(k[l_k + 1])$ can be written as a linear equation of type (11.43):

$$x(k[l_k + 1]) = H(k[l_k + 1])x(k[l_k + 1]) \oplus G((k[l_k])x(k[l_k]), \qquad (11.47)$$

where the matrices of operators H and G are the following:

$$h^i_j(k[l_k + 1]) = \delta^i_{j+1}\tau_j(k[l_k + 1]), \quad g^i_j(k[l_k]) = \delta^i_j\tau_j(k[l_k]), \qquad (11.48)$$

where $\delta^i_j = 1$ for $i = j$ and 0 otherwise. For instance, for $m = 4$ equation (11.47) takes the form

$$x(k[l_k + 1]) = \begin{pmatrix} 0 & 0 & 0 & 0 \\ \tau_1(k[l_k + 1]) & 0 & 0 & 0 \\ 0 & \tau_2(k[l_k + 1]) & 0 & 0 \\ 0 & 0 & \tau_3(k[l_k + 1]) & 0 \end{pmatrix} x(k[l_k + 1])$$

$$\oplus \begin{pmatrix} \tau_1(k[l_k]) & 0 & 0 & 0 \\ 0 & \tau_2(k[l_k]) & 0 & 0 \\ 0 & 0 & \tau_3(k[l_k]) & 0 \\ 0 & 0 & 0 & \tau_4(k[l_k]) \end{pmatrix} x(k[l_k]).$$

$$(11.49)$$

According to Proposition 11.25, the minimal (just which we need) solution to (11.47) is

$$x(k[l_k + 1]) = (H(k[l_k + 1]))^*(G(k[l_k])x(k[l_k]).$$

For the matrices H of type (11.47) with nonzero elements only on the second diagonal, the limit of iterations H^* is easy to calculate explicitly. Namely $H(k[l_k+1])^*$ is a lower diagonal matrix (i.e. has $\mathbf{0}$ above the main diagonal) with $\mathbf{1}$ on the diagonal and with elements

$$(H(k[l_k + 1])^*)^i_j = \prod_{p=j}^{i-1} \tau_p(k[l_k + 1])$$

below the main diagonal. For instance, for $m = 4$, $H(k[l_k + 1])^*$ equals

$$\begin{pmatrix} 1 & 0 & 0 & 0 \\ \tau_1(k[l_k + 1]) & 1 & 0 & 0 \\ \tau_1(k[l_k + 1])\tau_2(k[l_k + 1]) & \tau_2(k[l_k + 1]) & 1 & 0 \\ \tau_1(k[l_k + 1])\tau_2(k[l_k + 1])\tau_3(k[l_k + 1]) & \tau_2(k[l_k + 1])\tau_3(k[l_k + 1]) & \tau_3(k[l_k + 1]) & 1 \end{pmatrix}.$$

Suppose that the k-th customer losses are f^0_k for a unit of time waiting in the buffer. Assume also that during the service time the losses are f^i_k, $i = 1, \ldots, m$, for the time unit. Arriving at the buffer, the customers form a queue $\sigma_j = (l_1, \ldots, l_N)$ according to a certain principle of optimality. Total loss of the customer k subject to the queue σ can be computed by the formula:

$$F^\sigma_k = x_1(k[l_k])h^0_k + \sum_{i=1}^{m-1}(x_{i+1}(k[l_k]) - x_i(k[l_k]))f^i_k + \tau_m(k[l_k])h^m_k. \quad (11.50)$$

One can take these losses as functionals, for which one can calculate a compromise solution (or the Shapley value, etc.). For instance the *compromise solution* is found by the following procedure.

(1) Calculate all $x_i(k)$, $i = 1, \ldots, m$, $k = 1, \ldots, N$ — the instances when S_i starts serving the customer k for all possible queues $\sigma = (l_1, \ldots, l_N)$ (i.e. permutations of the customers);
(2) For each k and a permutation σ calculate the full incomes F^σ_k, (i.e. the negation of the (11.50);
(3) Determine the vector $M = \{M_1, \ldots, M_N\}$, where $M_k = \max_\sigma F^\sigma_k$;
(4) For each σ find the vector of the differences of the values of each functional and its maximal value: $M^\sigma_{otkl} = \{M^\sigma_{otkl,\,1}, \ldots, M^\sigma_{otkl,\,N}\}$, where $M^\sigma_{otkl,\,k} = M_k - F^\sigma_k$;

(5) For each functional determine its maximal difference: $M_{otklMax} = \{M_{otklMax,1}, \ldots, M_{otklMax,N}\}$, where $M_{otklMax,k} = \max_\sigma M^\sigma_{otkl,k}$;
(6) From the maximal differences of all functionals choose the minimal one:

$$M_{otklMinMax} = \min_{k=1,\ldots,N} M_{otklMax,k}.$$

The value σ, where $M_{otklMinMax}$ is attained is the *compromise value* that was looked for.

11.8 The first order partial differential equations in multi-criteria optimization problems

We shall begin this section by recalling briefly how first-order differential equations (such as the Hamilton–Jacobi-Bellman and the Isaacs equations) arise in optimal control and games thus giving an introduction to dynamic programming in continuous time and differential antagonistic games. The main goal of this section (achieved at the end) is to use the methods of idempotent mathematics (see previous section) to introduce a Bellman type equation that governs the evolution of the Pareto sets in the dynamic multi-criteria optimization problems (non zero-sum games).

Consider a controlled object with dynamics described by the equation $\dot{x} = f(x, u)$, when u is a control parameter ranging over a compact subset U in Euclidean space and f is a certain smooth function of two variables. A standard problem of optimal control theory consists in finding the minimum

$$S(t,x) = \min\{\int_0^t g(x,u)\, dt + S_0(x(0))\}, \qquad (11.51)$$

where g and S_0 are given functions and the minimum is sought for in the class of curves $x(\tau)$ with endpoint $x = x(t)$ that satisfy the dynamic equations $\dot{x} = f(x, u(t))$ for some choice of the control $u(t) \in U$. The dynamic programming approach to the solution of this problem consists in observing that if $S(t,x)$ from (11.51) is smooth, then it necessarily solves the Cauchy problem for the *Hamilton-Jacobi-Bellman (evolutionary differential) equation*

$$\frac{\partial S}{\partial t} + \max_{u \in U} \left(\left\langle \frac{\partial S}{\partial t}, f(x,u) \right\rangle - g(x,u) \right) = 0 \qquad (11.52)$$

with the initial condition $S(t,x)|_{t=0} = S_0(x)$.

Let us recall (on the heuristical level), how this equation comes into the play. Applying the obvious *optimality principle*, according to which an optimal trajectory remains optimal after deleting an arbitrary initial segment, yields the approximate equation

$$S(t, x) = \min_u (S(t - \Delta t, x - \Delta x(u)) + g(x, u)\Delta t). \qquad (11.53)$$

Let us expand $S(t - \Delta t, x - \Delta x)$ in the Taylor series, neglect the terms of the order of $(\Delta t)^2$, and take into account the fact that $\Delta x = f(x, u)\Delta t$ in this approximation. Then we obtain

$$S(t, x) = \min_u \left\{ S(t, x) - \Delta t \left(\frac{\partial S}{\partial t}(t, x) + \left\langle f(x, u), \frac{\partial S}{\partial x}(t, x) \right\rangle - g(x, u) \right) \right\},$$

which is obviously equivalent to

$$\min_u \left(-\frac{\partial S}{\partial t}(t, x) - \left\langle f(x, u), \frac{\partial S}{\partial x}(t, x) \right\rangle + g(x, u) \right) = 0.$$

On replacing min by max and changing the sign, we obtain (11.52). Connection with mechanics is supplied by the observation that equation (11.52) can be written in the form

$$\frac{\partial S}{\partial t} + H\left(x, \frac{\partial S}{\partial x} \right) = 0, \qquad (11.54)$$

i.e. as the *Hamilton-Jacobi equation* with the *Hamilton function*

$$H(x, p) = \max_{u \in U} \left(\left\langle p, f(x, u) \right\rangle - g(x, u) \right). \qquad (11.55)$$

The above considerations suggest that (now possibly non-smooth) solutions of minimization problems in optimal control theory are, in some sense, generalized solutions of the Cauchy problem for equations with non-smooth Hamiltonians. For the needs of optimal control theory one should envisage the situations, when the initial functions can be non-smooth and even discontinuous. A distinguishing feature of the Hamiltonians (11.55) occurring in optimal control theory is that they are convex (possibly, non-strictly) in p; in particular, they are continuous in p. We refer to [102], [103] for the theory of the solutions of equations (11.52) based on the ideas of Idempotent Mathematics.

The Hamiltonians of a different form occur in differential games. Let, for instance, $S(t, x_0, y_0)$ be the guaranteed approach in time t for an object P with dynamics $\dot{x} = f(x, u)$ pursuing an object E with dynamics $\dot{y} = g(y, v)$. Let $u \in U$ and $v \in V$ be the control parameters and $x_0, y_0 \in \mathbb{R}^3$ the initial states of the objects P and E, respectively. Just as above, we apply the optimality principle and find that $S(t, x, y)$ satisfies the Hamilton–Jacobi equation

$$\frac{\partial S}{\partial t} + \max_u \left\langle \frac{\partial S}{\partial x}, f(x, u) \right\rangle + \min_v \left\langle \frac{\partial S}{\partial y}, g(y, v) \right\rangle = 0 \qquad (11.56)$$

with the initial condition $S_0(x,y)$ that equals to the distance between x and y in $\mathbb{R}^3$. Equation (11.56) with the Hamiltonian

$$H(x,y,p_x,p_y) = \max_u \langle p_x, f(x,u) \rangle + \min_v \langle p_y, g(y,v) \rangle \qquad (11.57)$$

is known as the *Isaacs equation* in the theory of differential (in our example pursuit) games. The Hamiltonian (11.57), in contrast with (11.55), is convex-concave with respect to the momenta p_x and p_y. The theory of the Cauchy problem for equations with convex-concave Hamiltonians (by the way, any Hamilton–Jacobi equation, in a sense, can be reduced to such a problem) is much more complicated than the theory of the Cauchy problem for equations with convex Hamiltonians, see e.g. [180] for a game-theoretic approach to the analysis of these equations.

As we mentioned at the beginning we aim here at deducing a remarkable generalization of (11.52) obtained in [101] that arises in the multi-criteria optimization. We refer to [103] for the problems of existence and uniqueness of its solution as well as for the examples of its calculations.

Idempotent structures adequate to multi-criteria optimization are various semirings of sets or (in another representation) semirings of functions with an idempotent analog of convolution playing the role of multiplication.

Let $\leq$ denote the Pareto partial order on $\mathbb{R}^k$, i.e. $x = \{x_1, \ldots, x_k\} \leq y = \{y_1, \ldots, y_k\}$ whenever $x_i \leq y_i$ for all i. For any subset $M \subset \mathbb{R}^k$, we denote the set of minimal elements of the closure of M in $\mathbb{R}^k$ by $\text{Min}(M)$. Following [171], we introduce the class $P(\mathbb{R}^k)$ of subsets $M \subset \mathbb{R}^k$ whose elements are pairwise incomparable,

$$P(\mathbb{R}^k) = \{M \subset \mathbb{R}^k : \text{Min}(M) = M\}.$$

Obviously, $P(\mathbb{R}^k)$ is a semiring with respect to the operations $M_1 \oplus M_2 = \text{Min}(M_1 \cup M_2)$ and $M_1 \odot M_2 = \text{Min}(M_1 + M_2)$; the neutral element $\mathbf{0}$ with respect to addition in this semiring is the empty set, and the neutral element with respect to multiplication is the set whose sole element is the zero vector in $\mathbb{R}^k$. It is also clear that $P(\mathbb{R}^k)$ is isomorphic to the semiring of *normal sets*, that is, closed subsets $N \subset \mathbb{R}^k$ such that $b \in N$ implies $a \in N$ for any $a \geq b$; the sum and the product of normal sets are defined as their usual union and sum, respectively. Indeed, if N is normal, then $\text{Min}(N) \in P(\mathbb{R}^k)$; conversely, with each $M \in P(\mathbb{R}^k)$ we can associate the normalization $\text{Norm}(M) = \{a \in \mathbb{R}^k \mid \exists\, b \in M : a \geq b\}$.

For an arbitrary set X, by $B(X, \mathbb{R}^k)$ (respectively, by $B(X, P(\mathbb{R}^k))$) we shall denote here the set of mappings $X \to \mathbb{R}^k$ (respectively, $X \to P(\mathbb{R}^k)$) bounded below with respect to the order. Then $B(X, P(\mathbb{R}^k))$ is a semimodule with respect to pointwise addition $\oplus$ and multiplication $\odot$ by elements of $P(\mathbb{R}^k)$.

These semirings and semimodules naturally arise in multicriteria dynamic programming problems. Let a mapping $f \colon X \times U \to X$ specify a controlled dynamical system on X, so that any choice of $x_0 \in X$ and of a sequence of controls $\{u_1, \ldots, u_k\}$, $u_j \in U$, determines an admissible trajectory $\{x_j\}_{j=0}^k$ in X, where $x_j = f(x_{j-1}, u_j)$. Let $\varphi \in B(X \times U, \mathbb{R}^k)$, and let $\Phi(\{x_j\}_{j=0}^k) = \sum_{j=1}^k \varphi(x_{j-1}, u_j)$ be the corresponding vector criterion on the set of admissible trajectories. The element $\mathrm{Min}(\bigcup_{\{x_j\}} \Phi(\{x_j\}_{j=0}^k))$, where $\{x_j\}_{j=0}^k$ are all possible k-step trajectories issuing from x_0, is denoted by $\omega_k(x_0)$ and is called the *Pareto set* for the criterion Φ and the initial point x_0. Let us define the *Bellman operator* $\mathcal{B}$ on the semimodule $B(X, P(\mathbb{R}^k))$ by setting

$$(\mathcal{B}\omega)(x) = \mathrm{Min}\left(\bigcup_{u \in U} (\varphi(x, u) \odot \omega(f(x, u))) \right). \tag{11.58}$$

Obviously, $\mathcal{B}$ is linear in the semimodule $B(X, P(\mathbb{R}^k))$, and it follows from Bellman's optimality principle that the Pareto sets in k-step optimization problems satisfy the recursion relation $\omega_k(x) = \mathcal{B}\omega_{k-1}(x)$.

Sometimes, it is convenient to use a different representation of the set $P(\mathbb{R}^k)$. Proposition 11.26 below, which describes this representation, is a specialization of a more general result stated in [171]. Let L denote the hyperplane in $\mathbb{R}^k$ determined by the equation

$$L = \left\{ (a^j) \in \mathbb{R}^k : \sum a^j = 0 \right\},$$

and let $CS(L)$ denote the semiring of functions $L \to \mathbb{R} \cup \{+\infty\}$ with pointwise minimum as addition and the *idempotent convolution*

$$(g \circledast h)(a) = \inf_{b \in L} (g(a - b) + h(b))$$

as multiplication. Let us define a function $n \in CS(L)$ by setting $n(a) = \max_j(-a^j)$. Obviously, $n \circledast n = n$; that is, n is a multiplicatively idempotent element of $CS(L)$. Let $CS_n(L) \subset CS(L)$ be the subsemiring of functions h such that $n \circledast h = h \circledast n = h$. It is easy to see that $CS_n(L)$ contains the function identically equal to $\mathbf{0} = \infty$ and that the other elements of $CS_n(L)$ are just the functions that take the value $\mathbf{0}$ nowhere and satisfy the inequality $h(a) - h(b) \leq n(a - b)$ for all $a, b \in L$. In particular, for each $h \in CS_n(L)$ we have

$$|h(a) - h(b)| \leq \max_j |a^j - b^j| = \|a - b\|,$$

which implies that h is differentiable almost everywhere.

Proposition 11.26. *The semirings $CS_n(L)$ and $P(\mathbb{R}^k)$ are isomorphic.*

Proof. The main idea is that the boundary of each normal set in $\mathbb{R}^k$ is the graph of some real-valued function on L, and vice versa. More precisely, let us consider the vector $e = (1, \ldots, 1) \in \mathbb{R}^k$ normal to L and assign a function $h_M \colon L \to \mathbb{R}$ to each set $M \in P(\mathbb{R}^k)$ as follows. For $a \in L$, let $h_M(a)$ be the greatest lower bound of the set of $\lambda \in \mathbb{R}$ for which $a + \lambda e \in \mathrm{Norm}(M)$. Then the functions corresponding to singletons $\{\varphi\} \subset \mathbb{R}^k$ have the form

$$h_\varphi(a) = \max_j(\varphi^j - a^j) = \overline{\varphi} + n(a - \varphi_L), \qquad (11.59)$$

where $\overline{\varphi} = k^{-1} \sum_j \varphi^j$ and $\varphi_L = \varphi - \overline{\varphi} e$ is the projection of φ on L. Since idempotent sums $\oplus$ of singletons in $P(\mathbb{R}^k)$ and of functions (11.59) in $CS_n(L)$ generate $P(\mathbb{R}^k)$ and $CS_n(L)$, respectively, we can prove the proposition by verifying that the $\odot$-multiplication of vectors in $\mathbb{R}^k$ passes into the convolution of the corresponding functions (11.59). Namely, let us show that

$$h_\varphi \circledast h_\psi = h_{\varphi + \psi}.$$

Indeed, by virtue of (11.59), it suffices to show that

$$n_\varphi \circledast n_\psi = n_{\varphi + \psi},$$

where $n_\varphi(a) = n(a - \varphi_L)$, and the latter identity is valid since

$$n_\varphi \circledast n_\psi = n_0 \circledast n_{\varphi + \psi} = n \circledast n_{\varphi + \psi} = n_{\varphi + \psi}.$$

Finally, let us consider the controlled process in $\mathbb{R}^n$ specified by a controlled differential equation $\dot{x} = f(x, u)$ (where u belongs to a metric control space U) and by a continuous function $\varphi \in B(\mathbb{R}^n \times U, \mathbb{R}^k)$, which determines a vector-valued integral criterion

$$\Phi(x(\,\cdot\,)) = \int_0^t \varphi(x(\tau), u(\tau)) \, d\tau$$

on the trajectories. Let us pose the problem of finding the Pareto set $\omega_t(x)$ for a process of duration t issuing from x with terminal income determined by some function $\omega_0 \in B(\mathbb{R}^n, \mathbb{R}^k)$, that is,

$$\omega_t(x) = \mathrm{Min} \bigcup_{x(\cdot)} (\Phi(x(\,\cdot\,)) \odot \omega_0(x(t))), \qquad (11.60)$$

where $x(\,\cdot\,)$ ranges over all admissible trajectories issuing from x. By Proposition 11.26, we can encode the functions $\omega_t \in B(\mathbb{R}^n, P\mathbb{R}^k)$ by the corresponding functions

$$S(t, x, a) \colon \mathbb{R}_+ \times \mathbb{R}^n \times L \to \mathbb{R}.$$

The optimality principle permits us to write out the following equation, which is valid modulo $O(\tau^2)$ for small τ:

$$S(t, x, a) = \operatorname*{Min}_{u}(h_{\tau\varphi(x,u)} \circledast S(t - \tau, x + \Delta x(u)))(a). \tag{11.61}$$

It follows from the representation (11.59) of $h_{\tau\varphi(x,u)}$ and from the fact that n is, by definition, the multiplicative unit in $CS_n(L)$ that

$$S(t, x, a) = \min_{u}(\tau\overline{\varphi}(x, u) + S(t - \tau, x + \Delta x(u), a - \tau\varphi_L(x, u))).$$

Let us substitute $\Delta x = \tau f(x, u)$ into this equation, expand S in a series modulo $O(\tau^2)$, and collect similar terms. Then we obtain the equation

$$\frac{\partial S}{\partial t} + \max_{u} \left(\varphi_L(x, u) \frac{\partial S}{\partial a} - f(x, u) \frac{\partial S}{\partial x} - \overline{\varphi}(x, u) \right) = 0. \tag{11.62}$$

Although the presence of a vector criterion has resulted in a larger dimension, this equation coincides in form with the usual Bellman differential equation. Consequently, the generalized solutions can be defined on the basis of the idempotent superposition principle, and constructed e.g. by means of Pontryagin's maximum principle, see [103].

11.9 General flows of deterministic and stochastic replicator dynamics

This section is an introduction to the methods and results of the series of papers [93]-[97] (see full exposition in [98]) from the point of view of the evolutionary game theory. It is devoted to the dynamic law of large numbers for the Markov models of binary or more generally k-ary interacting particles that are described by nonlinear measure-valued equations or measure-valued stochastic processes that generalize the classical kinetic equations of statistical mechanics and the replicator dynamics from population biology and evolutionary games. Applications of these result in a more elementary setting of finite state spaces are addressed in Chapter 12.

1. Some notations.

We list here a few (rather) standard notations to be used in what follows.

For a locally compact topological space Y, $C(Y)$ denotes the Banach space of real bounded continuous functions f on Y equipped with the usual sup-norm $\|f\| = \sup_y |f(y)|$, $C_\infty(Y)$ denotes the closed subspace of $C(Y)$ consisting of continuous functions f vanishing at infinity (i.e. $\forall \epsilon > 0$ there exists a compact subset K of Y such that $|f(y)| \leq \epsilon$ for $y \notin K$). In case of a compact Y the spaces $C_\infty(Y)$ and $C(Y)$ coincide. By $\mathcal{M}(Y)$ is denoted the space of finite (signed) Borel measures on Y, considered as a Banach space with the total variation norm. The upper subscript "+" for all these spaces (e.g. $C_0^+(X)$, $\mathcal{M}^+(X)$) will denote the corresponding cones of non-negative elements.

A symmetric function of n variables is a function, which is symmetric with respect to all permutations of these variables, and by a symmetric operator on the space of functions of n variables we shall understand an operator that preserves the set of symmetric functions.

A continuous transitional kernel from X to Y is a continuous function $\nu(x, .)$ from $x \in X$ to the cone of positive finite measures on Y equipped with its $\star$-weak topology. Each continuous transition kernel from X to X specifies an integral operator

$$(G_\nu f)(x) = \int (f(y) - f(x))\nu(x, dy) \qquad (11.63)$$

on the space of continuous functions on X. A strongly continuous semigroup T_t of bounded linear operators on some Banach space H of continuous functions on X is said to be generated by G_f, if there exists a dense subspace $K \subset H$ that is invariant under T_t and such that $T_t f$ solves the evolution equation $\dot{f}_t = G_\nu f_t$ (with derivatives in the sense of Banach space H) for all $f \in K$. A Markov process Z_t on X is said to be a pure jump Markov process generated by G_ν if $\mathbf{E}_x f(Z_t) = (T_t f)(x)$ ($\mathbf{E}$ denotes the expectation) for f from $C_\infty(Y)$ (or its dense subspace). Usually it is not an obvious question to find out whether a given operator G_ν generates a unique Markov process (only in case of Markov chain, where X is a finite set, this becomes trivial).

2. State space of systems of interacting particles.

Further we shall denote by X a locally compact metric space. Denoting by X^0 a one-point space and by X^j the powers $X \times \ldots \times X$ (j-times) considered with their product topologies, we shall denote by $\mathcal{X}$ their disjoint union $\mathcal{X} = \cup_{j=0}^\infty X^j$, which is again a locally compact space. In applications, X specifies the state space of one particle and $\mathcal{X} = \cup_{j=0}^\infty X^j$ stands for the state space of a random number of similar particles. We shall denote by $C_{sym}(\mathcal{X})$ the Banach spaces of symmetric bounded continuous functions on $\mathcal{X}$ and by $C_{sym}(X^k)$ the corresponding spaces of functions on the finite power X^k. The space of symmetric measures will be denoted by $\mathcal{M}_{sym}(\mathcal{X})$. The elements of $\mathcal{M}^+_{sym}(\mathcal{X})$ and $C_{sym}(\mathcal{X})$ are respectively the *(mixed) states and observables* for a Markov process on $\mathcal{X}$. We shall denote the elements of $\mathcal{X}$ by bold letters, e.g. **x**, **y**. For a finite subset $I = \{i_1, \ldots, i_k\}$ of a finite set $J = \{1, \ldots, n\}$, we denote by $|I|$ the number of elements in I, by $\bar{I}$ its complement $J \setminus I$, by $\mathbf{x}_I$ the collection of the variables $x_{i_1}, \ldots, x_{i_k}$ and by $d\mathbf{x}_I$ the measure $dx_{i_1} \ldots dx_{i_k}$.

We shall denote by **1** the function on X that equals one identically and by $\mathbf{1}_A$ the indicator function of a set A.

Sometimes it is convenient to consider the factor spaces SX^k and $S\mathcal{X}$ obtained by the factorization of X^k and $\mathcal{X}$ with respect to all permutations, which allows for the identifications $C_{sym}(\mathcal{X}) = C(S\mathcal{X})$ and likewise. Clearly $S\mathcal{X}$ can be identified with the set of all finite subsets of X.

A key role in the theory of measure-valued limits of interacting particle systems is played by the inclusion $S\mathcal{X}$ to $\mathcal{M}^+(X)$ given by

$$\mathbf{x} = (x_1, ..., x_l) \mapsto \delta_{x_1} + ... + \delta_{x_l}, \tag{11.64}$$

which defines a bijection between $S\mathcal{X}$ and the space $\mathcal{M}^+_\delta(X)$ of finite linear combinations of δ-measures.

Clearly each $f \in C_{sym}(\mathcal{X})$ is defined by its components (restrictions) f^k on X^k so that for $\mathbf{x} = (x_1, \ldots, x_k) \in X^k \subset \mathcal{X}$, say, one can write $f(\mathbf{x}) = f(x_1, \ldots, x_k) = f^k(x_1, \ldots, x_k)$. Similar notations are for measures. In particular, the pairing between $C_{sym}(\mathcal{X})$ and $\mathcal{M}(\mathcal{X})$ can be written as

$$(f, \rho) = \int f(\mathbf{x})\rho(d\mathbf{x}) = f^0 \rho_0 + \sum_{n=1}^{\infty} \int f(x_1, \ldots, x_n)\rho(dx_1 \ldots dx_n),$$

$$f \in C_{sym}(\mathcal{X}), \ \rho \in \mathcal{M}(\mathcal{X})$$

so that $\|\rho\| = (\mathbf{1}, \rho)$ for $\rho \in \mathcal{M}^+(\mathcal{X})$.

A useful class of measures (and mixed states) on $\mathcal{X}$ is given by the decomposable measures of the form $Y^\otimes$, which are defined for an arbitrary finite measure $Y(dx)$ on X by their components

$$(Y^\otimes)_n(dx_1 \ldots dx_n) = Y^{\otimes n}(dx_1 \ldots dx_n) = Y(dx_1)...Y(dx_n)$$

Similarly the decomposable observables (multiplicative or additive)) are defined for an arbitrary $Q \in C(X)$ as

$$(Q^\otimes)^n(x_1, \ldots, x_n) = Q^{\otimes n}(x_1, \ldots, x_n) = Q(x_1)...Q(x_n)$$

and

$$(Q^\oplus)(x_1, \ldots, x_n) = Q(x_1) + ... + Q(x_n)$$

($Q^{\oplus}$ vanishes on X^0). In particular, if $Q = 1$, then $Q^{\oplus} = 1^{\oplus}$ is the number of particles: $1^{\oplus}(x_1, \ldots, x_n) = n$.

In the future, we shall be interested in pure jump processes on $\mathcal{X}$, whose semigroup and the generator preserves the space C_{sym} of continuous symmetric functions and hence are given by symmetric transition kernels $q(\mathbf{x}; d\mathbf{y})$ that could be thus considered as kernels on the factor space $S\mathcal{X}$. Occasionally we shall need the obvious formula

$$\sum_{I \subset \{1,\ldots,n\}, |I|=2} f(\mathbf{x}_I) = \frac{1}{2} \int \int f(z_1, z_2) \delta_{\mathbf{x}}(dz_1) \delta_{\mathbf{x}}(dz_2) - \frac{1}{2} \int f(z, z) \delta_{\mathbf{x}}(dz),$$

$$(11.65)$$

valid for any $f \in C^{sym}(X^2)$ and $\mathbf{x} = (x_1, \ldots, x_n) \in X^n$.

3. Pure jump Markov models of interacting particles.

To specify a binary particle interaction of a pure jump type one has to specify a continuous transition kernel

$$P^2(x_1, x_2; d\mathbf{y}) = \{P_m^2(x_1, x_2; dy_1 \ldots dy_m)\}$$

from SX^2 to $S\mathcal{X}$ such that $P^2(\mathbf{x}; \{\mathbf{x}\}) = 0$ for all $\mathbf{x} \in X^2$, the intensity being

$$P^2(x_1, x_2) = \int_{\mathcal{X}} P^2(x_1, x_2; d\mathbf{y}) = \sum_{m=0}^{\infty} \int_{X^m} P_m^2(x_1, x_2; dy_1 \ldots dy_m).$$

The intensity defines the rate of decay of any pair of particles x_1, x_2 and the measure $P^k(x_1, x_2; d\mathbf{y})$ defines the distribution of possible outcomes. Supposing that any randomly chosen pair of particles from a given set of n particles can interact, leads to the following generator of type (11.63) of binary interacting particles defined by the kernel P^2:

$$(G_2 f)(x_1, \ldots, x_n) = \sum_{I \subset \{1,\ldots,n\}, |I|=2} \int (f(\mathbf{x}_{\bar{I}}, \mathbf{y}) - f(x_1, \ldots, x_n)) P^2(x_I, d\mathbf{y}),$$

$$= \sum_{m=0}^{\infty} \sum_{I \subset \{1,\ldots,n\}, |I|=2} \int (f(\mathbf{x}_{\bar{I}}, y_1, \ldots, y_m) - f(x_1, \ldots, x_n)) P_m^k(x_I; dy_1 \ldots dy_m).$$

Though we shall not use this interpretation further, let us note here that the probabilistic description of the evolution of a pure jump Markov process Z_t on $\mathcal{X}$ specified by this generator (if this process is well defined!) is the following. Any two particles x_1, x_2 (randomly and uniformly chosen from exiting n particle) wait for the interaction an exponential random time with the parameter being the intensity $P^2(x_1, x_2)$. The first pair that manage to interact produce on its place a collection of particles $(y_1, \ldots, y_m)$ according to the distribution $P_m^2(x_1, x_2; dy_1 \ldots dy_m)/P^2(x_1, x_2)$. Then everything starts again from thus obtained new collection of particles.

Some sufficient conditions under which G_2 generates a unique Markov process will be specified later on.

Similarly, a k-ary interaction of a pure jump type is specified by a transition kernel

$$P^k(x_1, \ldots, x_k; d\mathbf{y}) = \{P_m^k(x_1, \ldots, x_k; dy_1 \ldots dy_m)\} \tag{11.66}$$

from SX^k to $S\mathcal{X}$ such that $P^k(\mathbf{x}; \{\mathbf{x}\}) = 0$ for all $\mathbf{x} \in \mathcal{X}$, having the intensity

$$P^k(x_1, \ldots, x_k) = \int P^k(x_1, \ldots, x_k; d\mathbf{y}) = \sum_{m=0}^{\infty} \int P_m^k(x_1, \ldots, x_k; dy_1 \ldots dy_m).$$
$$\tag{11.67}$$

This kernel defines the following generator of k-ary interacting particles:

$$(G_k f)(x_1, \ldots, x_n) = \sum_{I \subset \{1, \ldots, n\}, |I| = k} \int (f(\mathbf{x}_{\bar{I}}, \mathbf{y}) - f(x_1, \ldots, x_n)) P^k(x_I, d\mathbf{y}). \tag{11.68}$$

Changing the state space by (11.64) yields the corresponding Markov process on $\mathcal{M}_\delta^+(X)$. Choosing a positive parameter h, we shall perform now the following scaling: we scale the empirical measures $\delta_{x_1} + \ldots + \delta_{x_n}$ by a factor h and the operator of k-ary interactions by a factor h^{k-1} (similar to the scaling used in the theory of superprocesses, which in our notations corresponds to the 'interaction free' case of $k = 1$). This leads to the operator

$$\Lambda_k^h f(h\delta_{\mathbf{x}}) = h^{k-1} \sum_{I \subset \{1, \ldots, n\}, |I| = k} \int [f(h\delta_{\mathbf{x}} - \sum_{i \in I} h\delta_{x_i} + h\delta_{\mathbf{y}}) - f(h\nu)] P(\mathbf{x}_I; d\mathbf{y}) \tag{11.69}$$

(where we denoted $\delta_{\mathbf{y}} = \delta_{y_1} + \ldots + \delta_{y_m}$ for $\mathbf{y} = (y_1, \ldots, y_m)$), acting on the space of continuous functions on the set $\mathcal{M}_{h\delta}^+(X)$ of measures of the form $h\nu = h\delta_{\mathbf{x}} = h\delta_{x_1} + \ldots + h\delta_{x_n}$. This generator defines our basic Markov model of exchangeable particles with (h-scaled) k-ary interaction of pure jump type.

The above scaling (usually applied in statistical mechanics) is not the only reasonable one. For the theory of evolutionary games (or other biological model) a more natural scaling is by normalizing on the number of particles, i.e. by division of k-ary interaction by $n^{k-1} = (\|h\nu\|/h)^{k-1}$. This leads (instead of (11.69)) to the operator

$$\tilde{\Lambda}_k^h f(h\delta_{\mathbf{x}}) = h^{k-1} \sum_{I \subset \{1,\dots,n\}, |I|=k} \int [f(h\nu - \sum_{i \in I} h\delta_{x_i} + h\delta_{\mathbf{y}}) - f(h\nu)] \frac{P(\mathbf{x}_I; d\mathbf{y})}{\|h\delta_{\mathbf{x}}\|^{k-1}}.$$

(11.70)

4. Heuristic derivation of kinetic equations.

By (11.65) the operator Λ_2^h can be written in the form

$$\Lambda_2^h f(h\delta_{\mathbf{x}}) = -\frac{1}{2} \int_{\mathcal{X}} \int_X [f(h\delta_{\mathbf{x}} - 2h\delta_z + h\delta_{\mathbf{y}}) - f(h\delta_{\mathbf{x}})] P(z, z; d\mathbf{y})(h\delta_{\mathbf{x}})(dz)$$

$$+ \frac{1}{2h} \int_{\mathcal{X}} \int_{X^2} [f(h\delta_{\mathbf{x}} - h\delta_{z_1} - h\delta_{z_2} + h\delta_{\mathbf{y}}) - f(h\delta_{\mathbf{x}})]$$

$$P(z_1, z_2; d\mathbf{y})(h\delta_{\mathbf{x}})(dz_1)(h\delta_{\mathbf{x}})(dz_2).$$

Note that for the linear functions

$$f_g(\mu) = \int g(y)\mu(dy)$$

this operator acts as

$$\Lambda_2^h f_g(h\delta_{\mathbf{x}}) = \frac{1}{2} \int_{\mathcal{X}} \int_{X^2} [g^\oplus(z_1, z_2) - g^\oplus(\mathbf{y})] P(z_1, z_2; d\mathbf{y})(h\delta_{\mathbf{x}})(dz_1)(h\delta_{\mathbf{x}})(dz_2)$$

$$- \frac{1}{2} h \int_{\mathcal{X}} \int_X [g^\oplus(z, z) - g^\oplus(\mathbf{y})] P(z, z; d\mathbf{y})(h\delta_{\mathbf{x}})(dz).$$

It follows that if $h \to 0$ and $h\delta_{\mathbf{x}}$ tends to some finite measure μ (i.e. the number of particles tends to infinity, but the "whole mass" remains finite due to the scaling of each atom), the corresponding evolution equation $\dot{f} = \Lambda_2^h f$ for linear $f = f_g$ tends to the equation

$$\frac{d}{dt} \int_X g(z)\mu_t(dz) = \frac{1}{2} \int_{\mathcal{X}} \int_{X^2} (g^\oplus(\mathbf{y}) - g^\oplus(\mathbf{z})) P^2(\mathbf{z}; d\mathbf{y}) \mu_t^{\otimes 2}(d\mathbf{z}), \mathbf{z} = (z_1, z_2),$$

(11.71)

which represents the *general kinetic equation for binary interactions of pure jump type* in the *weak form*. The latter means that it has to hold for all $g \in C_\infty(X)$ (or at least its dense subspace). The famous Boltzman equation and the Smoluchovski coagulation equation are particular cases of (11.71).

Similar procedure with k-ary interactions leads to the *general kinetic equation for k-ary interactions of pure jump type in the weak form*:

$$\frac{d}{dt} \int_X g(z)\mu_t(dz) = \frac{1}{k} \int_{\mathcal{X}} \int_{X^k} (g^\oplus(\mathbf{y}) - g^\oplus(\mathbf{z})) P^k(\mathbf{z}; d\mathbf{y}) \mu_t^{\otimes k}(d\mathbf{z}), \mathbf{z} = (z_1, \dots, z_k).$$

(11.72)

On the other hand, similar procedure with the operator (11.70) leads to the equation

$$\frac{d}{dt}\int_X g(z)\mu_t(dz) = \frac{1}{k}\int_{\mathcal{X}}\int_{X^k}(g^{\oplus}(\mathbf{y})-g^{\oplus}(\mathbf{z}))P^k(\mathbf{z};d\mathbf{y})\left(\frac{\mu_t}{\|\mu_t\|}\right)^{\otimes k}(d\mathbf{z})\|\mu_t\|.$$
(11.73)

In biological context one traditionally writes the dynamics in terms of normalized (probability) measures. As for positive μ the norm equals $\|\mu\| = \int_X \mu(dx)$, one sees that for positive solutions μ_t of (11.73)

$$\frac{d}{dt}\|\mu_t\| = -\frac{1}{k}\int_{X^k}Q(\mathbf{z})\left(\frac{\mu_t}{\|\mu_t\|}\right)^{\otimes k}(d\mathbf{z})\|\mu_t\|,$$
(11.74)

where

$$Q(\mathbf{z}) = -\int_{\mathcal{X}}\int_{X^k}(\mathbf{1}^{\oplus}(\mathbf{y}) - \mathbf{1}^{\oplus}(\mathbf{z}))P^k(\mathbf{z};d\mathbf{y}).$$
(11.75)

Consequently, rewriting equation (11.73) in terms of normalized measure $\nu_t = \mu_t/\|\mu_t\|$ leads to the equation

$$\frac{d}{dt}\int_X g(z)\nu_t(dz) = \frac{1}{k}\int_{\mathcal{X}}\int_{X^k}(g^{\oplus}(\mathbf{y}) - g^{\oplus}(\mathbf{z}))P^k(\mathbf{z};d\mathbf{y})(\nu_t)^{\otimes k}(d\mathbf{z})$$

$$+ \frac{1}{k}\int_X g(z)\nu_t(dz)\int_{\mathcal{X}}\int_{X^k}Q(\mathbf{z})(\nu_t)^{\otimes k}(d\mathbf{z}).$$
(11.76)

It is worth noting that this different re-scaling of interactions leading to (11.73) is equivalent to a time change in (11.72) (a particular performance of this reduction in evolutionary biology is the well known trajectory-wise equivalence of Lotka-Volterra model and replicator dynamics, see [72]).

5. Well-posedness of the Cauchy problem for a class of kinetic equations and convergence of Markov approximation.
In this subsection we shall establish existence and uniqueness results for kinetic equations (11.72), (11.73) under a strong simplifying assumption of bounded coefficients (for unbounded coefficients we refer to the original papers cited at the beginning). This constitutes the first step in the justification of the above heuristics. The proof of the convergence itself will not be given here, we shall only formulate the result referring for a proof to the papers cited at the beginning.

As is the case with superprocesses, the central notion for the theory of kinetic equations is the criticality. The transition kernel $P = P^k(\mathbf{x};d\mathbf{y})$ from (11.66) is called *subcritical* (respectively *critical*), if

$$\int_{\mathcal{X}}(\mathbf{1}^{\oplus}(\mathbf{y}) - \mathbf{1}^{\oplus}(\mathbf{x}))P^k(\mathbf{x};d\mathbf{y}) \leq 0$$

for all $\mathbf{x} \in X^k$ (respectively if the equality holds).

Theorem 11.14. *Suppose the transition kernel* (11.66) *is subcritical and its intensity* (11.66) *is uniformly bounded. Then for any non-negative measure* $\mu \in \mathcal{M}^+(X)$ *there exists a unique global solution to* (11.72) *and* (11.73), *i.e. a unique continuous mapping* $t \mapsto \mu_t \in \mathcal{M}(X)$ *(with measures considered with their $\star$-weak topology) such that* $\mu_0 = \mu$ *and* (11.72) *holds for all* $g \in C_\infty(X)$. *Moreover this solution is also positive,* $\|\mu_t\| \le \|\mu\|$ *for all t, and the mapping* $t \mapsto \mu_t$ *is strongly differentiable (i.e. in the sense of the norm topology of $\mathcal{M}(X)$ and satisfies the corresponding strong version*

$$\frac{d}{dt}\mu_t(.) = \frac{1}{k} \int_{\mathcal{X}} \int_{X^k} (1^{\oplus}_{(.)}(\mathbf{y}) - 1^{\oplus}_{(.)}(\mathbf{z}))P(\mathbf{z}; d\mathbf{y})\mu_t^{\otimes k}(d\mathbf{z}), \quad \mathbf{z} = (z_1, \dots, z_k)$$

(11.77)

(where (.) denotes an arbitrary Borel set) of (11.72) *(respectively of* (11.73)).*

Proof. Step 1. First observe that the r.h.s. of (11.72) is a polynomial with bounded coefficients in μ_t and hence is Lipshitz continuous with respect to μ_t in the sense of the norm, as long as $\|\mu_t\|$ remains bounded. Hence for any $\mu \in \mathcal{M}(X)$ there exists a $T = T(\|\mu\|)$ such that the strong solution to (11.72) exists and is unique on the time interval $[0, T]$.

Step 2. The key moment is now to show that if μ is positive, then the above solution also has to be positive. There are two natural (and instructive) ways to see it. One is by comparison. Namely one observes that as long as μ stays positive the evolution μ_t described by (11.77) is bounded from below by the solution to the equation

$$\frac{d}{dt}\mu_t(.) = -\frac{1}{k} \int_{\mathcal{X}} \int_{X^k} 1^{\oplus}_{(.)}(\mathbf{z})P(\mathbf{z}; d\mathbf{y})\mu_t^{\otimes k}(d\mathbf{z}), \quad \mathbf{z} = (z_1, \dots, z_k) \quad (11.78)$$

which in turn is bounded from below by the solution to the equation

$$\frac{d}{dt}\mu_t(.) = -K\frac{1}{k} \int_{X^k} 1^{\oplus}_{(.)}(\mathbf{z})\mu_t^{\otimes k}(d\mathbf{z}), \quad \mathbf{z} = (z_1, \dots, z_k),$$

which is explicitly solvable and preserves positivity (where K is any upper bound for the intensity of the kernel P^k). Another way (seemingly first applied in the context of the Boltzmann equation) is to rewrite (11.72) first as

$$\frac{d}{dt} \int g(z)\mu_t(dz) = -K \int g(z)\mu_t(dz)$$

$$+ \frac{1}{k} \int_{\mathcal{X}} \int_{X^k} (g^{\oplus}(\mathbf{y}) - g^{\oplus}(\mathbf{z}))P(\mathbf{z}; d\mathbf{y})\mu_t^{\otimes k}(d\mathbf{z}) + K \int g(z)\mu_t(dz),$$

and then to rewrite this in the integral (so called *mild* or *interaction representation*) form

$$\mu_t(.) = e^{-Kt}\mu(.)$$

$$+ \int_0^t e^{-K(t-s)}\left[\frac{1}{k}\int\int(1_{(.)}^{\oplus}(\mathbf{y}) - 1_{(.)}^{\oplus}(\mathbf{z}))P(\mathbf{z};d\mathbf{y})\mu_s^{\otimes k}(d\mathbf{z}) + K\mu_s(.)\right]ds$$

and then observe that the natural perturbation series approximations to its solution preserve positivity.

Step 3. At last from sub-criticality it follows that $\int \mu_t(dx) \leq \int \mu(dx)$ for all t, which together with positivity implies $\|\mu_t\| \leq \|\mu\|$. Hence after a period T we can extend the solution uniquely for another period of length T and so on yielding a unique global (strong) solution.

Step 4. The simplest way to settle the problem with equation (11.73) one observes that its solution is obtained uniquely (by time change) from the solution μ_t of (11.72) as long as $\mu_t \neq 0$. But this has to be true, which follows from either comparison arguments of Step 2 or from the uniqueness. The proof is complete.

A worth noting feature of equation (11.72) is that its evolution preserves L_1 spaces. Namely, the following holds.

Theorem 11.15. *Under assumptions of the previous Theorem suppose we are given a reference (not necessarily finite, but σ-finite and positive) measure M on X. Let μ has a non-negative density f with respect to M, i.e. $f \in L_1(X, M)$. Then the solution μ_t will also have a non-negative density $f_t \in L_1(X, M)$ satisfying the weak equation*

$$\frac{d}{dt}\int_X g(z)f_t(z)M(dz) = \frac{1}{k}\int_{\mathcal{X}}\int_{X^k}(g^{\oplus}(\mathbf{y}) - g^{\oplus}(\mathbf{z}))P^k(\mathbf{z};d\mathbf{y})\prod_{i=1}^k f_t(z_i)M^{\otimes k}(d\mathbf{z}).$$

$$(11.79)$$

Proof. It repeats literarily the proof of the previous theorem. Namely, one first get the local existence and uniqueness and then extends to all times by positivity.

We have discussed above a formal deduction of (11.72) from a Markov process of interacting particle. Lots of work is required to make these arguments rigorous. This is done under various assumptions on the transition kernels. We shall formulate only a simple result for bounded coefficients.

Theorem 11.16. *Under the assumptions of the previous Theorem assume additionally that X is compact. Then the operator (11.69) (respectively (11.70)) generates a uniquely defined nonexplosive Markov process Z_t^h*

(respectively $\tilde{Z}_t^h$) on $\mathcal{M}_{h\delta}^+$ for any $h > 0$. Moreover, if the initial conditions Z_0^h converge weakly to a finite measure μ on X as $h \to 0$, then the processes Z_t^h (respectively $\tilde{Z}_t^h$) converge in the sense of distribution to the solution μ_t (a deterministic process) of the equation (11.72) (respectively (11.73)) constructed in Theorem 11.14.

Proof. We omit a proof of this result, as its full exposition would require a too long excursion to either functional analysis or stochastic processes). It is done analogously to more sophisticated facts obtained for instance in [96] by probabilistic methods, or in [97] by pure analytic tools.

6. Evolutionary games and Replicator dynamics.

Consider (the mixed strategy extension of) a symmetric k-person game with strategies of each players belonging to a compact space X and with payoffs given by a continuous function $H(x; y_1, \ldots, y_{k-1})$ on X^k symmetric with respect to the last variables $y_1, \ldots, y_{k-1}$ (see details of this definition in Section 11.2). A natural generalization of the *replicator dynamics* (RD) of two person finite game (discussed in Chapter 9) represents the measure-valued evolution described by the weak equation

$$\frac{d}{dt} \int_X g(x)\nu_t(dx) = \int_X (H^\star(\nu_t \| x) - H^\star(\nu_t))g(x)\nu_t(dx), \qquad (11.80)$$

which has to hold for all $g \in C(X)$ (the notation used are again from Section 11.2).

We like to show that this equation describes the law of large numbers of a Markov model of interacting particles obtained as a particular case of a general scheme presented above.

In biological context particles becomes species of a certain large population, a position of a particle $x \in X$ denoting its strategy. A key feature distinguishing the evolutionary game setting in the general model of the previous subsection is that the species produce new species of their own kind (with inherited behavioral patterns). In usual model of evolutionary game theory it is assumed that any k randomly chosen species can occasionally meet and play a k-person symmetric game specified by a payoff function $H(x; y_1, \ldots, y_{k-1})$ on X^k, payoff measuring fitness expressed in terms of expected number of offspring.

Remark. In basic evolutionary models a binary interaction is considered, i.e. the games of two players only with $k = 2$.

To specify a Markov model one needs to specify the game a bit further. Namely, we shall assume that the result of the game for player x playing against $(y_1, \ldots, y_{k-1})$ is given by the probability distribution

$H^m(x; y_1, \ldots, y_{k-1})$, $m = 0, 1, \ldots$ of the number of offspring it is going to get with the average given by the original function H, i.e.

$$H(x; y_1, \ldots, y_{k-1}) = \sum_{m=0}^{\infty} (m - 1) H^m(x; y_1, \ldots, y_{k-1}). \qquad (11.81)$$

Now we have a model of subsection 3 specified by the transition kernels (11.66) of the form

$$P_m^k(z_1, \ldots, z_k; d\mathbf{y}) = H^m(z_1; z_2, \ldots, z_k)\delta_{z_1}(dy_1) \ldots \delta_{z_1}(dy_m)$$
$$+ H^m(z_2; z_1, \ldots, z_k)\delta_{z_2}(dy_1) \ldots \delta_{z_2}(dy_m) + \ldots$$
$$+ H^m(z_k; z_1, \ldots, z_{k-1})\delta_{z_k}(dy_1) \ldots \delta_{z_k}(dy_m) \qquad (11.82)$$

so that

$$\int_{\mathcal{X}} (g^{\oplus}(\mathbf{y}) - g^{\oplus}(\mathbf{z})) P^k(\mathbf{z}; d\mathbf{y})$$

$$= \sum_{m=0}^{\infty} (m - 1)[g(z_1)H^m(z_1; z_2, \ldots, z_k) + \ldots + g(z_k)H^m(z_k; z_1, \ldots, z_{k-1})]$$

$$= g(z_1)H(z_1; z_2, \ldots, z_k) + \ldots + g(z_k)H(z_k; z_1, \ldots, z_{k-1}),$$

and (due to the symmetry of H) equation (11.73) takes the form

$$\frac{d}{dt} \int_X g(x)\mu_t(dx) = \|\mu_t\| \int_{X^k} g(x)H(z_1; z_2, \ldots, z_k) \left(\frac{\mu_t}{\|\mu_t\|}\right)^{\otimes k} (dz_1 \ldots dz_k), \qquad (11.83)$$

and hence for the normalized measure $\nu_t = \mu_t/\|\mu_t\|$ one gets precisely (11.80), which is thus a simple particular case of (11.76). (Of course, the well-posedness results of the previous subsection yield the corresponding well posedness results for (11.80).)

It is worth noting that if a reference measure M on X is chosen, equation (11.80) can be written in terms of the densities f_t of ν_t with respect to M as

$$\dot{f}_t(x) = f_t(x)(H^\star(f_t M \| x) - H^\star(f_t M)). \qquad (11.84)$$

7. Examples of equilibria.

For completeness of the picture we shall briefly note here how the connection of Nash equilibria with the replicator dynamics extends from finite game to the present more general setting.

Proposition 11.27. *(i) If ν defines a symmetric Nash equilibrium for symmetric k-person game (its mixed strategy extension) specified by payoff $H(x; y_1, \ldots, y_{k-1})$ on X^k (X again a compact space), then ν is a fixed point for RD (11.80). If ν is such that any open set in X has a positive ν measure (pure mixed profile), then the inverse statement holds.*

Proof. (i) By definition, ν defines a symmetric Nash equilibrium if and only if

$$H^\star(\nu\|x) \le H^\star(\nu) \tag{11.85}$$

for all $x \in X$. But the set $M = \{x : H^\star(\nu\|x) < H^\star(\nu)\}$ should have ν-measure zero (otherwise integrating (11.85) would lead to a contradiction). This implies that

$$\int_X (H^\star(\nu\|x) - H^\star(\nu))g(x)\nu_t(dx) = 0 \tag{11.86}$$

for all g. (ii) Conversely assuming (11.86) holds for all g implies (taking into account here that ν is purely mixed profile)

$$H^\star(\nu\|x) = H^\star(\nu)$$

on a open dense subset of X and hence everywhere, due to the continuity of H.

We shall conclude this subsection with a partial extension of Theorem 9.8 to the games with compact spaces of strategies.

Proposition 11.28. *Consider a mixed strategy extension of a two-person symmetric game with a compact space of pure strategies X of each player and a payoff matrix being an antisymmetric function H on X^2, i.e. $H(x,y) = -H(y,x)$. Assume there exists a positive finite measure M on X such that $\int H(x,y)M(dy) = 0$ for all x. Then M specifies a symmetric Nash equilibrium. Moreover, the function*

$$L(f) = \int \ln f_t(x)M(dx)$$

is the first integral (i.e. it is constant on the trajectories) of the system (11.84) (RD on densities with respect to M).

Proof. Is a straightforward generalization of the proof of Theorem 9.8. Note that the existence of the first integral is not enough to make a conclusion on the stability in this infinite-dimensional setting.

8. Stochastic replicator dynamics.

Kinetic equations and RD described above specify the law of large numbers limits for Markov processes of interacting particles when this limit is a deterministic process. However there are natural situations when the law of large numbers is itself random, see e.g. [44] in the context of evolutionary games. Studying such random limits leads to general measure-valued processes with (possibly infinite dimensional) pseudo-differential generators.

The simplest case of these limits obtained from branching (noninteracting particle systems) leads to a very popular in modern probability class of processes called superprocesses. For k-ary interacting particle systems (for instance for k-person evolutionary games) these random limits comprise a rather large class of processes with polynomial (in position) pseudo-differential generators, whose analysis was initiated in [93], [94]. We shall give here only a very brief introduction to those limiting processes that are connected with evolutionary games, which, as we noted above, correspond to a class of interactions where particles (or species) create the offspring of its own kind. For simplicity we return to two-person games with only a finite set $X = \{1, \dots, d\}$ of pure strategies.

Denoting by N_j the number of individuals playing the strategy j and by $N = \sum_{j=1}^{d} N_j$ the whole size of the population, assuming that the outcome of a game between players with strategies i and j is a probability distribution $A_{ij} = \{A_{ij}^m\}$ of the number of offsprings $m \geq -1$ of the players ($\sum_{m=-1}^{\infty} A_{ij}^m = 1$) and the intensity a_{ij} of the reproduction per time unit ($m = -1$ means the death of the individual) yields the Markov chain on $\mathbf{Z}_+^d$ with the generator

$$Gf(N) = \sum_{j=1}^{d} N_j \sum_{m=-1}^{\infty} \left(B_j^m + \sum_{k=1}^{d} a_{jk} A_{jk}^m \frac{N_k}{|N|} \right) (f(N + me_j) - f(N))$$

(11.87)

(where B_j^m describe the background reproduction process), which is a version of the generators of binary interactions introduced above. One can show that scaling combined with certain natural dependence of probabilities A_{jk}^m on a small parameter h (which roughly speaking is the inverse of the average number of particles) lead to the process on $\mathbf{R}_+^d$ with the generator of type

$$\Lambda = \sum_{j=1}^{d} x_j \left(\phi_j + \sum_{k=1}^{d} \frac{x_k}{\mu(x)} \phi_{jk} \right)$$

(11.88)

(quadratic in x), where all ϕ_j and ϕ_{jk} are the generators of one-dimensional Lévy processes, more precisely

$$\phi_{jk} f(x) = g_{jk} \frac{\partial^2 f}{\partial x_j^2}(x) + \beta_{jk} \frac{\partial f}{\partial x_j}(x)$$

$$+ \int \left(f(x + ye_j) - f(x) - \mathbf{1}_{y \leq 1}(y) \frac{\partial f}{\partial x_j}(x) y_j \right) \nu_{jk}(dy),$$

(11.89)

$$\phi_j f(x) = g_j \frac{\partial^2 f}{\partial x_j^2}(x) + \beta_j \frac{\partial f}{\partial x_j}(x)$$

$$+ \int \left(f(x + ye_j) - f(x) - \mathbf{1}_{y \leq 1}(y)\frac{\partial f}{\partial x_j}(x)y \right) \nu_j(dy),$$

where all ν_{jk}, ν_j are Borel measures on $(0, \infty)$ such that the function $\min(y, y^2)$ is integrable with respect to these measures, g_j and g_{jk} are non-negative. For the precise form of the approximation of the generators (11.88) by the generators of type (11.87) together with rigorous convergence results we refer to [94], noting here only the different interpretation of the approximations of the three terms in (11.89) (diffusion, drift and integral part): the first term stands for a quick game that can be called "death or birth" game, which describes some sort of fighting for reproduction, whose outcome is that an individual either dies or produces an offspring; the second term (approximating drift) describes games for death or for life depending on the sign of β_{jk}; third term corresponds to games for a slow reproduction of the large number of offsprings.

Chapter 12

Games of many agents in statistical limit

A very popular trend in modern game theory analyses the competitive interaction of a set of big players, referred to as principals or major agents, acting in the background of large pools of small players, where the impact on the overall evolution of the behavior of each small player in a group decreases as the size of the group increases. The methods of statistical physics are used to describe when the size of groups tends to infinity. This chapter is a brief introduction to authors' monograph [263], where this trend is more fully developed for the simplest case of finite state spaces of individual players and where an extensive bibliography can be found. We introduce the basic model of controlled mean-field interacting particle systems, its main performance in the pressure and resistance framework and the fundamental paradigm of the scaling limit (dynamic law of large numbers, with the origins in statistical physics) allowing one to reduce the analysis of these chaotic systems with exponentially large state spaces to a low dimensional dynamic system expressed in terms of simple ordinary differential equations (ODEs), so called kinetic equations. We describe various classes of real life socio-economic processes fitting to the general scheme: inspection, corruption, cyber-security, counterterrorism, coalition building, merging and splitting, evolutionary games with controlled mean-field dependent payoffs and many other. In Section 12.6 we shall briefly introduce the mean field games (MFGs), which represents one of the major trends in game theory of the 21st century. Our exposition will be mostly elementary. Some mathematical results for the case of compact state spaces are proved in Section 11.9.

12.1 Controlled mean-field interacting particle systems and their LLN

Let us introduce the Markov chains, which form the cornerstone of our analysis, with the minimal usage of technical terms.

One distinguishes Markov chains in discrete and continuous time.

The Markov chains in discrete time are characterized by a finite set $\{1, \ldots, d\}$, called the *state space*, of possible positions of an object under investigation (particle or agent), and a collection of non-negative numbers $P = \{P_{ij}\}$, i, j from $\{1, \ldots, d\}$, called the *transition probabilities* from i to j, such that $P_{i1} + \cdots + P_{id} = 1$ for any i. Collection of numbers $P = \{P_{ij}\}$ satisfying these conditions is called a *matrix of transition probabilities*, or a *transition matrix* or a *stochastic matrix*. The *Markov chain* specified by the collection P is the process of random transitions between the states $\{1, \ldots, d\}$ evolving in discrete times $t = 0, 1, 2, \ldots$ by the following rule. If at some time t the object is in a state i, in the next moment $t + 1$ it migrates from state i to some other state j, choosing j with the probability P_{ij}. Then the process repeats starting from j at time $t + 1$, and the same procedure continues ad infinitum.

The Markov chains in continuous time are characterized by a finite set $\{1, \ldots, d\}$, called the *state space*, of possible positions of an object under investigation (particle or agent), and a collection of non-negative numbers $Q = \{Q_{ij}\}$, $i \neq j$, i, j from $\{1, \ldots, d\}$, called the *transition rates* from i to j. To circumvent the restriction $i \neq j$ it is convenient to introduces the negative numbers

$$Q_{ii} = -\sum_{j \neq i} Q_{ij}. \tag{12.1}$$

The positive numbers $|Q_{ii}| = -Q_{ii}$ are called the *intensities of jumps* at state i. The collection $Q = \{Q_{ij}\}$, $i, j \in \{1, \ldots, d\}$, is then called a *Q-matrix* or a *Kolmogorov's matrix*. The *Markov chain* specified by the collection Q is the process of random transitions between the states $\{1, \ldots, d\}$ evolving by the following rule. If at some time t the object is in a state i, the object sits at i a random time τ characterized by the following waiting probability:

$$\mathbf{P}(\tau > s) = \exp\{-s|Q_{ii}|\} = \exp\{sQ_{ii}\},$$

such random time being referred to as $|Q_{ii}|$-*exponential waiting time*. At time τ the object instantaneously migrates from state i to some other state j, choosing j with the probability $Q_{ij}/|Q_{ii}|$. Then the process repeats starting from j, and the same procedure continues ad infinitum.

Models with discrete or continuous time are chosen from practical convenience. If we observe a process with some fixed frequency (say, measure a temperature in the sea every morning), discrete setting is more appropriate. If we observe a process steadily, continuous time setting is more appropriate.

The state spaces of Markov chains that we will be mostly concern with here are the strategies of individuals (say, corrupted or honest inspector, level of illegal activity, etc.) or the working conditions of agents or devices (say, infected or susceptible computer or individual, level of defence, etc.).

Let us turn to the basic setting of mean-field interacting particle systems with a finite number of types. We wish to model a large number N of particles or agents evolving each according to a Markov chain on some common state space $\{1, \ldots, d\}$, but with the transition rates depending on the overall distribution of agents among the states. Thus the overall state of the system of such agents is given by a collection of integers $n = (n_1, \ldots, n_d)$, where n_j is the number of agents in the state i, so that $N = n_1 + \cdots + n_d$.

Usually it is more convenient to work with frequencies, that is, instead of n, to use the normalized quantities $x = (x_1, \ldots, x_d) = n/N$, $x_j = n_j/N$. All such vectors x belong to the unit simplex Σ_d, defined as the collections of all d non-negative numbers summing up to unity:

$$\Sigma_d = \{x = (x_1, \ldots, x_d) : x_j \geq 0 \text{ for all } j \text{ and } \sum_{j=1}^{d} x_j = 1\}.$$

To describe transitions of individuals depending on the overall distribution x, we have to have a family $\{Q(x)\} = \{(Q_{ij})(x)\}$ of Kolmogorov's matrices depending on a vector x from Σ_d.

Thus, for any x, the family $\{Q(x)\}$ specifies a Markov chain on the state space $\{1, \ldots, d\}$. But this is not what we are looking for. We are interested in a system with many agents, i.e. a Markov chain on the states $n = (n_1, \ldots, n_d)$. Namely, the *mean-field interacting particle system* with N agents specified by the family $\{Q(x)\}$ is defined as the Markov chain on the set of collections $n = (n_1, \ldots, n_d)$ with $N = n_1 + \cdots + n_d$ evolving according to the following rule. Starting from any time and current state n one attaches to this state a $\sum_i n_i |Q_{ii}(n/N)|$-exponential random waiting time. Once the bell rings one chooses the transition $i \to j$, $i \neq j$, with the probability

$$\mathbf{P}(i \to j) = \frac{n_i Q_{ij}(n/N)}{\sum_k n_k |Q_{kk}(n/N)|}. \tag{12.2}$$

These are well defined probabilities, since

$$\sum_i \sum_{j \neq i} \mathbf{P}(i \to j) = 1.$$

As the number of particles N becomes large, the usual analysis of such Markov chain becomes unfeasible, the situation often referred to as the 'curse of dimensionality'. To overcome this problem one is searching for a simpler limiting evolution as $N \to \infty$. The situation is quite similar to the basic approaches used in physics to study the dynamics of gases and liquids: instead of following the random behavior of immense number of individual molecules, one looks for the evolution of the main mean characteristics, like bulk velocities or temperatures that often turns out to be described by some deterministic differential equations.

As can be shown (see e.g. [256], [257], [263] or a more advances convergence result in Section 11.9), the limiting evolution of the frequencies $x = (x_1, \ldots, x_d)$ evolving according to the Markov chain described above is governed by the system of ordinary differential equations

$$\dot{x}_k = \sum_{i \neq k} (x_i Q_{ik}(x) - x_k Q_{ki}(x)) = \sum_{i=1}^{d} x_i Q_{ik}(x), \quad k = 1, \ldots, d \quad (12.3)$$

(the last equation arising from (12.1)), called the *kinetic equations* for the process of interaction described above.

More precisely, under some continuity assumptions on the family $\{Q(x)\}$ one shows that if $X_x^N(t)$ is the position of our mean-field interacting particle system at time t when started at x at the initial time and $X_x(t)$ is the solution of system (12.3) at time t also started at x at the initial time, then the probability for the deviation $|X_x^N(t) - X_x(t)|$ to be larger than any number ϵ tends to zero as $N \to \infty$:

$$\mathbf{P}(|X_x^N(t) - X_x(t)| > \epsilon) \to 0, \quad N \to \infty. \quad (12.4)$$

In this most elementary setting the result (12.4) is well known. We are interested in more general situations when the transition rates Q are controlled by one or several big players that can influence the dynamics of the chain to their advantages. Moreover, we are concerned with the rates of convergence under mild regularity assumptions on Q. The rates are better characterized in terms of the convergence of some bulk characteristics (so-called weak convergence), rather than directly via the trajectories.

The convergence rates are crucial for any practical calculations. For instance, if the difference of some characteristics of Markov chain $X_x^N(t)$

and its limit $X_x(t)$ is of order $1/N$, then applying the approximation $X_x(t)$ to the chain $X_x^N(t)$ yields the error of order 1% for $N = 100$ players and the error of order 10% for $N = 10$ players. Thus the number N does not have to be very large for the approximation (12.3) of 'infinitely many players' to give reasonable predictions.

An important particular case of the scheme above occurs from the setting of *binary interactions*. Namely, let us assume that any player from any state i can be influenced by any player in any other state j to migrate from i to j, the random events occurring with some given intensity $T_{ij}(n/N)/N$ (the appearance of the multiplier $1/N$ being the mark of the general approach, according to which each individual's contribution to the overall evolution becomes negligible in the limit of large number of players N). More precisely, to any ordered pair of agents A, B in different states, say $i \neq j$, one attaches random $T_{ij}(n/N)/N$-exponential waiting time. The minimum of all these waiting time is again a random waiting time. If the minimum occurs on a pair (A, B) from some states i and j respectively, then A migrates from the state i to the state j, and then the process continues from the state n^{ij}. Equivalently, this process can be described by one clock with $\sum_{i \neq j} n_i n_j T_{ij}(n/N)/N$-exponential waiting time, attached to a state n. When such clock rings, the transition $i \to j$, $i \neq j$, is chosen with probability

$$\mathbf{P}(i \to j) = \frac{n_i n_j T_{ij}(n/N)/N}{\sum_{k \neq l} n_k n_l T_{kl}(n/N)/N} = \frac{n_i x_j T_{ij}(x)}{\sum_{k \neq l} n_k x_l T_{kl}(x)}.$$

As is seen directly, the process is the same as the one described above with the matrix $Q_{ij}(x)$ of the type

$$Q_{ij}(x) = x_j T_{ij}(x), \quad i \neq j.$$

The next step is to include a major player or principal that can influence the evolution of our mean-field interacting particle system. As examples, one can think about the management of complex stochastic systems consisting of large number of interacting components: agents, mechanisms, robots, vehicles, subsidiaries, species, police units, etc. To take this into account, our family of $Q(x)$ matrices should become dependent on some parameter b controlled by the principle: $\{Q = Q(x, b)\}$. This can be a real parameter (say, the budget used by the principle for management) or a vector-valued parameter, for instance, specifying resources for various directions of development. In any case, b is supposed to belong to some domain in a Euclidean space, usually bounded, as the resources can not be infinite.

We shall also assume that the principal has some objectives in this process, aiming to maximize some profit or minimize certain costs arising from the functioning of the Markov chain. In the simplest situation we can assume the principal to be a *best response principal*, which instantaneously chooses the value of b^* maximizing some current profit $B(x, b, N)$ for given x, N:

$$b^* = b^*(x, N) = argmax B(x, \ldots, N). \tag{12.5}$$

This procedure reduces the dynamics to the previous case but with $Q^N(x) = Q(x, b^*(x, N))$, which can now depend on N. Convergence (12.4) can still be shown if b^* stabilizes as $N \to \infty$, that is, $b^*(x, N) \to b^*(x)$, as $N \to \infty$, for some function $b^*(x)$. The limiting evolution (12.3) turns to the evolution

$$\dot{x}_k = \sum_{i=1}^{d} x_i Q_{ik}(x, b^*(x)), \quad k = 1, \ldots, d. \tag{12.6}$$

More realistic modeling would require the principal to choose the control parameter strategically with some planning horizon T and a final goal at the end. Moreover, the model should include costs for changing strategies. This leads to a dynamic control problem, where the principal chooses controls at fixed moments of time (discrete control setting) or in continuous time (continuous time setting).

Further extensions deal with several principals competing on the background of small players, where the pool of small players can be common for the major agents (think about advertising models, or several states fighting together a terrorist group) or be specific to each major agent (think about generals controlling armies, big banks controlling their subsidaries, etc.).

Since the limiting behavior of the controlled Markov chain is given by kinetic equations (12.12), one can expect that eventually the evolution will settle down near some stable equilibrium points of this dynamic systems. This remark motivates the analysis of these equilibria.

Moreover, these equilibria have an important game-theoretic interpretation, which is independent of the myopic hypothesis defining the Markov evolution. Namely, let us consider explicitly the following game Γ_N of $N+1$ players (that was tacitly borne in mind when discussing dynamics). When the major player chooses the strategy b and each of N small players chooses the state i, the major player receives the payoff $B(x, b, N)$ and each player in the state i receives $R_i(x, b)$, $i = 1, \ldots, d$ (as above, with $x = n/N$ and $n = (n_1, \ldots, n_d)$ the realized occupation numbers of all the states). Thus a strategy profile of small players in this game can be specified either by a sequence of N numbers (expressing the choice of the state by each agent), or more succinctly, by the resulting collection of frequencies $x = n/N$.

For the game Γ_N, the Nash equilibrium is a profile of strategies (x_N, b_N^*) such that

$$b_N^* = b_N^*(x_N, N) = argmax\, B(x_N, b, N)$$

and for any $i, j \in \{1, \ldots, d\}$

$$R_j(x - e_i/N + e_j/N, b_N^*) \leq R_i(x, b_N^*). \tag{12.7}$$

As finding Nash equilibria is a difficult task, one often preforms approximate calculations via some numeric schemes leading to approximate Nash equilibria or ϵ-Nash equilibria A profile of strategies is an ϵ-*Nash equilibrium* if the above conditions hold up to an additive correction term not exceeding ϵ.

As is shown in [257], the rest points of (12.12) describe all approximate Nash equilibria for Γ_N. These approximate Nash equilibria are ϵ-Nash equilibria with ϵ of order $1/N$.

Let us discuss the main concrete examples of the pressure and resistance framework.

12.2 Pressure and resistance framework

Let us present now a key particular performance of the above general scheme introduced by one of the authors in [257] and called there the *pressure and resistance game*. In this setting the interests of the major player P and the small players are essentially opposite, and the mean-field dependence of the actions of small players arises from their binary interactions. Namely, as in the above general setting, the strategies of player P are given by the choices of controls $b(x, N)$, from a given subset of Euclidean space, based on the number of small players N and the overall distribution x of the strategies of small players. The evolution of the states of small players goes as above with $T_{ij}(x, b)$ depending also on b, so that

$$Q_{ij}(x, b) = x_j T_{ij}(x, b), \quad i \neq j. \tag{12.8}$$

The resulting limiting process described by ODE (12.6) becomes

$$\dot{x}_j = x_j \sum_i x_i [T_{ij}(x, b^*(x)) - T_{ji}(x, b^*(x))], \quad j = 1, \ldots, d. \tag{12.9}$$

More concrete performance of this model is the following *pressure and resistance model*. Let us assume that each small player enters certain interaction with P (small players try to resist the pressure exerted on them by P) defined by the strategy $j \in \{1, \ldots, d\}$ and that the result of this interaction is the payoff $R_j(x, b)$ to the small player. Moreover, with rate $\varkappa/N$ ($\varkappa$ being some constant) any pair of agents can meet and discuss their payoffs, which may result in the player with lesser payoff R_i switching to the strategy with the better payoff R_j, which may occur with a probability proportional to $(R_j - R_i)$.

In other words, this is exactly the scheme described above by (12.8) with

$$T_{ij}(x, b) = \varkappa(R_j(x, b) - R_i(x, b))^+, \qquad (12.10)$$

where we used the standard notation $y^+ = \max(0, y)$. Therefore, this model is also a performance of the general model above with

$$Q_{ij}(x, b) = \varkappa x_j(R_j(x, b) - R_i(x, b))^+. \qquad (12.11)$$

The resulting limiting process described by ODE (12.9) becomes

$$\dot{x}_j = \sum_i \varkappa x_i x_j[R_j(x, b^*(x)) - R_i(x, b^*(x))], \quad j = 1, \ldots, d. \qquad (12.12)$$

Thus we are dealing with a class of models, where a distinguished 'big' player exerts certain level b of pressure on (or interference into the affairs of) a large group of N 'small' players that can respond with a strategy j. Often these j are ordered reflecting the strength or the level of the resistance. The term 'small' reflects the idea that the influence of each particular player becomes negligible as $N \to \infty$. As an example of this general setting one can mention the interference of humans on the environment (say, by hunting or fishing) or the use of medications to fight with infectious bacteria in a human body, with resisting species having the choice of occupying the areas of ample foraging but more dangerous interaction with the big player (large resistance levels r) or less beneficial but also less dangerous areas (low resistance level). Another example can be the level of resistance of the population on a territory occupied by military forces.

In many situations, the members of the pool of small players have an alternative class of strategies of collaborating with the big player on various levels. The creation of such possibilities can be considered as a strategic action of the major player (who can thus exert some control on the rules of the game). In biological setting this is, for instance, the strategy of dogs joining humans in hunting their 'relatives' wolves or foxes (nicely described recruitment. Thus, the model of [293] includes the recruitment benefits of terrorists as a positively correlated function of preemption efforts. A direct extension of this mode in the line indicated above (large number of players and the levels of attacks) suggests to write down the reward of a terrorist, or a terrorist group, considered as a representative of a large number of small players, using one of the levels of attack $j = 1, \ldots, d$ (in [293] there are two levels, normal and spectacular only), to be

$$R_j(x,b) = (1 - p_j(x,b))r_j^{fail}(b) + p_j(x,b)(S_j + r_j^{succ}(b)), \qquad (12.16)$$

where $p_j(x,b)$ is the probability of a successful attack (which depends on the level b of preemptive efforts of the principal b and the total distribution of terrorists playing different strategies), S_j is the direct benefits in case of a success and $r_j^{fail}(b)$, $r_j^{succ}(b)$ are the recruitment benefits in the cases of failure or success respectively. The costs of principal are given by

$$B(x,b) = \sum_j x_j \left[(1 - p_j(x,b))b + p_j(b)(b + S_j) \right].$$

It is seen directly that we are again in the same situation as described by (12.14) (up to constants and notations). The model extends naturally to account for possibility of the actions of two types, preemption and deterrence. Also very natural can be the extension to several major players For instance, USA and EU cooperation in fighting terrorism was considered in [198].

In a large number of problems it is essential to work with an infinite state-space of small players, in particular, with the state-space being the set of all natural numbers. Mathematical results are much rare for this case, as compared with finite state-spaces. This infinite-dimensional setting is crucial for the analysis of models with growth, like merging banks or firms or the evolution of species and the development of networks with preferential attachment, for instance scientific citation networks or the network of internet links (see a detailed discussion in [267]). Apart from the obvious economic examples mentioned above, similar process of the growth of coalitions under pressure can be possibly used for modeling the development of human cooperation (forming coalitions under the 'pressure' exerted by the nature) or the creation of liberation armies (from the initially small guerillas groups) by the population of the territories oppressed by an external military force. Yet another example can represent the analysis of the financial support of various projects (developmental or scientific), when the funding body has tools to merge or split projects and transfer funds between them in certain periods of time.

These processes of growth have a clear physical analogs, say the formation of dimers and trimers by the molecules of gas with eventual condensation under (now real physical) pressure.

12.4   Minority game, its strategic enhancement and sex ratio game

An important example of the pressure and resistance framework represents the famous minority game designed to model the evolution of financial markets. The remarkable properties of the minority game were revealed partially by the extensive computer simulations and partially by the application of the methods of statistical mechanics.

Recall that in the *minority game* each of N players declares (or adopts) independently and simultaneously one of the two possible strategies, denoted 0 and 1. The payoff 1 is paid to all players that declared the strategy that turned out to be in minority, and other players pay fine of amount 1. In a rare event of equal amounts of declared two strategies, all payoffs can be set to zero. The game is supposed to be repeated many times, so that players can learn to apply better strategies.

Let n_0, n_1 denote the amount of players that have chosen 0 or 1 respectively in one step of the game. Then $n_0 + n_1 = N$. Let $x = n_0/N$. We are exactly in the situation described in equation (12.10) (without the major player) with $d = 2$ and the payoffs

$$
R_1(x) = -R_0(x) = \operatorname{sgn}(x - 1/2) = \begin{cases} 1, & x > 1/2 \\ 0, & x = 1/2 \\ -1, & x < 1/2. \end{cases} \tag{12.17}
$$

The only specific feature of this setting is the obvious discontinuity of the payoffs $R_j(x)$. However, for large N the chance of a draw is really negligible, so that the model will be not changed essentially if one takes a smooth version of the payoffs,

$$
R_1(x) = -R_0(x) = \theta(x - 1/2), \tag{12.18}
$$

where $\theta(x)$ is a smooth non-decreasing function such that $\theta(x) = \operatorname{sgn}(x)$ for $|x| \geq \epsilon$ with some small ϵ. The kinetic equations (12.3) can be written down in terms of a single variable x as

$$
\dot{x} = -x(1 - x)\theta(x - 1/2). \tag{12.19}
$$

Placing the minority game in this general setting suggests various reasonable extensions of this market model, for instance, by including major players (say, a market makers) that have tools to control the interactions or the payoffs of minor players, or by a more subtle grouping of players. Moreover, it reveals a nice link with the so-called *sex-ratio game* of evolutionary biology. The latter game is designed to explain the common $1/2$ probability for births of males and females in basically any population. The rules of the sex-ratio game arise from the observation that, if there are n_0 males and n_1 females in a population (its current generation), then producing a male (resp. female) offspring would yield to this offspring an expected number of n_1/n_0 (resp. n_0/n_1) of its own offsprings. Thus we find ourselves in the situation described by the rules above with payoff (12.17) having the form

$$R_1(x) = \theta(x - 1/2) = x/(1 - x). \tag{12.20}$$

12.5 Replicator dynamics with mean-field dependent payoffs: Controlled hawk and dove evolution

As a final example let us touch upon the applications to the well established theory of evolutionary games and evolutionary dynamics reducing the discussion to the variations on most popular concrete model.

Recall the celebrated *Hawk-and-Dove game* (see Chapter 4), which is a symmetric two-player game with the matrix of payoffs (the letters in the table denote the payoffs to the row player)

	hawk	dove
hawk	H	V
dove	0	D

The *replicator dynamics* is given by the equation

$$\dot{x} = x(1 - x)[x(H + D - V) - (D - V)],$$

which is of course a particular case of dynamics (12.3), (12.11). This evolution may have only one internal rest point (point of internal equilibria),

$$x^* = \frac{V - D}{V - D - H}.$$

If $V > D$, as is usually assumed, then $x^* \in (0, 1)$ if and only if $H < 0$. In this case, this point is known (and easily checked) to be a stable equilibrium of the replicator dynamics.

The replicator dynamics can be obtained as the dynamic LLN (law of large number) for a Markov model of N interacting species, which can use one of two strategies, h or d (hawk or dove), with each pair meeting randomly and playing the game described by the Table above. The payoff for each player in this game of N players, the fraction x (resp. $1 - x$) of which playing h (resp. d), is $Hx + V(1 - x)$ for being a hawk and $D(1 - x)$ for being a dove.

If there are only a few hawks and plenty of dove to rob, the hawks get less impetus to fight between themselves meaning that H may increase with the decay of x, the fraction of species playing h. Similarly, if there are too many doves on the restricted resource area, they may start behaving more aggressively meaning that D may decrease with the decay of x. Choosing linear functions to express this dependence in the simplest form, the modified table of this *nonlinear Hawk-and-Dove game* becomes

	hawk	dove
hawk	$H + ax$	V
dove	0	$D - bx$

with some constants $a, b > 0$.

In this setting the game of two players stops being defined independently, but the game of N players choosing one of the strategies h or d remains perfectly well defined with the payoffs of hawks and doves becoming

$$\Pi_h(x) = (H + ax)x + V(1 - x), \quad \Pi_d(x) = (D - bx)(1 - x).$$

The replicator dynamics generalizes to the equation

$$\dot{x} = x(1 - x)[x(H + ax + D - bx - V) - (D - bx - V)], \qquad (12.21)$$

which remains to be a particular case of dynamics (12.3), (12.11). The internal fixed points of this dynamics are the solutions of the quadratic equation

$$(a - b)x^2 - (V - D - H - b)x + (V - D) = 0. \qquad (12.22)$$

It is seen that, say for $a = b$, the internal equilibrium becomes

$$x^* = \frac{V - D}{V - D - H - b},$$

which moves to the left with the increase of b. When the denominator in this expression becomes negative, the internal equilibrium gets lost and the pure strategy d becomes the stable equilibrium. On the other hand, if $b = 0$, then the roots of the quadratic equation (12.22) have asymptotic expressions

$$x_1 = \frac{V-D}{V-D-H}\left(1 + a\frac{V-D}{(V-D-H)^2} + O(a^2)\right), x_2 = \frac{1}{a}(V-D-H) + O(1)$$

for small a. Therefore, for small a, the internal rest point starts moving go the right remaining the only internal rest point. When a increases various bifurcations can occur. For instance, if $-H < a < (V - D - H)/2$, the internal rest points disappear. If

$$\frac{V-D-H}{2} < a < \frac{(V-D-H)^2}{4(V-D)},$$

two internal fixed points appear (both solutions of (12.22) belong to $(0, 1)$), the left being stable, and the right unstable, so that in total we get two stable equilibria, one mixed and one pure hawk.

Similar shifts of equilibria and bifurcations occur when mean-field dependence is introduced in other popular two-player games. It is interesting to see whether this kind of bifurcations could be observed in real field experiments.

The model can be used to describe the transformations of behavioral patterns and social norms in society, how doves (social behavior) develop themselves into hawks (aggressive behavior, violence) and vice versa. Moreover, making the parameter a, b depending on the some control parameter u of a major player allows one to introduce the model of controlled hawk and dove evolution, with the principal having a tool to drive the equilibria in the direction desired.

12.6 Mean field games (MFG) for finite-state models

We introduce now the MFG framework for discrete state-spaces. As was already mentioned, the difference between the present MFG setting and the previous models is that now the small agents themselves become rational optimizers and are not supposed just to follow some prescribed deterministic or stochastic strategies (like myopic behavior).

Let us start by recalling briefly the theory of controlled Markov chains. Let $Q = \{Q_{ij}(t, u)\}$ be a family of Q-matrices, $i, j \in \{1, \ldots, d\}$, depending continuously on parameter $u \in U \subset \mathbf{R}^n$ and piecewise continuously on time t. By a (Markov) strategy we mean any piecewise continuous functions $\hat{u}_j(t)$. A controlled Markov chain specified by such strategy is the Markov chain $X_{s,i}^{\hat{u}}(t)$ with the Q-matrices $Q_{ij}(t, \hat{u}_i(t))$. Suppose one gets a payoff $J(t, j, u)$ per unit of time staying at j with control u around time t and the terminal payoff $S_T(j)$, paid at time T if the process terminates in j at time T. The corresponding *Markov control problem* is to find the maximal total payoff

$$S(t, j) = \max_{\hat{u}} \mathbf{E}[\int_t^T J(s, j(s), \hat{u}_{j(s)}(s)) ds + S_T(X_{t,x}^{\hat{u}}(T))],$$

where $j(s) = X_{t,j}^{\hat{u}}(s)$, and a strategy, called an *optimal strategy*, where this maximum is attained.

It is known that $S(t, j)$ solves the backward (evolving from the terminal time T backward) *Bellman equation*:

$$\frac{\partial S(t, j)}{\partial t} + \max_u [J(t, j, u) + \sum_{k=1}^d Q_{jk}(t, u) S(t, k)] = 0, \qquad (12.23)$$

with the terminal condition $S(T, j) = S_T^j$.

The standard heuristic derivation of this equation goes as follows. Assuming S is smooth, and taking into account that our Markov chain can have only one jump during a small period of time τ, we can write approximately

$$S(t, j) = \max_u [J(t, j, u)\tau + \tau|Q_{jj}| \sum_{k \neq j} \frac{Q_{jk}}{|Q_{jj}|} S(t+\tau, k) + (1 - \tau|Q_{jj}|) S(t+\tau, j)].$$

This way of expressing optimal payoff at time t via its values at future times is often referred to in general as the *principle of optimality*. Expanding S in the Taylor series and keeping only terms linear in τ (free terms $S(t, j)$ cancel) yields (12.23).

Rigorous justification is usually performed by first proving the well-posedness of the Bellman equation and then showing that its solution supplies the required maximum (the so-called *verification theorem* arguments).

Suppose now we have a family $Q = \{Q_{ij}(x, u)\}$ of Q-matrices, $i, j \in \{1, \ldots, d\}$, depending continuously on parameter $u \in U$ and Lipshitz continuously on $x \in \Sigma_d$. Suppose there are N players, each moving according to Q and aiming at maximizing the payoff

$$\mathbf{E}[\int_t^T J(s, j(s), x(s), u(s)) ds + S_T(j(T))], \qquad (12.24)$$

where $j(s)$ is the position of the player at time s. Here the motions of all players are coupled, as all transitions depend on the total distribution of players $x = (n_1, \ldots, n_d)/N$, where n_j denote the number of players in states j, at any given time.

The *MFG methodology* suggests the following solution concept for this problem for large N. Let an evolution of the distributions $x(t)$, $t \in [0, T]$, is a known continuous curve. Then any given player should search for the maximal payoff

$$S(t, j) = \max_{\hat{u}} \mathbf{E}[\int_t^T J(s, j(s), x(s), \hat{u}_{j(s)}(s)) \, ds + S_T(j(T))],$$

where $j(s) = X_{s,j}^{\hat{u}}(s)$ is the Markov chain with the Q-matrices $Q_{ij}(x(t), \hat{u}_i(t))$. As follows from (12.23), $S(t, i)$ should satisfy the backward Bellman equation

$$\frac{\partial S(t, j)}{\partial t} + \max_u [J(t, j, x(t), u) + \sum_k Q_{jk}(x(t), u)S(t, k)] = 0, \qquad (12.25)$$

with the terminal condition $S(T, j) = S_T(j)$. Assume we can find a solution $S(t, x)$ and the corresponding optimal strategy $\hat{u}_j(t) = \hat{u}_j(t, x(t))$ providing max in this equation at any time t. Now, if all players are using this optimal strategy, the mean-field interacting particle system of N players given by the Q-matrices

$$\hat{Q}_{ij}(t, x) = Q_{ij}(x, \hat{u}_i(t))$$

converge (see discussion around (12.3)), to the solutions $X_{0,x(0)}(t)$ of the system of (forward) kinetic equations

$$\dot{x}_k = \sum_{i \neq k} [x_i Q_{ik}(x, \hat{u}_i(t))) - x_k Q_{ki}(x, \hat{u}_k(t)))] = \sum_{i=1}^d x_i Q_{ik}(x, \hat{u}_i(t))), \, k = 1, \ldots, d.$$
$$(12.26)$$

Let $\hat{x}(t)$ be the solution of this system with the initial condition $\hat{x}(0) = x(0)$. The consistency between the controlled dynamics and the mean-field evolution can be naturally described by the requirement that $\hat{x}(t) = x(t)$. This is exactly the forward-backward *MFG consistency problem* or MFG consistency condition, also referred by some authors as *Nash-MFG equilibrium*. Equivalently starting with a control $u_j(t)$ we can solve the corresponding kinetic equation finding the distribution $x(t)$ and then find the corresponding optimal control $\hat{u}_j(t)$ fitting to the solution of the HJB (12.25). The MFG consistency condition can then be expressed by the equation $u(t) = \hat{u}(t)$. In any case it can be expressed by saying that the pair $(\hat{x}(t), \hat{u}_j(t))$ provides a solution to the coupled *forward-backward system*

(12.25)-(12.26), more precisely, to its *initial-terminal value problem* (initial x_0 is given for (12.26) and terminal S_T for (12.25)).

One can expect that solutions to MFG consistency problem should provide some approximations to the solutions of the games of N players, where each player is trying to maximize (12.24) (for a very simple version of such result see [263]). Thus we are led to two basic problems of MFG theory: describe the solutions to the MFG consistency problem (say, prove existence and/or uniqueness theorem) and provide exact link between these solutions and the corresponding original game of a finite number of players. The latter task can be discussed with two approaches (often requiring different techniques): showing that Nash equilibria of games with N players converge to a solution of the MFG consistency problem, or showing that the solutions of the MFG consistency problem yield approximate Nash equilibria for finite-player games.

An alternative approach to the MFG consistency problem that arises from looking directly at the limiting evolution of the pair $(j(t), x(t))$, where $x(t)$ evolves according to the kinetic equations,

$$\dot{x}_k = \sum_{i=1}^{d} x_i Q_{ik}(x(t), \hat{u}_i(t)), \quad k = 1, \ldots, d,$$

and $j(t) = X_{s,x}^{\hat{u}}(t)$ is the Markov chain with the Q-matrices $Q_{ij}(x(t), \hat{u}_i(t))$, as at the controlled Markov process in the continuous state space $\{1, \ldots, d\} \times \Sigma_d$. This is, strictly speaking, not a chain anymore, as it evolves by jumps and continuous displacements in the continuous state space. Nevertheless, controlling the process with the objective to maximize (12.24) we find for the optimal payoff the Bellman equation in the form

$$\frac{\partial S(t, j, x)}{\partial t} + \max_{u_1, \ldots, u_d} \big[J(t, j, x, u_j)$$

$$+ \sum_k Q_{jk}(x, u_j) S(t, k, x) + \sum_{k,i} \frac{\partial S(t, j, x)}{\partial x_k} x_i Q_{ik}(x, u_i) \big] = 0. \quad (12.27)$$

This is obtained analogously to (12.23) from the approximate equation

$$S(t, j, x) = \max_{u_1, \ldots, u_d} \big[\tau J(t, j, x, u_j) + \tau |Q_{jj}| \sum_{k \neq j} \frac{Q_{jk}(x, u_j)}{|Q_{jj}|} S(t+\tau, k, X_{x,t}(t+\tau))$$

$$+ (1 - \tau |Q_{jj}|) S(t + \tau, j, X_{x,t}(t + \tau)) \big],$$

or, discarding the higher terms in τ,

$$S(t, j, x) = \max_{u_1, \ldots, u_d} \big[J(t, j, x, u) \tau$$

$$+\tau \sum_{k \neq j} Q_{jk} S(t,k,x) + (1 - \tau |Q_{jj}|) S(t+\tau, j, X_{x,t}(t+\tau)))],$$

by using the first order Taylor approximation

$$S(t+\tau, j, X_{x,t}(t+\tau)) = S(t,j,x) + \tau \frac{\partial S}{\partial t}(t,j,x) + \tau \sum_{k,i} \frac{\partial S}{\partial x_k}(t,j,x) x_i Q_{ik}(x, u_i).$$

Equation (12.27) is called the *master equation* (in backward form). The next statement shows that this equation provides (at least in case it is reasonably well-posed) an alternative approach to the analysis of the MFG consistency problem, which selects the most effective solutions to backward-forward systems, thus forming an envelop for various solutions of the MFG consistency problem.

Proposition 12.1. *Let $S(t,j,x)$ be a smooth function solving equation (12.27) with the terminal condition $S_T(j)$ and giving the optimal payoff in the Markov decision problem on $\{1,\ldots,d\} \times \Sigma_d$ used to derive this equation. Let it be possible to choose Lipschitz continuous (in x) selector $\tilde{u}_j(t,x)$ giving maximum in (12.27) (for this solution S) and hence to build the corresponding trajectories $X_{x_0}(t)$ solving the kinetic equations*

$$\dot{x}_j = \sum_i x_i Q_{ij}(x, \tilde{u}_i(t,x))$$

with any initial x_0.

Then the pair $(X_{x_0}(t), \tilde{u}_j(t, X_{x_0}(t)))$ is a solution to the MFG consistency problem and

$$\tilde{S}(t,j) \leq S(t,j,x_0), \tag{12.28}$$

for the payoff $\tilde{S}(t,j)$ on any other solution $(\hat{x}(t), \hat{u}_j(t))$ to the forward-backward MFG consistency problem with $\hat{x}(t) = X_{x_0}(t)$.

Proof. By the definition of $S(t,j,x)$ as the solution to a Markov decision problem in $\{1,\ldots,d\} \times \Sigma_d$, $S(t,j,x_0)$ is not less than the payoff that can be obtained by any player using any symmetric strategy (the same as all other players) given the dynamics $X_{x_0}(t)$ of the total distributions, implying (12.28). It follows that the payoff $S(t,j,x_0)$ cannot be improved by changing the strategies inside the class of symmetric strategies. Consequently, $\tilde{u}_i(t, x(t))$ provides the maximum payoff in the class of these strategies and hence provides a solution to the forward-backward MFG consistency problem. ∎

An important extension of the above MFGs includes a major player. Namely, let us assume that the transition matrices $Q = Q(x, u, b)$ and the payoffs $J(t, j, x, u, b)$ depend additionally on a parameter b controlled by the major player (a principal). If the principal is playing just the best response $b^*(x)$ we are directly back to the original problem with $Q = Q(x, u, b^*(x))$. However, if the major player choose b strategically aiming at maximizing some payoff of the general type

$$\int_t^T B(x(s), b(s))\, ds + V_T(x(T)),$$

the situation becomes different.

In this case the *MFG methodology* works as follows. If the evolution of the distributions $x(t)$, $t \in [0, T]$, is a known continuous curve, the major player finds the optimal strategy $\hat{b}(t)$ and, based on this strategy, any given player should search for the maximal payoff

$$S(t, i) = \max_{\hat{u}} \mathbf{E}\Big[\int_t^T J(s, j(s), x(s), \hat{u}_{j(s)}(s), \hat{b}(s))dt + S_T(j(T))\Big],$$

where $j(t) = X^{\hat{u}}_{s,x}(t)$ is the Markov chain with the Q-matrices $Q_{ij}(x(t), \hat{u}_i(t), \hat{b}(t))$. The Bellman equation for the optimal payoff of each small player $S(t, i)$ takes the form

$$\frac{\partial S(t, j)}{\partial t} + \max_u [J(t, j, x(t), u, \hat{b}(t)) + \sum_k Q_{jk}(x(t), u, b(t))S(t, k)] = 0.$$

$$(12.29)$$

After finding a solution $S(t, x)$ and the corresponding optimal strategy $\hat{u}_j(t) = \hat{u}_j(t, x(t))$ providing max in this equation at any time t we can solve the corresponding kinetic equations

$$\dot{x}_k = \sum_{i=1}^d x_i Q_{ik}(x, \hat{u}_i(t)), \hat{b}(t)), \quad k = 1, \ldots, d. \qquad (12.30)$$

Let $\hat{x}(t)$ be the solution of this system with the initial condition $\hat{x}(0) = x(0)$. The *MFG consistency problem* or *MFG consistency condition with major player* can be expressed by the equation $\hat{x}(t) = x(t)$.

MFG were introduced by Lasry-Lions [269] and Huang-Malhame-Caines [241], [242]. A deep theory developed so far is reflected already in several books and surveys including [210], [238], [236], [217], and the fundamental 2 volume monograph [218]. Many concrete models with finite state spaces are analysed in [263].

Chapter 13

Quantum games

Some basic knowledge of quantum mechanics is a necessary prerequisite
for studying quantum games. The most fundamental facts are collected
in Section 13.1 describing closed finite-dimensional systems and comple-
mented by the rules of quantum measurements in Section 13.2. These facts
are sufficient for the main trend of our introductory exposition. Essentially, they can be looked at as a linear algebra recast in physics terms and
physics notations. However, for further developments of quantum games
one needs some notions of open quantum systems. They are presented in
Section 13.9, to which we refer occasionally, and which supplies the background that is necessary for reading modern research papers. The main
Sections 13.3-13.7 build the foundations of quantum games. The main papers on this topic are written by physicists for physicists, and the inevitable
exploitation of physics jargon may create difficulties for mathematicians or
economists. Our main goal here is to make clear the physical content and to
stress the new features of the games that may be revealed in their quantum
versions, in particular, the new cooperative equilibria. We omit sometimes
the lengthy calculations (referring to the original papers) once the physical
part is sorted out and the problem is reformulated as pure game-theoretic
problem of calculating the Nash or dominated equilibria. Section 13.8 is devoted to quantum games arising from the classical games with infinite state
space (like the classical Cournot duopoly). Section 13.10 touches upon general theory of finite quantum static games. Further links and references are
provided in Section 13.11.

13.1   On finite-dimensional quantum mechanics

Finite-dimensional quantum systems are described by finite-dimensional complex Euclidean or Hilbert spaces $H = \mathbf{C}^n$, equipped with the standard scalar product $(.,.)$, which is usually considered to be *anti-linear* or *conjugate linear* with respect to the first argument, $(au, v) = \bar{a}(u, v)$ for $a \in \mathbf{C}$, and linear with respect to the second one: $(v, au) = a(v, u)$. For any orthonormal basis $e_1, \ldots, e_n$ in H, the scalar product writes down in coordinates as $(u, v) = \sum_j \bar{u}_j v_j$, where $u = \sum_j u_j e_j$, $v = \sum_j v_j e_j$. The usual Euclidean norm is $\|x\| = \sqrt{(x, x)}$. These systems are referred to as *qubits*, *qutrits* in case of $n = 2$ or 3, and *qunits* for general n (or rather *qudits* with general d).

Pure states of a quantum system described by such space H are unit vectors in H.

Remark 13.1. More precisely, two vectors that differ by a multiplier are considered to describe the same state, so that the state space is the projective space $\mathbf{C}P^{k-1}$ of the equivalence classes of vectors with equivalence defined as proportionality.

The space of all $n \times n$ matrices $A = (A_{ij})$ can be considered as the space of all linear operators $\mathcal{L}(\mathbf{C}^n)$ in $\mathbf{C}^n$. This correspondence is described equivalently either via the action on the basis vectors as $Ae_j = \sum_k A_{kj} e_k$, or in coordinates as $(Au)_j = \sum_k A_{jk} u_k$. $\mathcal{L}(\mathbf{C}^n)$ is a space of dimension n^2 with the usual operator norm defined as

$$\|A\| = \sup_{\|x\|=1} \|Ax\|. \tag{13.1}$$

Similarly the space $\mathcal{L}(H_1, H_2)$, $H_2 = \mathbf{C}^m$, of linear operators $H_1 \to H_2$ can be identified with the space of $m \times n$ matrices. It becomes a norm space under the standard operator norm (13.1), where $\|x\|$ is the norm in H_1 and $\|Ax\|$ is the norm in H_2.

The subspace of $\mathcal{L}(\mathbf{C}^n)$ consisting of self-adjoint or Hermitian matrices, defined by the equation $A^* = A$, where $A^* = \bar{A}^T$ (T for transpose and the bar for complex conjugation) will be denoted $\mathcal{L}_s(\mathbf{C}^n)$. The trace of $A \in \mathcal{L}(\mathbf{C}^n)$ is defined as $\operatorname{tr} A = \sum_j A_{jj}$. It implies that $\operatorname{tr}(AB) = \operatorname{tr}(BA)$, which in turn implies that $\operatorname{tr}(C^{-1}AC) = \operatorname{tr} A$ for any invertible C. Consequently one obtains two other equivalent expressions for the trace of $A \in \mathcal{L}_s(\mathbf{C}^n)$: $\operatorname{tr} A = \sum_j \lambda_j$, where $\{\lambda_j\}$ is the collection of all eigenvalues of A and

$$\operatorname{tr} A = \sum_j (e_j, Ae_j),$$

where $\{e_j\}$ is an arbitrary orthonormal basis in $\mathbf{C}^n$.

The space $\mathcal{L}(\mathbf{C}^n)$ is a Hilbert space with respect to the scalar product

$$(A, B)_{HS} = \operatorname{tr}(A^* B) = \sum_{i,j} \bar{A}_{ij} B_{ij}, \tag{13.2}$$

with the corresponding norm called the *Hilbert-Schmidt norm*

$$\|A\|_{HS} = [\operatorname{tr}(A^* A)]^{1/2}.$$

For Hermitian operators it simplifies to $\operatorname{tr}(A^* B) = \operatorname{tr}(AB)$.

Any $A \in \mathcal{L}_s(\mathbf{C}^n)$ is diagonalizable, meaning that there exists a unitary U such that $A = U^* D U$, where $D = D(\lambda_1, \ldots, \lambda_n)$ is diagonal with the eigenvalues $\{\lambda_j\}$ of A on the diagonal. Then $|A|$ is defined as the positive operator $|A| = U^* |D| U$ with $|D| = D(|\lambda_1|, \ldots, |\lambda_n|)$, the diagonal operator with the numbers $\{|\lambda_j|\}$ on the diagonal. The functional $A \mapsto \operatorname{tr}(|A|)$ defines yet another norm on $\mathcal{L}_s(\mathbf{C}^n)$, the *trace norm*:

$$\|A\|_{tr} = \operatorname{tr}(|A|) = \sum_j |\lambda_j|.$$

The key point is that this norm is dual to the usual operator norm with respect to the duality provided by the trace: for $A \in \mathcal{L}_s(\mathbf{C}^n)$,

$$\|A\|_{tr} = \sup_{\|B\|=1} |\operatorname{tr}(AB)|, \quad \|A\| = \sup_{\|B\|_{tr}=1} |\operatorname{tr}(AB)|. \tag{13.3}$$

To show these equations it is handy to work in the basis where A is diagonal: $A = D(\lambda_1, \ldots, \lambda_n)$. Then $\operatorname{tr}(AB) = \sum_j \lambda_j B_{jj}$. Choosing B to be diagonal with B_{ij} equal the sign of λ_j it follows that

$$\sup_{\|B\|=1} |\operatorname{tr}(AB)| \geq \sum_j |\lambda_j| = \|A\|_{tr}.$$

On the other hand,

$$\sup_{\|B\|=1} |\operatorname{tr}(AB)| = \sup_{\|B\|=1} |\sum_j \lambda_j B_{jj}| \leq \sum_j |\lambda_j| \max_j |B_{jj}| \leq \sum_j |\lambda_j|.$$

Therefore the first equation in (13.3) is proved. The second equation is proved similarly.

For two spaces $H_1 = \mathbf{C}^n$ and $H_2 = \mathbf{C}^m$ with orthonormal bases $e_1, \ldots, e_n$ and $f_1, \ldots, f_m$ the *tensor product space* $H_1 \otimes H_2$ can be defined as the nm-dimensional space generated by vectors denoted $e_j \otimes f_k$, so that any $\psi \in H_1 \otimes H_2$ can be represented as

$$\psi = \sum_{j=1}^{n} \sum_{k=1}^{m} a_{jk} e_j \otimes f_k. \tag{13.4}$$

The *tensor product* of any two-vectors $u = \sum_j u_j e_j$, $v = \sum_k v_k f_k$ is defined as the vector

$$u \otimes v = \sum_{j,k} u_j v_k e_j \otimes f_k.$$

Similarly, the tensor product is defined for several Hilbert spaces. Namely, the product $H = H_1 \otimes H_2 \otimes \cdots \otimes H_K$ of spaces of dimensions $n_1, \ldots, n_K$ can be described as the $n = \prod n_j$-dimensional space generated by vectors denoted $e_{j_1}^1 \otimes \cdots \otimes e_{j_K}^K$ and called the tensor products of $e_{j_1}^1, \ldots, e_{j_K}^K$, where $\{e_1^m, \ldots, e_{n_m}^m\}$ is an orthonormal basis in H_m.

Any $\psi \in H_1 \otimes H_2$ can be represented by the so-called *Schmidt decomposition*

$$\psi = \sum_{j=1}^{\min(n,m)} \lambda_j \xi_j \otimes \eta_j, \tag{13.5}$$

where $\lambda_j \geq 0$, and $\{\xi_j\}$ and $\{\eta_j\}$ are some orthonormal bases in H_1 and H_2. In fact, one just has to write the singular decomposition of the matrix $A = (a_{jk})$ from (13.4), namely to represent it as $A = UDV^T$, where $D = D(\lambda_1, \lambda_2, \ldots)$ is diagonal of dimension $m \times n$ and U, V are unitary matrices. Here λ_j^2 are the (common) eigenvalues of the matrices A^*A and AA^* with $\lambda_j \geq 0$. Then the vectors $\xi_j = U_{lj} e_l$ and $\eta_k = V_{lj} f_l$ form orthonormal bases in H_1 and H_2 and

$$\psi = \sum_{j,k,l} U_{jl} \lambda_l V_{kl} e_j \otimes f_k$$

equals (13.5).

Pure states of the tensor product $\psi \in H_1 \otimes H_2$ are called *entangled* (the term introduced by E. Schrödinger in 1935) if they cannot be written in the product form $\psi = u \otimes v$ with some u, v. It is seen that ψ is not entangled if and only if its Schmidt decomposition has only one nonzero term. On the other hand, a pure state $\psi \in H_1 \otimes H_2$ is called *maximally entangled*, if its Schmidt decomposition has the maximal number of nonvanishing λ_j and they all are equal, and thus they equal $1/\sqrt{\min(n,m)}$.

In physics one usually works in *Dirac's notations*. In these notations usual vectors $u = \sum_j u_j e_j \in H$ are referred to as *ket-vectors*, are denoted $|u\rangle$ and are considered to be column vectors with coordinates u_j. The corresponding *bra-vectors* are denoted $\langle u|$ and are considered to be row vectors with coordinates $\bar{u}_j$. These notations are convenient, because they allow to represent both scalar and tensor products as usual matrix multiplications: for two ket-vectors $|u\rangle$ and $|v\rangle$ we have

$$\langle u|v\rangle = \langle u|.|v\rangle = (u,v) = \sum_j \bar{u}_j v_j,$$

$$|v\rangle\langle u| = v \otimes \bar{u} = \sum_{j,k} v_j \bar{u}_k e_j \otimes e_k.$$

Therefore the latter product is often identified with the $n \times n$-matrix ρ with the entries $\rho_{jk} = v_j \bar{u}_k$. As matrices, they act on vectors $w = |w\rangle$ in the natural way:

$$\rho w = |v\rangle\langle u|w\rangle, \quad (\rho w)_j = \sum_k \rho_{jk} w_k = v_j \sum_k \bar{u}_k w_k = v_j \langle u|w\rangle.$$

On the other hand, the bra-vectors form the space H^* of linear functionals on H specified via the scalar product.

Remark 13.2. It is worth stressing that the operation of conjugation (usually denoted by bar or a star) in a Hilbert space H (defined as an anti-linear convolution map $A : H \to H$, the latter meaning that $A^2 = \mathbf{1}$) is not unique and depends on a chosen 'real' basis. For instance, in $\mathbf{C} = \mathbf{R}^2$, a reflection with respect to any real line in $\mathbf{C}$ (chosen to be real in $\mathbf{C}$), or analytically, any map of type $v \to \bar{v}e^{i\phi}$ with any real ϕ defines such a convolution.

Continuing the analogy we see that the tensor product $H_1 \otimes H_2^*$ is naturally isomorphic to the space of linear operators $\mathcal{L}(H_2, H_1)$. Namely, for orthnormal bases $\{e_i\}, \{f_j\}$ in H_1 and H_2 any

$$X = \sum_{i,j} X_{ij} e_i \otimes \bar{f}_j \in H_1 \otimes H_2^*$$

can be identified with the operator

$$X = \sum_{ij} X_{ij} |e_i\rangle\langle f_j| : f_k \mapsto \sum_i X_{ik} |e_i\rangle \tag{13.6}$$

with the matrix X_{ij}. These matrix elements can be written in two equivalent forms:

$$X_{ij} = \langle e_i|Xf_j\rangle = \text{tr}(X|f_j\rangle\langle e_i|). \tag{13.7}$$

General or mixed states, also referred to as *density matrices* or *density operators* of a qunit are defined to be non-negative $n \times n$-matrices ρ with unit trace: $\text{tr}\,\rho = 1$. The *state space of a qunit* is usually defined either as the set of all density matrices (which is not a linear space) or the linear space

Since for any operator A such that $A^2 = \mathbf{1}$, we have

$$\exp\{ixA\} = \cos x \mathbf{1} + i \sin x\, A,$$

for any real x, it follows that

$$e^{ibZ} = \begin{pmatrix} e^{ia} & 0 \\ 0 & e^{-ia} \end{pmatrix}, \quad e^{icY} = \begin{pmatrix} \cos c & \sin c \\ -\sin c & \cos c \end{pmatrix}, \qquad (13.13)$$

so that

$$e^{ibZ}e^{icY}e^{idZ} = \begin{pmatrix} e^{i(b+d)}\cos c & e^{i(b-d)}\sin c \\ -e^{-i(b-d)}\sin c & e^{-i(b-d)}\cos c \end{pmatrix}. \qquad (13.14)$$

Comparing with (13.12) we see that (13.14) is yet another way to represent arbitrary element of $SU(2)$. This way is referred to as the $Z - Y$ decomposition or the *Cartan decomposition* of the elements of $SU(2)$.

Finally, (13.11) can be rewritten as

$$U = \cos\phi\cos\theta\mathbf{1} + i\sin\psi\sin\theta\sigma_x + i\cos\psi\sin\theta\sigma_y + i\sin\phi\cos\theta\sigma_z, \quad (13.15)$$

and thus

$$U = u_0\mathbf{1} + iu_1\sigma_1 + iu_2\sigma_2 + iu_3\sigma_3, \qquad (13.16)$$

with real u_k satisfying the condition $u_0^2 + u_1^2 + u_2^2 + u_3^2 = 1$. this representation shows that topologically $SU(2)$ is a unit sphere in $\mathbf{R}^4$.

Any element A of $\mathcal{T}_s(\mathbf{C}^2)$ can be written in unique way as

$$A = x_0 I + x_1\sigma_1 + x_2\sigma_2 + x_3\sigma_3 \qquad (13.17)$$

with real x_j.

Let us see how $u \in SU(2)$ act on $\mathcal{T}_s(\mathbf{C}^2)$ via the dressing $A \mapsto UAU^*$: $e^{i\phi\sigma_3}$ leaves σ_3 invariant and acts on σ_1, σ_2 as rotation with the matrix

$$\begin{pmatrix} \cos(2\phi) & -\sin(2\phi) \\ -\sin(2\phi) & \cos(2\phi) \end{pmatrix};$$

and similarly, $e^{i\psi\sigma_2}$ leaves σ_2 invariant and acts on σ_1, σ_3 as rotation with the matrix

$$\begin{pmatrix} \cos(2\psi) & -\sin(2\psi) \\ -\sin(2\psi) & \cos(2\psi) \end{pmatrix}.$$

But these rotations generate the group $SO(3)$ in $\mathbf{R}^3$ with the basis $\sigma_1, \sigma_2, \sigma_3$. Hence any rotation in this space can be achieved via dressing with certain $u \in SU(2)$.

How to represent elements of $O(3) \setminus SO(3)$ via dressing? Here one needs anti-unitary operators.

A mapping $A : H \to H$ is called *ani-linear* or *conjugate-linear* if $A(v + w) = Av + Aw$ for $v, w \in H$ and $A(av) = \bar{a}v$, $a \in \mathbf{C}$. The simplest example is the equivalence between bra and ket vectors: $|x\rangle \mapsto \langle x|$, or equivalently, just the mapping $\psi \to \bar{\psi}$.

To any linear map $A : H \to H$ there corresponds an anti-linear map $\tilde{A} : v \mapsto A \circ s(v)$, where s is the conjugation: $s(v) = \bar{v}$. If A, B and ρ are linear operators, then $\tilde{A}\rho\tilde{B}$ is also a linear operator given by the matrix $A\bar{\rho}\bar{B}$.

An anti-linear map A is called *anti-unitary* if $(Av, Aw) = (w, v) = (\bar{v}, \bar{w})$. It is seen that U is unitary if and only if $\tilde{U} = U \circ s$ is anti-unitary, and for any two unitary operators U, V the operators $\tilde{U} \circ V$ and $U \circ \tilde{V}$ are anti-unitary. The famous *Wigner theorem* states that any mapping U in a Hilbert space of dimension $d > 2$ such that $|(Uu, Uv)| = |(u, v)|$ for all vectors u, v is either unitary or anti-unitary.

Exercise 13.2. As an elementary version of the Wigner theorem show that any mapping $M : \mathbf{R}^n \to \mathbf{R}^n$, which either preserves the scalar product or is continuous and preserves the magnitude of the scalar product, is necessarily a linear orthogonal operator.

If $\tilde{U} = U \circ s$ is an anti-unitary operator in $\mathbf{C}^2$, then the mapping

$$\rho \mapsto \tilde{U}\rho\tilde{U}^{-1} = U\bar{\rho}\overline{U^{-1}} = U\bar{\rho}U^T \qquad (13.18)$$

acting on matrices (13.17) preserves x_0 and transforms $x = (x_1, x_2, x_3)$ to Ox with some orthogonal operator O with $\det O = -1$; and vice versa, any such transformation O can be obtained in this way. To prove this claim it is sufficient to show that such representation is available for the reflection $R : (x_1, x_2, x_3) \mapsto (x_1, -x_2, x_3)$, because any orthogonal transformation O with $\det O = -1$ can be written as $O = R \circ S$ with $S \in SO(3)$. The reflection R can be obtained from the anti-unitary operator $\tilde{U}$ with $U = e^{ibZ}$ from (13.13). In fact, we see by (13.18) that

$$\tilde{U}\rho\tilde{U}^{-1} = \begin{pmatrix} e^{ia} & 0 \\ 0 & e^{-ia} \end{pmatrix} \bar{\rho} \begin{pmatrix} e^{ia} & 0 \\ 0 & e^{-ia} \end{pmatrix}$$

and hence

$$\tilde{U}\sigma_1\tilde{U}^{-1} = \sigma_1, \quad \tilde{U}\sigma_2\tilde{U}^{-1} = -\sigma_2, \quad \tilde{U}\sigma_3\tilde{U}^{-1} = \sigma_3.$$

For two qubits one has a straightforward criterion for states to be entangled. In fact, if $\psi = \psi_0|0\rangle + \psi_1|1\rangle$, $\phi = \phi_0|0\rangle + \phi_1|1\rangle$, then

$$\psi \otimes \phi = \psi_0\phi_0|00\rangle + \psi_0\phi_1|01\rangle + \psi_1\phi_0|10\rangle + \psi_1\phi_1|11\rangle, \qquad (13.19)$$

where we use the standard notation for the products of the basis vectors:

$$|00\rangle = |0\rangle \otimes |0\rangle, \ |01\rangle = |0\rangle \otimes |1\rangle, \ |10\rangle = |1\rangle \otimes |0\rangle, \ |11\rangle = |1\rangle \otimes |1\rangle. \ (13.20)$$

An arbitrary state in $\mathbf{C}^2 \otimes \mathbf{C}^2$ can be written as

$$\xi = \xi_{00}|00\rangle + \xi_{01}|01\rangle + \xi_{10}|10\rangle + \xi_{11}|11\rangle. \qquad (13.21)$$

Comparing with (13.19) it is seen that (13.21) is not entangled (is a product state) if and only if

$$\xi_{00}\xi_{11} = \xi_{10}\xi_{01}. \qquad (13.22)$$

Apart from the standard basis (13.20) a key role in application is played by the *Bell basis* consisting of fully entangled states:

$$\frac{1}{\sqrt{2}}(|00\rangle + |11\rangle), \ \frac{1}{\sqrt{2}}(|00\rangle - |11\rangle), \ \frac{1}{\sqrt{2}}(|01\rangle + |10\rangle), \ \frac{1}{\sqrt{2}}(|01\rangle - |10\rangle). \ (13.23)$$

Specific role belongs also to the last vector $(|01\rangle - |10\rangle)/\sqrt{2}$, referred to as the *singleton*, because it is rotationally invariant. Namely, the group $SU(2)$ acts naturally in $\mathbf{C}^2 \otimes \mathbf{C}^2$ as $u(\phi \otimes \psi) = u(\phi) \otimes u(\psi)$, and the vector $(|01\rangle - |10\rangle)/\sqrt{2}$ is invariant under this action.

Exercise 13.3. Check this invariance.

Remark 13.3. The above mentioned action of $SU(2)$ on $\mathbf{C}^2 \otimes \mathbf{C}^2$ decomposes into the direct sum of two irreducible representations, one-dimensional one generated by $(|01\rangle - |10\rangle)/\sqrt{2}$ and three-dimensional one, generated by other tree vectors of the Bell basis. Therefore these three vectors are referred to as a triplet.

Up to the phase shifts (that is, up to multiplications by numbers of unit magnitude), arbitrary orthonormal basis in a qubit can be given by the vectors

$$e_0^{\beta\phi} = \begin{pmatrix} \cos(\beta/2) \\ \sin(\beta/2)e^{i\phi} \end{pmatrix}, \quad e_1^{\beta\phi} = \begin{pmatrix} -\sin(\beta/2)e^{-i\phi} \\ \cos(\beta/2) \end{pmatrix}, \qquad (13.24)$$

which can be obtained by acting on the standard basis by the operator $U_{\beta\phi}$:

$$U_{\beta\phi} = \begin{pmatrix} \cos(\beta/2) & -\sin(\beta/2)e^{-i\phi} \\ \sin(\beta/2)e^{i\phi} & \cos(\beta/2) \end{pmatrix}, \quad U_{\beta\phi}e_j = e_j^{\beta\phi}, \ j = 0, 1.$$

These vectors are eigenvectors of the operator

$$S_{\beta\phi} = U_{\beta\phi}\sigma_3 U^*_{\beta\phi}$$

$$= \begin{pmatrix} \cos\beta & \sin\beta e^{-i\phi} \\ \sin\beta e^{i\phi} & -\cos\beta \end{pmatrix} = \sin\beta\cos\phi\sigma_x + \sin\beta\sin\phi\sigma_y + \cos\beta\sigma_z,$$

which can be considered as the projection of the matrix-valued *spin-vector* $\sigma = (\sigma_x, \sigma_y, \sigma_z)$ on the unit vector $n = (\sin\beta\cos\phi, \sin\beta\sin\phi, \cos\beta) \in \mathbf{R}^3$, and therefore referred to as the *component of the spin in the direction n*.

This basis can be used to demonstrate that the Schmidt decomposition (13.5) is not unique, but the notion of maximally entangled state is still well defined. In fact, one sees directly that maximally entangled vector $(|00\rangle + |11\rangle)/\sqrt{2}$ can be written analogously in other bases $e_j^{\beta\phi}$:

$$\frac{1}{\sqrt{2}}(|00\rangle + |11\rangle) = \frac{1}{\sqrt{2}}\left(e_0^{\beta\phi} \otimes \overline{e_0^{\beta\phi}} + e_1^{\beta\phi} \otimes \overline{e_1^{\beta\phi}}\right). \tag{13.25}$$

Also another way to express the rotation invariance of the singleton state is to observe that it has the same form when expressed in any pair (13.24):

$$\frac{1}{\sqrt{2}}(|01\rangle - |10\rangle) = \frac{1}{\sqrt{2}}\left(e_0^{\beta\phi} \otimes e_1^{\beta\phi} - e_1^{\beta\phi} \otimes e_0^{\beta\phi}\right). \tag{13.26}$$

Exercise 13.4. Check (13.25) and (13.26).

13.2 Measurement in quantum mechanics

Measurements in quantum mechanics occur via an interaction of the measured quantum system described by the Hilbert space H with another system, an apparatus, so that this interaction changes the state of the initial system. *Physical observables* are given by self-adjoint matrices $A \in \mathcal{L}(H)$. Such matrices A are known to have the spectral decomposition $A = \sum_j \lambda_j P_j$, where P_j are orthogonal projections on the eigenspaces of A corresponding to the eigenvalues λ_j. According to the *basic postulate of quantum measurement*, measuring observable A in a state ρ (often referred to as the *Stern-Gerlach experiment*) can yield each of the eigenvalue λ_j with the probability

$$\text{tr}\,(\rho P_j), \tag{13.27}$$

and, if the value λ_j was obtained, the state of the system changes to the reduced state

$$P_j \rho P_j / \operatorname{tr}(\rho P_j).$$

In particular, if the state ρ was pure, $\rho = |\psi\rangle\langle\psi|$, then the probability to get λ_j as the result of the measurement becomes $(\psi, P_j \psi)$ and the reduced state also remains pure and is given by the vector $P_j \psi$. If the interaction with the apparatus was preformed 'without reading the results', the state ρ is said to be subject to a *non-selective measurement* that changes ρ to the state $\sum_j P_j \rho P_j$.

Remark 13.4. The notion of a general state (a density matrix) arises naturally from the simple duality (13.3) (with a bit more nontrivial extension to the case of infinite-dimensional spaces). In fact, von Neumann introduced a state as a linear functional on the space of observables (Hermitian linear operators) that was supposed to show the results of possible measurements of all these observables. By (13.3), this led inevitably to the notion of a mixed state as given above.

Extended to all self-adjoint operators the transformation

$$E : B \mapsto \sum_j P_j B P_j \tag{13.28}$$

is sometimes called the *conditional expectation* from $\mathcal{L}(H)$ to the subalgebra $N \subset \mathcal{L}(H)$ of operators that commute with all P_j. The conditional expectation is seen to satisfy the following properties reminiscent to its classical counterpart: (i) $E(X^*) = [E(X)]^*$, (ii) $X \geq 0$ implies $E(X) \geq 0$, (iii) $E(X) = X$ if and only if $X \in N$, (iv) If $X_1, X_2 \in N$, then $E(X_1 B X_2) = X_1 E(B) X_2$ (take-out-what-we-know property). Another accepted term for (13.28) is the *pinching map*.

For instance, applying the above scheme to the Pauli operator σ_3 of a qubit, allows one to conclude that, assuming the state of a qubit is some ρ, the measurement would reveal the values 1 or -1 corresponding to the pure states e_0 or e_1, with the probabilities

$$p_0 = \operatorname{tr}(\rho|e_0\rangle\langle e_0|) = \langle e_0|\rho|e_0\rangle = \rho_{00}, \quad p_1 = \rho_{11}, \tag{13.29}$$

showing in particular that the condition for density matrices to have the unit trace is necessary to make the probabilistic postulate of quantum measurement consistent. This measurement also corroborates the interpretation of the states ρ being the mixture of the pure states $|e_0\rangle\langle e_0|$ and $|e_1\rangle\langle e_1|$ with probabilities p_1 and p_2:

$$\rho = p_0|e_0\rangle\langle e_0| + p_1|e_1\rangle\langle e_1|. \tag{13.30}$$

However, all ρ with the same diagonal elements as in (13.30) will give the same result under this measurement.

In representation (13.9) the probabilities p_0, p_1 from (13.29) take the values

$$p_0 = (1 + x_3)/2, \quad p_1 = (1 - x_3)/2. \tag{13.31}$$

More generally, if the state of n dimensional system is ρ and we performed the measurement of any observable A that is diagonal in the standard basis $e_1, \ldots, e_n$ of $\mathbf{C}^n$, the probability to obtain e_j as the result of the measurement will be

$$\langle e_j | \rho | e_j \rangle = \rho_{jj}. \tag{13.32}$$

As seen from this formula all A which are diagonal in the basis $e_1, \ldots, e_n$ and have different eigenvalues produce the same probabilities of finding e_j. Thus effectively we are measuring the operator that labels the elements of the basis. This calculation is often referred to as measurement in the *computational basis* $e_1, \ldots, e_n$. If we are working in the product of two qubits $\mathbf{C}^2 \otimes \mathbf{C}^2$, which is the most basic scene for two-player two-action quantum games, the simplest computational basis is (13.20). For a state $\rho \in \mathcal{T}(\mathbf{C}^2) \otimes \mathcal{T}(C^2)$ the measurement in this computational basis will produce any of this vectors with the probabilities

$$\langle jk | \rho | jk \rangle = \mathrm{tr}(\rho | jk \rangle \langle jk |). \tag{13.33}$$

For a tensor product of two spaces H_A and H_B, with bases $\{e_j^A\}$ and $\{e_j^B\}$, the matrix elements of the linear operators (for instance, density matrices) are defined as

$$\rho(e_i^A \otimes e_j^B) = \sum_{i', j'} \rho_{i'j', ij}(e_{i'}^A \otimes e_{j'}^B),$$

and the probability to obtain $e_i^A \otimes e_j^B$ with the measurement performed on ρ (in the computational basis $e_i^A \otimes e_j^B$) is the diagonal element $\rho_{ij, ij}$. In particular,

$$(\rho^A \otimes \rho^B)_{i'j', ij} = \rho_{i'i}^A \rho_{j'j}^B,$$

and the probability to obtain $e_i^A \otimes e_j^B$ with the measurement performed on $\rho^A \otimes \rho^B$ is the product

$$\rho_{ii}^A \rho_{jj}^B. \tag{13.34}$$

If we are in the pure state

$$\rho = |\psi\rangle\langle\psi|, \quad \psi = \sum_{i,j} \psi_{ij} e_i^A \otimes e_j^B,$$

then this probability reduces to $|\psi_{ij}|^2$.

As we shall see, most of the quantum games can be ultimately reformulated in classical terms. However, specific feature lies in the physical
realizability of the strategies involved. For instance, all unitary operators
on a single qubit can be realized by the *Mach-Zender interferometer*, which
manipulates with photons, whose two states are usually denoted $|R\rangle = |0\rangle$
and $|L\rangle = |1\rangle$ (for right and left polarization), and which is built from
the following three units (referred to as *passive optical devices* that form
the *standard linear optics toolbox*): (1) *beam splitter (BS)* preforming the
unitary transformation U_{BS}:

$$U_{BS} = \frac{1}{\sqrt{2}}\left(|R\rangle\langle R| + |L\rangle\langle L| + i|R\rangle\langle L| + i|L\rangle\langle R|\right) = \frac{1}{\sqrt{2}}(1 + i\sigma_1),$$

$$U_{BS}(|R\rangle) = \frac{1}{\sqrt{2}}(|R\rangle + i|L\rangle), \quad U_{BS}(|L\rangle) = \frac{1}{\sqrt{2}}(|L\rangle + i|R\rangle), \qquad (13.35)$$

(2) *mirror operator* $U_{mir} = -i\sigma_1$ that takes $|R\rangle$ to $i|L\rangle$ and $|L\rangle$ to $i|R\rangle$,
and (3) *phase shifters*

$$U_R(\phi) = |R\rangle e^{i\phi}\langle R| + |L\rangle\langle L| = \begin{pmatrix} e^{i\phi} & 0 \\ 0 & 0 \end{pmatrix},$$

$$U_L(\phi) = |R\rangle\langle R| + |L\rangle e^{i\phi}\langle L| = \begin{pmatrix} 0 & 0 \\ 0 & e^{i\phi} \end{pmatrix}. \qquad (13.36)$$

Remark 13.5. Sometimes also other optical devices realizing unitary operators are referred to as beam splitters, for instance the ideal BS is the
rotation $\cos\theta\sigma_1 + \sin\theta\sigma_3$ with $\cos\theta$ and $\sin\theta$ referred to as transmittance
and reflectivity parameters respectively.

The standard combination of these units in the *Mach-Zender setting*
acts as

$$U_{MZ} = U_R(\theta_2)U_{BS}U_R(\phi_1)U_L(\phi_2)U_{mir}U_{BS}U_R(\theta_1), \qquad (13.37)$$

which is easily seen to yield the full 4-parameter representations of the
group $U(2)$ (compare with representation (13.11) of $SU(2)$).

Another scheme of physically realizable units that reproduces directly
the Cartan decomposition (13.14) can be built from the so-called *quarter
wave plates* and *half wave plates* acting as

$$U_{qwp}(\theta) = e^{-i\theta\sigma_2}e^{-i\pi\sigma_3/4}e^{-i\theta\sigma_2},$$

$$U_{hwp}(\theta) = [U_{qwp}(\theta)]^2 = e^{-i\theta\sigma_2}e^{-i\pi\sigma_3/2}e^{-i\theta\sigma_2}. \qquad (13.38)$$

Combining these two schemes one can naturally build the *universal unitary gate for two-qubit states*, that is, the schemes realizing all possible unitary transformations of two-qubit states (see detail in [229]).

Yet another optical devise is the so-called filter, which can be oriented in different ways to make computations in the bases (13.24). It corresponds to the observable that is diagonal in this basis and gives values 0 and 1 for $e_0^{\beta\phi}$ and $e_1^{\beta\phi}$ respectively. Physically it detects a photon if finds it in $e_0^{\beta\phi}$ and absorbs it if finds it in the state $e_1^{\beta\phi}$. This devise is called a *filter* oriented along the vector $n = (\sin\beta\cos\phi, \sin\beta\sin\phi, \cos\beta) \in \mathbf{R}^3$.

A key property of the entangled states is that the entanglement is destroyed when a measurement is carried out on one of the two systems only. For instance, if, in the Schmidt state (13.5), we measure the observable A of the first system that is diagonal in the basis $\{\xi_j\}$, that is $A = \sum_j \lambda_j P_j$, where P_j is the projection (in the product space) to the subspace generated by $\xi_j \otimes H_2$, then we can obtained either of the (not entangled) vectors $\xi_j \otimes \eta_j$ with the equal probability $1/\min(n, m)$.

This leads one to the far reaching consequence of non locality of quantum interaction. Namely, suppose that two photons are emitted by some device in the maximally entangled state (13.24) and then move in two different directions. Suppose we measure one of them by a filter oriented along $n = (\sin\beta, 0, \cos\beta)$. According to (13.24), the result will be the states $e_0^{\beta 0} \otimes e_0^{\beta 0}/\sqrt{2}$ or $e_1^{\beta 0} \otimes e_1^{\beta 0}/\sqrt{2}$ with probabilities $\cos^2\beta/2$ and $\sin^2\beta/2$. The remarkable thing is that this measurement on the first particle affects the other particle, as it brings it to a well defined state (for any β) and, what is more important, is independent of the distance (non locality!) between the particles at the time of the measurement. Thus if after the measurement of the first particle we measure the second particle with the same filter (oriented along the same vector), we obtain the same result as for the first particle with probability one (correlation 100%). This situation is essentially the famous *Einstein-Podolskii-Rosen (EPR) paradox* of quantum theory (in its simplified version suggested by D. Bohm). Though Einstein considered such actions on arbitrary distances as something unnatural, the recent experiments fully confirmed the conclusions of the EPR thought experiment paving the path to the experimental work on quantum communication, quantum computation and quantum games.

Measurements arising from self-adjoint operators as described above do not exhaust the effects of possible measuring instruments. Therefore a more general formulation of the measurement postulate is needed. Let Ω be a finite or countable set. A collection of positive operators $\{M_\omega\}$ in a Hilbert space H parametrized by $\omega \in \Omega$ (or a mapping from Ω to $\mathcal{L}^+(H)$) is called a *positive operator valued measure (POVM)* if it satisfies the normalization condition

$$\sum_{\omega \in \Omega} M_\omega = 1. \tag{13.39}$$

The space Ω describes the set of possible outcomes of an experiment. The *measurement based on a POVM* performed in a state ρ produces an outcome ω with the probability

$$\mathrm{tr}(\rho M_\omega). \tag{13.40}$$

If all M_ω are orthogonal projections, the POVM is called the *projection valued measure (PVM)*. Only the PVMs arise from self-adjoint operators as described above.

13.3   Meyer's quantum penny-flip game

Let us now introduce the first quantum game proposed by D. Meyer in [288]. It is an example of a quantum *sequential games*, where players act in some order on one and the same devise, and represents a version of the penny flip-over game. The classical setup is as follows. The referee places a coin (penny) head up in a box. Then three moves are performed sequentially by the two players, P (assumed to play by the rule of classical probability) and Q (which next will be assumed to play by the quantum rules). First Q makes a move by either flipping a coin (action F) or not (action N). Then P, not seeing the result of the action of the Q, makes her move by either flipping it (action F) or not (action N), and finally Q (not seeing what P has done) has the right to flip it again (action F) or not (action N). Then the referee opens the box. If the coin is head up, Q wins and P pays Q a penny. Otherwise P wins and Q pays P a penny. This is a zero-sum game with the table (the numbers in the table are the payoffs of P)

Table 13.1

		Q			
		NN	NF	FN	FF
P	N	-1	1	1	-1
	F	1	-1	-1	1

Easy to see that under the usual rules the optimal minimax value of the game is 0 and the optimal minimax strategies of the players are to choose their strategies uniformly (with probability $1/2$ for P and probability $1/4$ for Q).

To construct a quantum version of the game, one augments the two-state classical system to the qubit by associating the basis vectors $e_0 = |0\rangle = |H\rangle$ and $e_1 = |1\rangle = |T\rangle$ of the qubit $\mathbf{C}^2$ with the states H and P. Thus pure quantum states are unit vectors $|\psi\rangle = a|H\rangle + b|T\rangle$, $|a|^2 + |b|^2 = 1$ (more precisely, the corresponding elements of the projective space $\mathbf{CP}^1$) and the mixed quantum states are given by the density matrices (13.9).

Possible (pure) transformations of quantum states are given by the unitary matrices (13.10). However, Meyer has chosen to work with unitary matrices

$$U = U(u, v) = \begin{pmatrix} u & \bar{v} \\ v & -\bar{u} \end{pmatrix}, \quad |u|^2 + |v|^2 = 1, \qquad (13.41)$$

that, unlike $SU(2)$, have the determinant -1.

The opening of a state means the act of its measurement. The result of such action on a state ρ will be $|H\rangle$ or $T\rangle$ with probabilities (13.31).

Flip and no-flip actions F and N are thus presented by the Pauli matrix $F = \sigma_1$ and the unit matrix $N = I$:

Let us now assume that P can play classical strategies, F and N, and their classical mixtures, i.e. apply F with some probability $p \in [0, 1]$ and N with the probability $1 - p$. The key point (or assumption) in the quantum setting is that classical mixtures randomize the actions on the density matrices, that is, the mixed p-strategy of P acts on a density matrix ρ by the rule

$$\rho \mapsto pF\rho F^* + (1 - p)N\rho N^*.$$

The quantum player Q is supposed to play by (pure) *quantum strategies*, that is, by applying arbitrary unitary operators $U(u, v)$. Thus, after the first move of Q the initial state

$$\rho_0 = |0\rangle\langle 0| = |H\rangle\langle H| = \begin{pmatrix} 1 & 0 \\ 0 & 0 \end{pmatrix}$$

turns to the state

$$\rho_1 = U(u, v)\rho_0 U^*(u, v) = \begin{pmatrix} u\bar{u} & u\bar{v} \\ v\bar{u} & v\bar{v} \end{pmatrix}.$$

After the move of P the state turns to

$$\rho_2 = pF\rho_1 F^* + (1-p)N\rho_1 N^* = \begin{pmatrix} pv\bar{v} + (1-p)u\bar{u} & pv\bar{u} + (1-p)u\bar{v} \\ pu\bar{v} + (1-p)v\bar{u} & pu\bar{u} + (1-p)v\bar{v} \end{pmatrix}.$$

If the game would stop here, the payoff to the player P would be

$$(-1)[pv\bar{v} + (1-p)u\bar{u}] + pu\bar{u} + (1-p)v\bar{v} = (2p-1)(u\bar{u} - v\bar{v}). \quad (13.42)$$

It is seen that the game with such payoff has the value:

$$\max_p \min_{u,v}[(2p-1)(u\bar{u} - v\bar{v})] = \min_{u,v}\max_p[(2p-1)(u\bar{u} - v\bar{v})] = 0,$$

and the minimax strategies of the players are $p = 1/2$, $u^2 = v^2 = 1/2$.

To see what happens if the third move is included, assume that Q plays both times with the *Hadamard matrix* or *Hadamard gate* $U = U(1/\sqrt{2}, 1/\sqrt{2})$. Then

$$\rho_1 = \begin{pmatrix} 1/2 & 1/2 \\ 1/2 & 1/2 \end{pmatrix} = |\psi\rangle\langle\psi|$$

with

$$\psi = (H\rangle + T\rangle)/\sqrt{2}.$$

This state can be thought of imaginatively as describing the coin standing on its side. Then $\rho_2 = \rho_1$ independently of the choice of p, and thus

$$\rho_3 = U(1/\sqrt{2}, 1/\sqrt{2}) \begin{pmatrix} 1/2 & 1/2 \\ 1/2 & 1/2 \end{pmatrix} U^*(1/\sqrt{2}, 1/\sqrt{2})$$

$$= |U(1/\sqrt{2}, 1/\sqrt{2})\psi\rangle\langle U(1/\sqrt{2}, 1/\sqrt{2})\psi| = |H\rangle\langle H| = \begin{pmatrix} 1 & 0 \\ 0 & 0 \end{pmatrix},$$

so that Q wins with probability 1 independently of the actions of P! The power of quantum strategies (or quantum communications) is thus explicitly revealed.

In paper [288] one can also find the discussion of what can happen if both players are allowed to use pure quantum or even mixed quantum strategies.

13.4 First sequential games: Quantum Prisoner's dilemma

In this and the next sections we present the two basic approaches proposed for the quantization of the simultaneous static games, firstly the so-called *EWL protocol* suggested in [228] on the example of Prisoner's dilemma and secondly the so-called *MW protocol* suggested in [133] on the example of the Battle of the Sexes.

The table of the Prisoner's dilemma worked with in [228] was as follows:

Table 13.2

		Bob	
		C	D
Alice	C	(3,3)	(0,5)
	D	(5,0)	(1,1)

which is a performance of a more general version

Table 13.3

		Bob	
		C	D
Alice	C	(r,r)	(s,t)
	D	(t,s)	(p,p)

with r for reward, p for punishment, s for sucker's payoff, where $t > r > p > s$.

In the quantum version each player can manipulate a qubit (rather than playing with two bits in the classical version) generated by the basis of two vectors that now are identified with the actions of cooperate or defect: $e_0 = |0\rangle = |C\rangle$, $e_1 = |1\rangle = |D\rangle$. Only the referee has access to the combine system of two qubits and she prepares the initial state $\psi_{in} = J|CC\rangle$ with some unitary operator J in $\mathbf{C}^2 \otimes \mathbf{C}^2$ (made known to both players), which is symmetric with respect to the interchange of the players. Physically J is said to act as the *entanglement* that mixes in some way the initial product form $|CC\rangle = |C\rangle \otimes |C\rangle$. Then Alice and Bob choose (simultaneously and independently) some unitary operators U_A and U_B in $\mathbf{C}^2$ to act on their qubits, which transform the state ψ_{in} into $(U_A \otimes U_B)\psi_{in}$. Finally the referee redoes the entanglement by applying $J^* = J^{-1}$ yielding the final state

$$|\psi_{fin}\rangle = J^*(U_A \otimes U_B)J|CC\rangle, \tag{13.43}$$

often referred to as *EWL protocol*. For

$$|\psi_{fin}\rangle = \psi_{CC}|CC\rangle + \psi_{CD}|CD\rangle + \psi_{DC}|DC\rangle + \psi_{DD}|DD\rangle,$$

the squares $|\psi_{CC}|^2$, $|\psi_{CD}|^2$, $|\psi_{DC}|^2$, $|\psi_{DD}|^2$ are the probabilities of the corresponding outcomes, so that the payoffs for Alice and Bob are

$$\Pi_A = r|\psi_{CC}|^2 + p|\psi_{DD}|^2 + t|\psi_{DC}|^2 + s|\psi_{CD}|^2,$$

$$\Pi_B = r|\psi_{CC}|^2 + p|\psi_{DD}|^2 + s|\psi_{DC}|^2 + t|\psi_{CD}|^2. \tag{13.44}$$

Clearly the game depends on both the choice of the entangling operator J and the set of allowed unitary operators, that is, the strategy spaces of Alice and Bob.

Concretely, in [228], the set of unitary operators used by both players was restricted, rather artificially in fact, to the two-parameter set

$$U(\theta, \phi) = \begin{pmatrix} e^{i\phi}\cos(\theta/2) & \sin(\theta/2) \\ -\sin(\theta/2) & e^{-i\phi}\cos(\theta/2) \end{pmatrix} \tag{13.45}$$

with $0 \le \theta \le \pi$, $0 \le \phi \le \pi/2$. In particular, the operators $\hat{C}$ and $\hat{D}$,

$$\hat{C} = U(0,0) = \begin{pmatrix} 1 & 0 \\ 0 & 1 \end{pmatrix} = \mathbf{1}, \quad \hat{\mathbf{D}} = \mathbf{U}(\pi,\mathbf{0}) = \begin{pmatrix} 0 & 1 \\ -1 & 0 \end{pmatrix},$$

were associated with the cooperative and defective classical strategies, because as the starting point was supposed to be $|CC\rangle$, the identity operator $\hat{C}$ preserves the cooperative behavior $|C\rangle$ and $\hat{D}$ flips it to the defective behavior $|D\rangle$.

Remark 13.6. In later publications the changing sign feature of $\hat{D}$ (looking a bit artificially) was mostly abandoned and one used the exact flipping operator $F = \sigma_x$ instead.

The assumptions on J made in [228] were introduced with a very clear interpretation, as those that would allow to reproduce the classical game. Namely, the commutativity conditions

$$[J, \hat{D} \otimes \hat{D}] = 0, \quad [J, \hat{D} \otimes \hat{C}] = 0, \quad [J, \hat{C} \otimes \hat{D}] = 0, \tag{13.46}$$

were assumed, implying that

$$[J, U(\theta, 0)] = 0 \tag{13.47}$$

for all θ. If this holds, then

$$\psi_{fin}\rangle = U_A \otimes U_B |CC\rangle$$
$$= [\cos(\theta_A/2)|C\rangle - \sin(\theta_A/2)|D\rangle] \otimes [\cos(\theta_B/2)|C\rangle - \sin(\theta_B/2)|D\rangle],$$

and all probabilities factorize, whenever U_A and U_B are restricted to $U(\theta, 0)$, in particular, if U_A and U_B are allowed to be only the 'classical actions', i.e. either $\hat{C}$ or $\hat{D}$. Thus, identifying $\cos^2(\theta_A/2)$ and $\cos^2(\theta_B/2)$ with classical probabilities p and q we reproduce the payoffs of the classical prisoners' dilemma played with the mixed strategies.

Exercise 13.5. (i) Check that in the basis $|CC\rangle, |CD\rangle, |DC\rangle, |DD\rangle$

$$\hat{C} \times \hat{D} = -\begin{pmatrix} \hat{D} & 0 \\ 0 & \hat{D} \end{pmatrix}, \quad \hat{D} \times \hat{C} = \begin{pmatrix} 0 & -1 \\ 1 & 0 \end{pmatrix},$$

$$\hat{D} \times \hat{D} = \begin{pmatrix} 0 & \hat{D} \\ -\hat{D} & 0 \end{pmatrix}, \quad \sigma_x \times \sigma_x = \begin{pmatrix} 0 & \sigma_x \\ \sigma_x & 0 \end{pmatrix}.$$

(ii) For $J = \begin{pmatrix} A & B \\ R & S \end{pmatrix}$, condition (13.46) is equivalent to

$$J = \begin{pmatrix} A & B \\ -B & A \end{pmatrix}, \quad A\hat{D} = \hat{D}A, \quad B\hat{D} = \hat{D}B$$

(and under $[J, C \otimes D] = 0$ the first two conditions of (13.46) are equivalent). Hence J is a linear combination of the matrices

$$\mathbf{1}, \hat{C} \times \hat{D}, \hat{D} \times \hat{C}, \hat{D} \times \hat{D}.$$

In [228] the operator J was chosen as

$$J_\gamma = \exp\{i\gamma \hat{D} \otimes \hat{D}/2\}, \quad \gamma \in [0, \pi/2],$$

so that

$$J_\gamma = \cos(\gamma/2)\hat{C} \otimes \hat{C} + i\sin(\gamma/2)\hat{D} \otimes \hat{D}.$$

(The choice of coefficients is also restricted by the requirement that J is unitary.)

Remark 13.7. This choice of J can be considered as the most general fully symmetric choice. In future publications, when the flipping $F = \sigma_x$ became standard substitute to $\hat{D}$, the version with

$$J_\gamma = \cos(\gamma/2)\mathbf{1} \otimes \mathbf{1} + i\sin(\gamma/2)\sigma_x \otimes \sigma_x$$

became a more or less canonical choice.

We have arrived now at the problem of finding Nash equilibria for a purely classical stationary game of two players with payoffs (13.44), where ψ_{fin} is calculated from (13.43) based on the players strategies, which are the pairs θ_A, ϕ_A of Alice and θ_B, ϕ_B of Bob, defining U_A and U_B via (13.45). The whole quantum content is encoded in the particular way the payoffs are calculated.

If $\gamma = 0$, J_0 is the identity operator and the probabilities factorize for all U_A, U_B. Hence $\hat{D} \otimes \hat{D}$ is equilibrium in dominated strategies, as in classical game. Therefore γ is considered as the 'entangling parameter', and the case with $\gamma = \pi/2$ as the 'maximally entangled game'. In this case

$$J_{\pi/2} = \frac{1}{\sqrt{2}}[\hat{C} \otimes \hat{C} + i\hat{D} \otimes \hat{D}], \quad J_{\pi/2}|CC\rangle = \frac{1}{\sqrt{2}}(|CC\rangle + i|TT\rangle),$$

and $\hat{D} \otimes \hat{D}$ is not a Nash equilibrium. The calculations show (see [228] and a very detailed presentation in review [239]) that in the maximally entangled game there is a unique Nash equilibrium $\hat{Q} \otimes \hat{Q}$ with

$$\hat{Q} = i\sigma_3 = \begin{pmatrix} i & 0 \\ 0 & -i \end{pmatrix}$$

with the payoff $(3,3)$, that is, the Pareto optimal (effective) solution became the unique Nash equilibrium!

As was noted in [207], this effect was achieved precisely by the artificial restriction of the strategy space to operators (13.45). The same effect can be achieved just with three strategies $I, \sigma_y = \sigma_2, \sigma_z = \sigma_3$ for each player with σ_z giving the desired equilibrium. In [239] detailed calculations are given reproducing the same effect from the three strategies $\mathbf{1}, \sigma_\mathbf{x}, \sigma_\mathbf{z}$. But if we allow the full discrete set $\mathbf{1}, \sigma_\mathbf{x}, \sigma_\mathbf{y}, \sigma_\mathbf{z}$ the equilibrium disappears. This is in fact a consequence of a general result, see below Proposition 13.1.

13.5 First sequential games: Quantum Battle of the Sexes

A slightly different approach to the 'quantization of games' was proposed in [133] on the example of the Battle of the Sexes, namely the game given by the table

Table 13.4

		Bob	
		O	T
Alice	O	(α, β)	(γ, γ)
	T	(γ, γ)	(β, α)

where $\alpha > \beta > \gamma$. Here O reflects the preferred activity of the wife (opera, ballet, etc.) and T that of the husband (television, football, etc.). Again each player has a quibit $\mathbf{C}^2$ at their disposal with the two basic states now denoted $e_0 = |O\rangle$, $e_1 = |T\rangle$.

Classical theory (see (9.16)) yields the conclusion that this game has two pure Nash equilibria (O, O), (T, T) and one mixed equilibrium (p and q denote the probabilities used by Alice and Bob to play O):

$$p^* = \frac{\alpha - \gamma}{\alpha + \beta - 2\gamma}, \quad q^* = 1 - p^* = \frac{\beta - \gamma}{\alpha + \beta - 2\gamma}, \qquad (13.48)$$

the payoff for both Alice and Bob in this equilibrium being

$$(\alpha + \beta)p^*q^* + \gamma(p^*p^* + q^*q^*) = \frac{\alpha\beta - \gamma^2}{\alpha + \beta - 2\gamma} < \min(\alpha, \beta). \qquad (13.49)$$

As the simplest possible quantum version one can suggest that instead of choosing probabilities p, q to play the classical strategies, Alice and Bob are allowed to choose quantum superpositions, that is, the pure quantum states $a|O\rangle + b|T\rangle$ and $c|O\rangle + d|T\rangle$ respectively, with $|a|^2 + |b|^2 = 1$ and $|c|^2 + |d|^2 = 1$ and the outcome is then measured according to the measurement rules of quantum mechanics (see the end of Section 13.1). Namely, their common pure state in $\mathbf{C}^2 \otimes \mathbf{C}^2$ becomes

$$\psi_{in} = (a|O\rangle + b|T\rangle) \otimes (c|O\rangle + d|T\rangle) = ac|OO\rangle + ad|OT\rangle + bc|TO\rangle + bd|TT\rangle,$$

so that after the measurements one gets $|OO\rangle$ with probability $|ac|^2$, $|OT\rangle$ with probability $|ad|^2$, $|TO\rangle$ with probability $|bc|^2$ and $|TT\rangle$ with probability $|bd|^2$. This is exactly the same result, as if they play classical mixed strategies choosing O with probabilities $p = |a|^2$ and $q = |c|^2$ respectively, so that this quantum version of the game reproduces the classical game.

Introducing more advanced quantum operations one can assume that the players start at some initial pure state ψ_{in} (the analog of $J|CC\rangle$ of the EWL protocol), or even mixed state ρ_{in}, and then Alice and Bob are allowed to perform on their parts of the product $\mathbf{C}^2 \otimes \mathbf{C}^2$ some quantum operations. In [133] it was suggested that the allowed strategies for Alice and Bob are either the identity operator $I = \mathbf{1}$ or the exchange (flip) operator $F = \sigma_1$, or their classical mixtures, that is, choosing I with probabilities p and q respectively (and thus F with probabilities $1 - p$ and $1 - q$). The point to stress is that, like in Meyer's penny flipping game, these probabilities are applied to the action of I and F on the density matrices by dressing (13.8).

As one can expect, the situation will be quite different depending on the initial ψ_{in} or ρ_{in}.

Suppose first that they start with a factorizable initial density matrix $\rho_{in} = \rho_A \otimes \rho_B$. Applying their eligible mixtures given by the probabilities p and q, Alice and Bob transform the initial state into the final $\rho_A^{fin} \otimes \rho_B^{fin}$ with

$$\rho_A^{fin} = pI\rho_A I + (1-p)F\rho_A F, \quad \rho_B^{fin} = qI\rho_B I + (1-p)F\rho_B F.$$

For a general initial state the final state becomes

$$\rho_{fin} = pq(I_A \otimes I_B)\rho_{in}(I_A \otimes I_B) + p(1-q)(I_A \otimes F_B)\rho_{in}(I_A \otimes F_B)$$

$$+ (1-p)q(F_A \otimes I_B)\rho_{in}(F_A \otimes I_B) + (1-p)(1-q)(F_A \otimes F_B)\rho_{in}(F_A \otimes F_B).$$
$$(13.50)$$

If they start with the initial density matrix ρ_{in} arising from the pure states $|OO\rangle$, $|OT\rangle$, $|TO\rangle$, $|TT\rangle$, then we get back the same classical outcome as in the first simplest scenario. In fact, starting for instance with $|OO\rangle$, one obtains

$$\rho_A^{fin} \otimes \rho_B^{fin} = \left[p \begin{pmatrix} 1 & 0 \\ 0 & 0 \end{pmatrix} + (1-p) \begin{pmatrix} 0 & 0 \\ 0 & 1 \end{pmatrix} \right] \otimes \left[q \begin{pmatrix} 1 & 0 \\ 0 & 0 \end{pmatrix} + (1-q) \begin{pmatrix} 0 & 0 \\ 0 & 1 \end{pmatrix} \right]$$

$$= (p|O\rangle\langle O| + (1-p)|T\rangle\langle T|) \otimes (q|O\rangle\langle O| + (1-q)|T\rangle\langle T|)$$

$$= pq|OO\rangle\langle OO| + p(1-q)|OT\rangle\langle OT| + (1-p)q|TO\rangle\langle TO| + (1-p)(1-q)|TT\rangle\langle TT|.$$

This density matrix yields outcomes $|OO\rangle$, $|OT\rangle$, $|TO\rangle$, $|TT\rangle$ with the probabilities pq, $p(1-q)$, $(1-p)q$ and $(1-p)(1-q)$, respectively.

The situation changes if we start with $\psi_A = a|O\rangle + b|T\rangle$ and $\psi_B = c|O\rangle + d|T\rangle$, so that

$$|\psi_{in}\rangle = \psi_A \otimes \psi_B = ac|OO\rangle + ad|OT\rangle + bc|TO\rangle + ad|TT\rangle$$

$$= c_{OO}|OO\rangle + c_{OT}|OT\rangle + c_{TO}|TO\rangle + c_{TT}|TT\rangle,$$

or, in terms of the density matrices,

$$\rho_{in} = \rho_A \otimes \rho_B,$$

$$\rho_A = |\psi_A\rangle\langle\psi_A| = \begin{pmatrix} a\bar{a} & a\bar{b} \\ b\bar{a} & b\bar{b} \end{pmatrix}, \quad \rho_B = |\psi_B\rangle\langle\psi_B| = \begin{pmatrix} c\bar{c} & c\bar{d} \\ d\bar{c} & d\bar{d} \end{pmatrix}.$$

Applying I with probabilities p and q yields

$$\rho_A^{fin} = p \begin{pmatrix} a\bar{a} & a\bar{b} \\ b\bar{a} & b\bar{b} \end{pmatrix} + (1-p) \begin{pmatrix} b\bar{b} & b\bar{a} \\ a\bar{b} & a\bar{a} \end{pmatrix}, \rho_B^{fin} = q \begin{pmatrix} c\bar{c} & c\bar{d} \\ d\bar{c} & d\bar{d} \end{pmatrix} + (1-q) \begin{pmatrix} d\bar{d} & d\bar{c} \\ c\bar{d} & c\bar{c} \end{pmatrix}.$$

According to (13.34), the probabilities to get $|OO\rangle$, $|OT\rangle$, etc., become

$$P_{OO} = [pa\bar{a} + (1-p)b\bar{b}][qc\bar{c} + (1-q)d\bar{d}]$$

$$= pq|c_{OO}|^2 + (1-p)q|c_{TO}|^2 + p(1-q)|c_{OT}|^2 + (1-p)(1-q)|c_{TT}|^2, \quad (13.51)$$

$$P_{TT} = [pb\bar{b} + (1-p)a\bar{a}][qd\bar{d} + (1-q)c\bar{c}]$$

$$= pq|c_{TT}|^2 + (1-p)q|c_{OT}|^2 + p(1-q)|c_{TO}|^2 + (1-p)(1-q)|c_{OO}|^2, \quad (13.52)$$

$$P_{OT} = [pa\bar{a} + (1-p)b\bar{b}][qd\bar{d} + (1-q)c\bar{c}]$$

$$= pq|c_{OT}|^2 + (1-p)q|c_{TT}|^2 + p(1-q)|c_{OO}|^2 + (1-p)(1-q)|c_{TO}|^2, \quad (13.53)$$

$$P_{TO} = [pb\bar{b} + (1-p)a\bar{a}][qc\bar{c} + (1-q)d\bar{d}]$$

$$= pq|c_{TO}|^2 + (1-p)q|c_{OO}|^2 + p(1-q)|c_{TT}|^2 + (1-p)(1-q)|c_{OT}|^2, \quad (13.54)$$

which are different from the classical outcomes.

One can expect to have the same outcomes for the general initial superposed state:

$$|\psi_{in}\rangle = \sum_{i,j=0}^{1} c_{ij}|ij\rangle = c_{OO}|OO\rangle + c_{OT}|OT\rangle + c_{TO}|TO\rangle + c_{TT}|TT\rangle. \quad (13.55)$$

Let us check it. The corresponding density matrix is

$$\rho_{in} = |\psi_{in}\rangle\langle\psi_{in}| = \sum_{i,j,k,l,} c_{ij}\bar{c}_{kl}|ij\rangle\langle kl| = \sum_{i,j,k,l,} c_{ij}\bar{c}_{kl}|i\rangle\langle k| \otimes |j\rangle\langle l|.$$

Hence,

$$I_A\rho_{in}I_A = I_B\rho_{in}I_B = \rho_{in},$$

$$F_A|ij\rangle\langle kl|F_A = F_A|i\rangle\langle k|F_A \otimes |j\rangle\langle l| = |i'\rangle\langle k'| \otimes |j\rangle\langle l| = |i'j\rangle\langle k'l|,$$

$$F_B|ij\rangle\langle kl|F_B = |ij'\rangle\langle kl'|, \quad (F_A \otimes F_B)|ij\rangle\langle kl|(F_A \otimes F_B) = |i'j'\rangle\langle k'l'|,$$

where prime denotes the complementary index.

Hence, by (13.50),

$$\rho_{fin} = \sum_{i,j,k,l,} c_{ij}\bar{c}_{kl}(pq|ij\rangle\langle kl| + p(1-q)|ij'\rangle\langle kl'|$$

$$+ (1-p)q|i'j\rangle\langle k'l| + (1-p)(1-q)|i'j'\rangle\langle k'l'|). \quad (13.56)$$

Probability to measure $|ij\rangle$ is the diagonal element

$$P_{ij} = (\rho_{fin})_{ij,ij} = pq|c_{ij}|^2 + (1-p)q|c_{i'j}|^2 + p(1-q)|c_{ij'}|^2 + (1-p)(1-q)|c_{i'j'}|^2,$$
(13.57)

which is a compact form of equations (13.51)-(13.54) above.

If we now consider the general bi-matrix game with the matrix

Table 13.5

		Bob	
		O	T
Alice	O	$(\alpha_{00}, \beta_{00})$	$(\alpha_{01}, \beta_{01})$
	T	$(\alpha_{10}, \beta_{10})$	$(\alpha_{11}, \beta_{11})$

we get the payoffs for the first player

$$\Pi^A = \sum_{ij} \alpha_{ij} P_{ij}$$

$$= \sum \alpha_{ij}[pq|c_{ij}|^2 + (1-p)q|c_{i'j}|^2 + p(1-q)|c_{ij'}|^2 + (1-p)(1-q)|c_{i'j'}|^2]$$

$$= pq \sum \alpha_{ij}|c_{ij}|^2 + (1-p)q \sum \alpha_{ij}|c_{i'j}|^2$$

$$+ p(1-q) \sum \alpha_{ij}|c_{ij'}|^2 + (1-p)(1-q) \sum \alpha_{ij}|c_{i'j'}|^2,$$
(13.58)

and the same (with β instead of α) for the second player.

Thus, as the result, applying quantum rules in this *MW protocol*, means effectively just applying the parametric family of transformations to the initial payoff matrix: $\alpha \mapsto \tilde{\alpha}$, $\beta \mapsto \tilde{\beta}$:

$$\tilde{\alpha} = \begin{pmatrix} \sum \alpha_{ij}|c_{ij}|^2 & \sum \alpha_{ij}|c_{i'j}|^2 \\ \sum \alpha_{ij}|c_{ij'}|^2 & \sum \alpha_{ij}|c_{i'j'}|^2 \end{pmatrix}, \tilde{\beta} = \begin{pmatrix} \sum \beta_{ij}|c_{ij}|^2 & \sum \beta_{ij}|c_{i'j}|^2 \\ \sum \beta_{ij}|c_{ij'}|^2 & \sum \beta_{ij}|c_{i'j'}|^2 \end{pmatrix}.$$
(13.59)

In case of the 'most entangled' initial state $\psi = a|OO\rangle + b|TT\rangle$, $|a|^2 + |b|^2 = 1$, it simplifies to

$$\tilde{\alpha} = \begin{pmatrix} \alpha_{00}|a|^2 + \alpha_{11}|b|^2 & \alpha_{10}|a|^2 + \alpha_{01}|b|^2 \\ \alpha_{01}|a|^2 + \alpha_{10}|b|^2 & \alpha_{11}|a|^2 + \alpha_{00}|b|^2 \end{pmatrix},$$

$$\tilde{\beta} = \begin{pmatrix} \beta_{00}|a|^2 + \beta_{11}|b|^2 & \beta_{10}|a|^2 + \beta_{01}|b|^2 \\ \beta_{01}|a|^2 + \beta_{10}|b|^2 & \beta_{11}|a|^2 + \beta_{00}|b|^2 \end{pmatrix}.$$
(13.60)

In particular, the matrix of Table 13.4 transforms to the matrix

Table 13.6

Bob

	O	T
Alice O	$(\tilde{\alpha}, \tilde{\beta})$	$(\tilde{\gamma}, \tilde{\gamma})$
T	$(\tilde{\gamma}, \tilde{\gamma})$	$(\tilde{\beta}, \tilde{\alpha})$

with

$$\tilde{\alpha} = \alpha|a|^2 + \beta|b|^2, \tilde{\beta} = \alpha|b|^2 + \beta|a|^2, \quad \tilde{\gamma} = \gamma.$$

Now the value of the mixed Nash equilibrium is still less than the payoffs at both pure equilibria, $p^* = q^* = 0$ and $p^* = q^* = 1$. These pure equilibria give payoffs $(\tilde{\alpha}, \tilde{\beta})$ and $(\tilde{\beta}, \tilde{\alpha})$ respectively. In the special case of $a = b = 1/\sqrt{2}$, the payoffs for Alice and Bob coincide in both pure equilibria. They equal $(\alpha + \beta)/2$, are efficient (Pareto optimal), and get better payoff than the third equilibrium payoff $(\alpha+\beta+2\gamma)/4$. Hence it was argued in [133] that both equilibria $p^* = q^* = 0$ and $p^* = q^* = 1$ represent somehow the unique solution and thus solve the dilemma of the Battle of the Sexes. This is of course arguable. As was commented in [206], there remain the possibility of mismatch (one chooses 0 and another 1) giving lower payoff, which leaves essentially the same dilemma as the initial classical one. Marinatto and Weber argued back that it was natural for players to stick to $p^* = q^* = 1$, which means doing nothing, rather than start flipping.

13.6 Variations on MW protocol

Using transformation (13.59) one can automatically transform any game to a new quantum version obtained by the MW protocol.

This transforms, of course, in a systematic way, all properties of the games: equilibria, their stability, etc. For instance, stability of the equilibria of the transformed RD for two-player two-action games was analyzed in [244], ESS stability for the transformed Rock-Paper-Scissors game in [245], and for 3-player games in [246]. The transformations of the simplest cooperative games were analyzed in [247].

Transformation (13.59) extends directly to games with arbitrary number of players and arbitrary number of strategies (in order to preserve the dimension under this transform, if we have n strategies for a player, then one must choose exactly n basic transformations allowing to reshuffle them (say, n transforms taking the first strategy to any of the n existing strategies).

One can extend the setting of MW protocol by allowing arbitrary unitary strategies of the players (rather than just I and F) and their classical mixtures.

If the same extension performed with the EWL protocol, the only real difference between MW and EWL approaches lies in the application by EWL protocol the disentangling operator J^* before the measurement, which is not the case in the MW protocol.

Let us review couple of the extensions performed along these lines and their conclusions.

In [223] the MW protocol is applied to the Battle of the Sexes starting from the 'maximally entangled' initial state

$$|\psi_{in}\rangle = J|OO\rangle = (|OO\rangle + |TT\rangle)\sqrt{2}, \quad \rho_{in} = |\psi_{in}\rangle\langle\psi_{in}|$$

while the strategy space of the players is taken to consist of arbitrary unitary U_A and U_B, given by the matrices (see (13.14))

$$U = \begin{pmatrix} e^{i(\phi+\psi)}\cos\theta & ie^{i\phi-\psi}\sin\theta \\ ie^{-i(\phi-\psi)}\sin\theta & e^{-i(\phi+\psi)}\cos\theta \end{pmatrix},$$

and their arbitrary mixtures. Namely, Alice and Bob are supposed to choose probability densities $f_A(U)$ and $f_B(U)$ such that

$$\int_{SU(2)} f_A(U)\,dU = 1, \quad \int_{SU(2)} f_B(U)\,dU = 1,$$

where the integration is with respect to the Haar measure on $SU(2)$, and the final state becomes

$$\rho_{fin} = \int\int f_A f_B (U_A \otimes U_B)\rho_{in}(U_A \otimes U_B)^* dU_A dU_B. \qquad (13.61)$$

Diagonal elements of this matrix defines the probabilities of the outcomes $|OO\rangle$, $|OT\rangle$, $|TO\rangle$, $|TT\rangle$. Hence, introducing the payoff operators for Alice and Bob by

$$\hat{S}_A = \alpha|OO\rangle\langle OO| + \beta|TT\rangle\langle TT| + \gamma(|OT\rangle\langle OT| + |TO\rangle\langle TO|),$$

$$\hat{S}_B = \beta|OO\rangle\langle OO| + \alpha|TT\rangle\langle TT| + \gamma(|OT\rangle\langle OT| + |TO\rangle\langle TO|),$$

it follows that the expected payoffs are

$$\mathbf{E}S_A(f_A, f_B) = \mathrm{tr}\,(\rho_{fin}\hat{S}_A) = \int\int f_A f_B S_A(U_A, U_B)\,dU_A dU_B,$$

$$\mathbf{E}S_B(f_A, f_B) = \mathrm{tr}\,(\rho_{fin}\hat{S}_B) = \int\int f_A f_B S_B(U_A, U_B)\,dU_A dU_B,$$

where

$$S_A(U_A, U_B) = \mathrm{tr}[(U_A \otimes U_B)\rho_{in}(U_A \otimes U_B)^*\hat{S}_A],$$

$$S_B(U_A, U_B) = \mathrm{tr}[(U_A \otimes U_B)\rho_{in}(U_A \otimes U_B)^*\hat{S}_B]$$

are their payoffs in the pure unitary strategies. Thus all the quantum content is encoded in the structure of these payoffs, and the problem to find Nash equilibria is now fully classical. The calculations can be simplified by noting that, since ρ_{in} is decomposable density matrix, so are also the matrices $(U_A \otimes U_B)\rho_{in}(U_A \otimes U_B)^*$, and hence probabilities to get an outcome $|\sigma\tau\rangle$ (here σ and τ are either O or T) in the game with pure strategies is just

$$P_{\sigma\tau} = |\langle\sigma\tau|(U_A \otimes U_B)|\psi_{in}\rangle|^2,$$

so that

$$S_A(U_A, U_B) = \alpha|\langle OO|(U_A \otimes U_B)|\psi_{in}\rangle|^2 + \beta|\langle TT|(U_A \otimes U_B)|\psi_{in}\rangle|^2$$

$$+ \gamma(|\langle OT|(U_A \otimes U_B)|\psi_{in}\rangle|^2 + |\langle TO|(U_A \otimes U_B)|\psi_{in}\rangle|^2),$$

and $S_B(U_A, U_B)$ the same with α, β interchanged.

Calculations show (see [223]) that in this game there are infinitely many Nash equilibria, but they all give the same payoff $S = (\alpha+\beta+2\gamma)/4$. Moreover, the problem of mismatch does not arise, because all Nash equilibria are of the form (f_A, f_B) with $f_A \in F_A$, $f_B \in F_B$ and some sets F_A, F_B, and any combination yields the same payoff.

In [235] another new version of MW protocol is introduced and analyzed, where players are allowed additional choice, to accept given initial entangled state or not. More precisely, both players declare independently whether they like to start with a suggested (by referee) quantum state, and this quantum state is actually prepared by the referee if both players declare their willingness for it, otherwise they start with the classical initial $|OO\rangle$.

In [286] the MW protocol and the transformation (13.59) are used for the quantization of the ultimatum game, where the first player is supposed to have two strategies: to offer some preassigned unfair division of the total sum of 100, say $99 + 1$ and the fair one: $50 + 50$. The second player can either accept the offer or reject. Thus the table is

Table 13.7

Bob

Alice		accept	not
	unfair	(99,1)	(0,0)
	fair	(50,50)	(0,0)

Paper [286] also analyses this game under the set of all unitary strategies, where it just reproduces for this concrete setting the general remark of [206] on the absence of Nash equilibria for general MW protocol extended to full unitary strategies. For mixed unitary strategies (like in (13.61)) it is shown the existence of Nash equilibria (the corresponding general result is given in Theorem 13.5).

13.7 Variations on EWL protocol

As in the case of MW protocol, various extensions of EWL protocol were analyzed by using more general strategy spaces and the games with more players and more initial classical actions. Let us review some of these contributions.

In [224] the EWL protocol is applied to the Battle of Sexes. It is shown the existence of infinitely many Nash equilibria when the strategies of players are restricted to a two-parameter set of unitary transformations (like in the original EWL protocol). What seems more important they show that for 'nontrivial' two action two player games, if the players are allowed to play the full set of $SU(2)$ strategies, the quantum EWL game has no Nash equilibria, when started in maximally entangled state (see exact formulation below in Proposition 13.1).

In [225] the analysis of equilibria for the general prisoner's dilemma with Table 13.3 above was provided. Under restricted set of unitary operators (EWL like), the phase transitions are found: the desired cooperative equilibria Q arises when the entanglement parameter γ crosses certain critical values expressed in terms of the parameters r, s, t. For the full unitary strategies there is a similar transition between the situation with infinitely many equilibria and no equilibria at all.

In [227] the *three-player quantum Prisoner's dilemma* is considered. The story behind the dilemma is the same as for two prisoners. The payoffs are chosen to reflect the idea that defection brings advantage to each player that is 'inversely proportional' to the number of other defecting players.

There are two natural equivalent ways to represent three player games, via two tables distinguished by a particular choice of the third player:

Table 13.8   Colin C

Bob

	C	D
Alice C	(3,3,3)	(2,5,2)
D	(5,2,2)	(4,4,0)

Table 13.9   Colin D

Bob

	C	D
Alice C	(2,2,5)	(0,4,4)
D	(4,0,4)	(1,1,1)

or equivalently by a two-row table showing payoffs of each player A, B, C obtained from each possible profile. For a symmetric game, this table can be reduced showing all entries with the equal numbers of D and C only once. For the general payoffs of a three-player symmetric game the table can be given as

profile	(C,C,C)	(D,D,D)	(D,C,C)	(D,D,C)
payoff	(r_{3c}, r_{3c}, r_{3c})	(r_{3d}, r_{3d}, r_{3d})	(r_{1d}, r_{2c}, r_{2c})	(r_{2d}, r_{2d}, r_{1c})

$$(13.62)$$

The above story of the prisoner's dilemma corresponds to the ordering

$$r_{1c} < r_{3d} < r_{2c} < r_{3c} < r_{2d} < r_{1d}.$$

Like its two-player counterpart, the classical version of this game is a symmetric game with defecting D being the dominating strategies, so that the profile (D, D, D) is a Nash equilibrium that is also the solution in dominating strategies.

Quantum scheme extends the two-player game by choosing $J = \exp\{i(\gamma/2)\sigma_x \otimes \sigma_x \times \sigma_x\}$, with $0 \leq \gamma \leq \pi/2$. The final state is

$$|\psi_{fin} = J^*(U_A \otimes U_B \otimes U_C)J|OOO\rangle.$$

The payoff for Alice, say, is

$$S_A = 5P_{DCC} + 4(P_{DDC} + P_{DCD}) + 3P_{CCC} + 2(P_{CCD} + P_{CDC}) + P_{DDD},$$

where

$$P_{\sigma\xi\eta} = |\langle \sigma\xi\eta|\psi_{fin}\rangle|^2.$$

The strategic space is chosen to be restricted to the two-parameter set:

$$U(\theta, \phi) = \begin{pmatrix} \cos\theta & e^{i\phi}\sin\theta \\ -e^{-i\phi}\sin\theta & \cos\theta \end{pmatrix}, \quad \theta \in [0, \pi/2], \ \phi \in [0, \pi/2].$$

Here $U(0,0) = \mathbf{1}$ represents the strategy 'cooperate' and $U(\pi/2, \pi/2) = i\sigma_x$ represents the flipping operator of the 'defecting' strategy.

The calculations show (see [227]) that $i\sigma_y \otimes i\sigma_y \otimes i\sigma_y$ is a Nash equilibrium for all γ (this is a new feature as compared to the two-player setting) with the payoff

$$S_A = S_B = S_C = 1 + 2\sin^2 \gamma.$$

For $\gamma = \pi/2$ this yields the desired cooperative and Pareto optimal payoffs of value 3. We see also that the symmetric equilibrium payoff increases monotonically and continuously with the entanglement parameter γ.

A straightforward extension of EWL scheme to arbitrary number N of players is as follows. The entangling operator is taken to be the 'maximally entangling' one:

$$J = \frac{1}{\sqrt{2}}(I^{\otimes N} + iF^{\otimes N}) = \frac{1}{\sqrt{2}}(I^{\otimes N} + i\sigma_x^{\otimes N}), \quad J^* = \frac{1}{\sqrt{2}}(I^{\otimes N} - i\sigma_x^{\otimes N}),$$

and thus the initial state is

$$\xi_{in} = J(|0\rangle^{\otimes N}) = J|0\cdots 0\rangle = \frac{1}{\sqrt{2}}(|0\rangle^{\otimes N} + i|1\rangle^{\otimes N}).$$

The final state is

$$\xi_{fin} = J^*(U_1 \otimes \cdots \otimes U_N)\xi_{in} = J^*(U_1 \otimes \cdots \otimes U_N)J|0\rangle^{\otimes N}.$$

Abandoning the artificial restrictions to the allowed unitary strategies, the game can be naturally considered with arbitrary unitary strategies $U \in SU(2)$ of all players, that is with U given by (13.15):

$$U = \cos\theta(\cos\phi\mathbf{1} + i\sin\phi\sigma_z) + i\sin\theta(\sin\psi\sigma_x + \cos\psi\sigma_y).$$

Two further extensions of the strategy spaces are natural. One can use classically mixed quantum strategies (like (13.61) for MW protocols), or one can allow to players to use the full set of $TP - CP$ operations (see Section 13.9), their Kraus representation being given in (13.98).

Remark 13.8. These strategies can be realized physically via the interaction with additional quantum systems, referred in this context to as ancillas (or ancillary qubits), see Theorem 13.3, where H_C is the ancilla. For this reason the authors of [208] point out that in quantum setting all strategies can be considered as 'deterministic'.

With the general unitary operators U as above,

$$U|0\rangle = \cos\theta\cos\phi|0\rangle + i\cos\theta\sin\phi|0\rangle + i\sin\theta\sin\psi|1\rangle - \sin\theta\cos\psi|1\rangle$$

$$= \cos\theta e^{i\phi}|0\rangle - \sin\theta e^{-i\psi}|1\rangle,$$

$$U|1\rangle = \cos\theta\cos\phi|1\rangle - i\cos\theta\sin\phi|1\rangle + i\sin\theta\sin\psi|0\rangle + \sin\theta\cos\psi|0\rangle$$

$$= \cos\theta e^{-i\phi}|1\rangle + \sin\theta e^{i\psi}|0\rangle,$$

and

$$(U_1 \otimes \cdots \otimes U_N)J|0\rangle^{\otimes N} = \frac{1}{\sqrt{2}}U_1|0\rangle \otimes \cdots \otimes U_N|0\rangle + \frac{i}{\sqrt{2}}U_1|1\rangle \otimes \cdots \otimes U_N|1\rangle.$$

For instance, in the case $N = 2$,

$$U_1 \otimes U_2 \xi_{in} = \frac{1}{\sqrt{2}}(\cos\theta_1 e^{i\phi_1}|0\rangle - \sin\theta_1 e^{-i\psi_1}|1\rangle) \otimes (\cos\theta_2 e^{i\phi_2}|0\rangle - \sin\theta_2 e^{-i\psi_2}|1\rangle)$$

$$+ \frac{i}{\sqrt{2}}(\cos\theta_1 e^{-i\phi_1}|1\rangle + \sin\theta_1 e^{i\psi_1}|0\rangle) \otimes (\cos\theta_2 e^{-i\phi_2}|1\rangle + \sin\theta_2 e^{i\psi_2}|0\rangle).$$

$$\xi_{fin} = \frac{1}{2}(\cos\theta_1 e^{i\phi_1}|0\rangle - \sin\theta_1 e^{-i\psi_1}|1\rangle) \otimes (\cos\theta_2 e^{i\phi_2}|0\rangle - \sin\theta_2 e^{-i\psi_2}|1\rangle)$$

$$+ \frac{i}{2}(\cos\theta_1 e^{-i\phi_1}|1\rangle + \sin\theta_1 e^{i\psi_1}|0\rangle) \otimes (\cos\theta_2 e^{-i\phi_2}|1\rangle + \sin\theta_2 e^{i\psi_2}|0\rangle).$$

$$- \frac{i}{2}(\cos\theta_1 e^{i\phi_1}|1\rangle - \sin\theta_1 e^{-i\psi_1}|0\rangle) \otimes (\cos\theta_2 e^{i\phi_2}|1\rangle - \sin\theta_2 e^{-i\psi_2}|0\rangle)$$

$$+ \frac{1}{2}(\cos\theta_1 e^{-i\phi_1}|0\rangle + \sin\theta_1 e^{i\psi_1}|1\rangle) \otimes (\cos\theta_2 e^{-i\phi_2}|0\rangle + \sin\theta_2 e^{i\psi_2}|1\rangle)$$

$$= \sum_{k,l=0}^{1} \xi_{kl}|kl\rangle,$$

$$\xi_{00} = \frac{1}{2}\cos\theta_1 e^{i\phi_1}\cos\theta_2 e^{i\phi_2} + \frac{1}{2}i\sin\theta_1 e^{i\psi_1}\sin\theta_2 e^{i\psi_2}$$

$$- \frac{1}{2}i\sin\theta_1 e^{-i\psi_1}\sin\theta_2 e^{-i\psi_2} + \cos\theta_1 e^{-i\phi_1}\cos\theta_2 e^{-i\phi_2}$$

$$= \cos\theta_1\cos\theta_2\cos(\phi_1 + \phi_2) - \sin\theta_1\sin\theta_2\sin(\psi_1 + \psi_2), \qquad (13.63)$$

$$\xi_{11} = \frac{1}{2}\sin\theta_1 e^{-i\psi_1}\sin\theta_2 e^{-i\psi_2} + \frac{1}{2}i\cos\theta_1 e^{-i\phi_1}\cos\theta_2 e^{-i\phi_2}$$

$$- \frac{1}{2}i\cos\theta_1 e^{i\phi_1}\cos\theta_2 e^{i\phi_2} + \frac{1}{2}\sin\theta_1 e^{i\psi_1}\sin\theta_2 e^{i\psi_2}$$

$$= \sin\theta_1 \sin\theta_2 \cos(\psi_1 + \psi_2) + \cos\theta_1 \cos\theta_2 \sin(\phi_1 + \phi_2), \qquad (13.64)$$

$$\xi_{01} = -\frac{1}{2}\cos\theta_1 e^{i\phi_1}\sin\theta_2 e^{-i\psi_2} + \frac{i}{2}\sin\theta_1 e^{i\psi_1}\cos\theta_2 e^{-i\phi_2}$$

$$\frac{i}{2}\sin\theta_1 e^{-i\psi_1}\cos\theta_2 e^{i\phi_2} \frac{1}{2}\cos\theta_1 e^{-i\phi_1}\sin\theta_2 e^{i\psi_2}$$

$$= i\cos\theta_1 \sin\theta_2 \sin(\psi_2 - \phi_1) + i\sin\theta_1 \cos\theta_2 \cos(\phi_2 - \psi_1), \qquad (13.65)$$

$$\xi_{10} = i\cos\theta_2 \sin\theta_1 \sin(\psi_1 - \phi_2) + i\sin\theta_2 \cos\theta_1 \cos(\phi_1 - \psi_2). \qquad (13.66)$$

For the probabilities of the four outcomes we thus have

$$P_{00} = |\xi_{00}|^2 = \cos^2\theta_1 \cos^2\theta_2 \cos^2(\phi_1 + \phi_2) + \sin^2\theta_1 \sin^2\theta_2 \sin^2(\psi_1 + \psi_2)$$

$$-2\sin\theta_1 \sin\theta_2 \cos\theta_1 \cos\theta_2 \cos(\phi_1 + \phi_2)\sin(\psi_1 + \psi_2),$$

$$P_{11} = |\xi_{11}|^2 = \cos^2\theta_1 \cos^2\theta_2 \sin^2(\phi_1 + \phi_2) + \sin^2\theta_1 \sin^2\theta_2 \cos^2(\psi_1 + \psi_2)$$

$$+2\sin\theta_1 \sin\theta_2 \cos\theta_1 \cos\theta_2 \cos(\psi_1 + \psi_2)\sin(\phi_1 + \phi_2),$$

$$P_{01} = |\xi_{01}|^2 = \cos^2\theta_1 \sin^2\theta_2 \sin^2(\psi_2 - \phi_1) + \sin^2\theta_1 \cos^2\theta_2 \cos^2(\phi_2 - \psi_1),$$

$$+2\sin\theta_1 \sin\theta_2 \cos\theta_1 \cos\theta_2 \sin(\psi_2 - \phi_1)\cos(\phi_2 - \psi_1),$$

$$P_{10} = |\xi_{10}|^2 = \cos^2\theta_2 \sin^2\theta_1 \sin^2(\psi_1 - \phi_2) + \sin^2\theta_2 \cos^2\theta_1 \cos^2(\phi_1 - \psi_2)$$

$$+2\sin\theta_1 \sin\theta_2 \cos\theta_1 \cos\theta_2 \sin(\psi_1 - \phi_2)\cos(\phi_1 - \psi_2).$$

Exercise 13.6. Check that these probabilities really sum up to 1.

Proposition 13.1. *Unless there is an outcome $|\sigma\tau\rangle$ such that it gives the best payoff to both Alice and Bob, there is no Nash equilibrium for full pure quantum strategies $SU(2)$ for a two-player two-action game.*

Proof. It is seen from formulas (13.63)-(13.66) that whatever choice of parameter θ_2, ϕ_2, ψ_2 is made by Bob, Alice can choose her θ_1, ϕ_1, ψ_1 in a way that would make any of the coefficients $\xi_{00}, \xi_{01}, \xi_{10}, \xi_{11}$ equal 1 in magnitude, and thus to ensure the corresponding outcome to occur with probability 1. For instance, in order to achieve $\xi_{00} = 1$, Alice can choose $\phi_1 = -\phi_2$, $\psi_1 = (\pi/2) - \psi_2$, which turns ξ_{00} to

$$\cos\theta_1 \cos\theta_2 - \sin\theta_1 \sin\theta_2 = \cos(\theta_1 + \theta_2)$$

and then $\theta_1 = -\theta_2$ converts this to 1. The same is possible for Bob under any strategy of Alice. Hence a Nash equilibrium can be only an outcome that gives the best payoff to both Alice and Bob.

Remarkably enough, for $N > 2$ the situation changes drastically. For $N = 3$ we have

$$U_1 \otimes U_2 \otimes U_3 \xi_{in} = \frac{1}{\sqrt{2}}(\cos\theta_1 e^{i\phi_1}|0\rangle - \sin\theta_1 e^{-i\psi_1}|1\rangle) \otimes (\cdots 2 \cdots) \otimes (\cdots 3 \cdots)$$

$$+ \frac{i}{\sqrt{2}}(\cos\theta_1 e^{-i\phi_1}|1\rangle + \sin\theta_1 e^{i\psi_1}|0\rangle) \otimes (\cdots 2 \cdots) \otimes (\cdots 3 \cdots).$$

$$\xi_{fin} = \frac{1}{2}(\cos\theta_1 e^{i\phi_1}|0\rangle - \sin\theta_1 e^{-i\psi_1}|1\rangle) \otimes (\cdots 2 \cdots) \otimes (\cdots 3 \cdots)$$

$$+ \frac{i}{2}(\cos\theta_1 e^{-i\phi_1}|1\rangle + \sin\theta_1 e^{i\psi_1}|0\rangle) \otimes (\cdots 2 \cdots) \otimes (\cdots 3 \cdots)$$

$$- \frac{i}{2}(\cos\theta_1 e^{i\phi_1}|1\rangle - \sin\theta_1 e^{-i\psi_1}|0\rangle) \otimes (\cdots 2 \cdots) \otimes (\cdots 3 \cdots)$$

$$+ \frac{1}{2}(\cos\theta_1 e^{-i\phi_1}|0\rangle + \sin\theta_1 e^{i\psi_1}|1\rangle) \otimes (\cdots 2 \cdots) \otimes (\cdots 3 \cdots)$$

$$= \sum_{k,l,m=0}^{1} \xi_{klm}|klm\rangle,$$

where the second and the third brackets reproduce the first one with all indices changed to 2 or 3 respectively. This is the linear combination of the 8 basis vectors. But only 4 need to be calculated, as the other are obtained by permutations. We have

$$\xi_{000} = \frac{1}{2}\cos\theta_1 e^{i\phi_1} \cos\theta_2 e^{i\phi_2} \cos\theta_3 e^{i\phi_3} + \frac{1}{2}i \sin\theta_1 e^{i\psi_1} \sin\theta_2 e^{i\psi_2} \sin\theta_3 e^{i\psi_3}$$

$$+ \frac{1}{2}i \sin\theta_1 e^{-i\psi_1} \sin\theta_2 e^{-i\psi_2} \sin\theta_3 e^{-i\psi_3} + \frac{1}{2}\cos\theta_1 e^{-i\phi_1} \cos\theta_2 e^{-i\phi_2} \cos\theta_3 e^{-i\phi_3}$$

$$= \cos\theta_1 \cos\theta_2 \cos\theta_3 \cos(\phi_1 + \phi_2 + \phi_3) + i \sin\theta_1 \sin\theta_2 \sin\theta_3 \cos(\psi_1 + \psi_2 + \psi_3),$$

$$\xi_{111} = -\frac{1}{2}\sin\theta_1 e^{-i\psi_1} \sin\theta_2 e^{-i\psi_2} \sin\theta_3 e^{-i\psi_3}$$

$$+ \frac{1}{2}i \cos\theta_1 e^{-i\phi_1} \cos\theta_2 e^{-i\phi_2} \cos\theta_3 e^{-i\phi_3}$$

$$- \frac{1}{2}i \cos\theta_1 e^{i\phi_1} \cos\theta_2 e^{i\phi_2} \cos\theta_3 e^{i\phi_3} + \frac{1}{2}\sin\theta_1 e^{i\psi_1} \sin\theta_2 e^{i\psi_2} \sin\theta_3 e^{i\psi_3}$$

$$= i \sin\theta_1 \sin\theta_2 \sin\theta_3 \sin(\psi_1 + \psi_2 + \psi_3) + \cos\theta_1 \cos\theta_2 \cos\theta_3 \sin(\phi_1 + \phi_2 + \phi_3),$$

$$\xi_{001} = -\frac{1}{2}\cos\theta_1 e^{i\phi_1} \cos\theta_2 e^{i\phi_2} \sin\theta_3 e^{-i\psi_3} + \frac{1}{2}i \sin\theta_1 e^{i\psi_1} \sin\theta_2 e^{i\psi_2} \cos\theta_3 e^{-i\phi_3}$$

$$-\frac{1}{2}i\sin\theta_1 e^{-i\psi_1}\sin\theta_2 e^{-i\psi_2}\cos\theta_3 e^{i\phi_3} + \frac{1}{2}\cos\theta_1 e^{-i\phi_1}\cos\theta_2 e^{-i\phi_2}\sin\theta_3 e^{i\psi_3}$$

$$= i\cos\theta_1\cos\theta_2\sin\theta_3\sin(\psi_3-\phi_1-\phi_2)+\sin\theta_1\sin\theta_2\cos\theta_3\sin(\phi_3-\psi_1-\psi_2),$$

$$\xi_{011} = \frac{1}{2}\cos\theta_1 e^{i\phi_1}\sin\theta_2 e^{-i\psi_2}\sin\theta_3 e^{-i\psi_3}$$

$$+\frac{1}{2}i\sin\theta_1 e^{i\psi_1}\cos\theta_2 e^{-i\phi_2}\cos\theta_3 e^{-i\phi_3}$$

$$+\frac{1}{2}i\sin\theta_1 e^{-i\psi_1}\cos\theta_2 e^{i\phi_2}\cos\theta_3 e^{i\phi_3} + \frac{1}{2}\cos\theta_1 e^{-i\phi_1}\sin\theta_2 e^{i\psi_2}\sin\theta_3 e^{i\psi_3}$$

$$= \cos\theta_1\sin\theta_2\sin\theta_3\cos(\psi_2+\psi_3-\phi_1)+i\sin\theta_1\cos\theta_2\cos\theta_3\cos(\phi_3+\phi_2-\psi_1).$$

In [208] a detailed discussion is devoted to the quantized version of the famous minority game. In its classical versions the players are supposed to choose 0 or 1 and submit to the referee. Those whose choice turns out to be in minority get one point reward each. If there is an even split, or all player made the same choice, no payments arise.

Probability for the 1st player to be in minority is

$$|\xi_{011}|^2 + |\xi_{100}|^2 = \cos^2\theta_1\sin^2\theta_2\sin^2\theta_3\cos^2(\phi_1-\psi_2-\psi_3)$$

$$+\sin^2\theta_1\cos^2\theta_2\cos^2\theta_3\cos^2(\psi_1-\phi_2-\phi_3)$$

$$+\cos^2\theta_1\sin^2\theta_2\sin^2\theta_3\sin^2(\phi_1-\psi_2-\psi_3)$$

$$+\sin^2\theta_1\cos^2\theta_2\cos^2\theta_3\sin^2(\psi_1-\phi_2-\phi_3)$$

$$= \cos^2\theta_1\sin^2\theta_2\sin^2\theta_3 + \sin^2\theta_1\cos^2\theta_2\cos^2\theta_3,$$

which is the same as in the classical game for $\cos^2\theta = p$ denoting the probability of flipping (or of choosing 1). Hence for $N=3$ player the quantum version of the minority game does not offer anything new.

The situation changes when the number of payers increases. The analysis of these cases exploits a simple observation that for minority games the result is not changed whether or not the final gate J^* is applied. In fact, J^* transforms any basis vectors $j_1\cdots j_k\rangle$ within the sub-space generated by $j_1\cdots j_k\rangle$ and $j'_1\cdots j'_k\rangle$ (prime denotes the complimentary index), but both these vectors yield the same payoff. Hence for these particular games EWL and MW schemes are equivalent. It is shown in [208] that new (and more profitable than classical) equilibria arise for the minority games with $N>3$. An example of such equilibrium for $N=4$ is (u,u,u,u), where

$$u = \frac{1}{\sqrt{2}}\cos\frac{\pi}{16}(I+i\sigma_x) + \frac{1}{\sqrt{2}}\sin\frac{\pi}{16}(i\sigma_y - i\sigma_z).$$

As also shown in [208], there exist games of 3 player where new profitable equilibria arise. For instance, the game with the table of type (13.62):

profile	(C,C,C)	(D,D,D)	(D,C,C)	(D,D,C)
payoff	(2,2,2)	(0,0,0)	(1,9,9)	(-9,-9,1)

have new profitable equilibria in its quantum version. The equilibria are given by unitary strategies, but represent equilibria even if considered among all $TP - CP$ strategies (which is proved using the Kraus representations for such maps). On the other hand, there are examples, for instance given by the table

profile	(C,C,C)	(D,D,D)	(D,C,C)	(D,D,C)
payoff	(-9,-9,-9)	(7,7,7)	(8,-9,-9)	(1,1,-9)

where classical rules produce outcomes (Nash equilibria) with better performance than their quantum counterparts.

13.8 Quantization of games with continuous strategy spaces

The extension of EWL protocol for games with initially continuous strategy space was first suggested in [271].

The underlying classical model was that of Cournot's duopoly. Recall that, for $Q = q_1 + q_2$ denoting the total amount of a product produced by two firms, one assumes that the price per unit of the product equals $P(Q) = (a - Q)^+ = \max(0, a - Q)$. If c is the cost of the production of a unit of the product, the profits of two firms are (for $a \geq Q$)

$$u_j(q_1, q_2) = q_j[P(Q) - c] = q_j[a - c - (q_1 + q_2)]. \tag{13.67}$$

Though the unique Nash equilibrium is $q_1^* = q_2^* = (a - c)/3$ with each firm getting $(a - c)^2/9$, the cooperative behavior would be to choose $q_1' = q_2' = (a - c)/4$ yielding to each firm the better profit $(a - c)^2/4$.

To quantise this game let us assume that each player is working with the Hilbert space $L^2(\mathbf{R})$. The simplest initial functions for both players are the Gaussian packets

$$\psi_j(x_j) = (\pi h)^{-1/4} \exp\left\{-\frac{x_j^2}{2h}\right\}$$

(normalized to $\int |\psi_j(x_j)|^2 dx_j = 1$, so that

$$\psi^{in}(x_1, x_2) = \psi_1(x_1)\psi_2(x_2) = (\pi h)^{-1/2} \exp\{-\frac{x_1^2 + x_2^2}{2h}\}. \tag{13.68}$$

The two basic operators in $L^2(\mathbf{R})$ are the operator X of multiplication by the variable x and the momentum operator $P = -ihd/dx$. The unitary shift operators

$$D(y)f(x) = \exp\{-iyP/h\}f(x) = \exp\{-yd/dx\}f(x) = f(x-y)$$

are the simplest possible operators allowing the players to manipulate their positions (the amount of product to produce). Therefore they are natural candidates to be chosen as possible actions of the players. Thus, copying the finite-dimensional EWL scheme, we can introduce a quantum version of Cournot's game by asserting that the final state of the system should be

$$\psi_{y_1,y_2}^{fin} = J^*[D_1(y_1) \otimes D_2(y_2)]J\psi_{in} \tag{13.69}$$

with an appropriately chosen unitary entangling operator J on $L^2(\mathbf{R}) \otimes L^2(\mathbf{R}) = L^2(\mathbf{R}^2)$. By the canonical interpretation of the wave mechanics, the probability distribution of finding a system described by the wave function $\psi(x_1, x_2)$ in a position (x_1, x_2) has the probability density $|\psi(x_1, x_2)|^2$. Hence the average positions and final payoffs to the players can be calculated by the formulas

$$q_j(y_1, y_2) = \mathbf{E}[x_j] = \int \int x_j |\psi_{y_1,y_2}^{fin}(x_1, x_2)|^2 \, dx_1 dx_2, \tag{13.70}$$

$$u_j(y_1, y_2) = \mathbf{E}[x_j(a - c - (x_1 + x_2))]$$

$$= \int \int [x_j(a - c - (x_1 + x_2))]|\psi_{y_1,y_2}^{fin}(x_1, x_2)|^2 \, dx_1 dx_2. \tag{13.71}$$

Looking at the simplest J that may mix up the variables, one can copy the unitary rotations of (13.13) and suggest to use J of the type

$$Jf(x_1, x_2) = f(U(x_1, x_2)), \quad U(x_1, x_2) = \begin{pmatrix} \cos c & \sin c \\ -\sin c & \cos c \end{pmatrix} \begin{pmatrix} x_1 \\ x_2 \end{pmatrix}. \tag{13.72}$$

However, by physical reasons (see some comments below) the authors of [274] suggest to use instead the 'Lorenz rotations':

$$J_\gamma f(x_1, x_2) = f(U_\gamma(x_1, x_2)),$$

$$U_\gamma(x_1, x_2) = \begin{pmatrix} \cosh\gamma & \sinh\gamma \\ \sinh\gamma & \cosh\gamma \end{pmatrix} \begin{pmatrix} x_1 \\ x_2 \end{pmatrix} = \begin{pmatrix} x_1\cosh\gamma + x_2\sinh\gamma \\ x_1\sinh\gamma x_1 + x_2\cosh\gamma \end{pmatrix}, \tag{13.73}$$

with the inverse operator

$$J_\gamma^* f(x_1, x_2) = J_\gamma^{-1} f(x_1, x_2) = f(U_\gamma^{-1}(x_1, x_2)),$$

$$U_\gamma^{-1}(x_1, x_2) = \begin{pmatrix} \cosh\gamma & -\sinh\gamma \\ -\sinh\gamma & \cosh\gamma \end{pmatrix} \begin{pmatrix} x_1 \\ x_2 \end{pmatrix}. \tag{13.74}$$

With this choice of J_γ and denoting $x = (x_1, x_2)$, $y = (y_1, y_2)$, we get for an arbitrary $\psi(x_1, x_2)$ that

$$(J\psi)(x_1, x_2) = (J_\gamma\psi)(x) = \psi(Ux),$$

$$[D_1(y_1) \otimes D_2(y_2)]J_\gamma\psi(x_1, x_2) = \psi(Ux - Uy),$$

$$J_\gamma^*[D_1(y_1) \otimes D_2(y_2)]J_\gamma\psi(x_1, x_2) = \psi(x - Uy).$$

Therefore, with ψ^{in} given by (13.68),

$$\psi_{y_1,y_2}^{fin} = J_\gamma^*[D_1(y_1) \otimes D_2(y_2)]J_\gamma\psi_{in}$$

$$= (\pi h)^{-1/2}\exp\{-\frac{1}{2h}[(x_1 - y_1\cosh\gamma - y_2\sinh\gamma)^2$$
$$+ (x_2 - y_1\sinh\gamma + y_2\cosh\gamma)^2]\}. \tag{13.75}$$

The average positions in this state, defined by (13.70), equal

$$q_1(y_1, y_2) = y_1\cosh\gamma + y_2\sinh\gamma, \quad q_1(y_1, y_2) = y_1\sinh\gamma + y_2\cosh\gamma. \tag{13.76}$$

Though in principle we are mostly interested in payoffs (13.71), the final simplification suggested in [271] is that before the measurement, the final state ψ_{y_1,y_2}^{fin} is squeezed in a way that it effectively becomes the δ-function centered at the mean position (q_1, q_2), and therefore, instead of (13.71), the payoffs simplify to (13.67) with (q_1, q_2) given by (13.76):

$$u_1^\gamma(y_1, y_2) = q_1[a - c - (q_1 + q_2)] = (y_1\cosh\gamma + y_2\sinh\gamma)[a - c - e^\gamma(y_1 + y_2)],$$
$$u_2^\gamma(y_1, y_2) = q_2[a - c - (q_1 + q_2)] = (y_1\sinh\gamma + y_2\cosh\gamma)[a - c - e^\gamma(y_1 + y_2)]. \tag{13.77}$$

Solving for the Nash equilibrium, that is, solving the equations

$$\frac{\partial u_1}{\partial y_1} = 0, \quad \frac{\partial u_2}{\partial y_2} = 0, \tag{13.78}$$

yields the equilibrium

$$y_1^* = y_2^* = \frac{(a - c)\cosh\gamma}{1 + 2e^{2\gamma}} \tag{13.79}$$

with the profit

$$u_1^\gamma(y_1^*, y_2^*) = u_2^\gamma(y_1^*, y_2^*) = \frac{(a - c)^2 e^\gamma \cosh\gamma}{(3\cosh\gamma + \sinh\gamma)^2}. \tag{13.80}$$

As $\gamma = 0$ we recover the classical game. But as $\gamma \to \infty$, we have

$$\lim_{\gamma \to \infty} u_j^\gamma(y_1^*, y_2^*) = (a - c)^2/8, \qquad (13.81)$$

which is the effective outcome. Thus in this limit the dilemma between the Nash equilibrium and the Pareto optimum disappears.

Exercise 13.7. Calculate the Nash equilibrium using the full formula (13.71) instead of its simplified version (13.77).

Remark 13.9. Physical realization of quantum games are usually performed via the methods of quantum optics. There the main role is played by the creation and annihilation operators $\hat{a}^\pm$ of quantum oscillators, which are given by the formulas

$$\hat{a}^\pm = \frac{1}{\sqrt{2\omega h}}(\omega X \mp iP),$$

or equivalently

$$X = \sqrt{\frac{\hbar}{2\omega}}(\hat{a}^- + \hat{a}^+), \quad P = -i\sqrt{\frac{\hbar\omega}{2}}(\hat{a}^- - \hat{a}^+).$$

In quantum optics the operators X and P are referred to as the quadratures (of a single mode of the electromagnetic field given by $\hat{a}^\pm$). In paper [271] the units with $\omega = 1$ and $h = 1$ are used, in which case it is seen that the operator J of (13.73) is given by the formula

$$J(\gamma) = \exp\{-\gamma(\hat{a}_1^+ \hat{a}_2^+ - \hat{a}_1^- \hat{a}_2^-)\} = \exp\{i\gamma(X_1 P_2 + X_2 P_1)\}, \qquad (13.82)$$

and $J_\gamma \psi_{in}$ turns out to represent the important two-mode squeezed vacuum state used in the theory of quantum teleportation.

In [274] the above scheme (again with the simplification (13.77)) was used to analyze the Stackelberg duopoly. The difference with the above game is that now the moves are sequential. Firstly the first firm makes the move by choosing y_1, and then the second firm makes the move choosing its y_2 that maximises its profit given y_1. Thus the optimal choice of the second firm arises from solving the second equation in (13.78) yielding

$$y_2(y_1) = \frac{(a - c)\cosh\gamma - y_1 e^{2\gamma}}{1 + e^{2\gamma}}.$$

Then the first firm should find y_1 maximising

$$u_1^\gamma(y_1, y_2(y_1)) = (y_1 \cosh\gamma + y_2(y_1)\sinh\gamma)[a - c - e^\gamma(y_1 + y_2(y_1))].$$

Simple analysis yields the optimal value

$$y_1^* = \frac{(a-c)(1+\cosh(2\gamma)}{2(\cosh\gamma + e^\gamma)},$$

with the corresponding optimal $y_2^* = y_2(y_1^*)$. Of course, the optimal profit of the second firm turns out to be lower than the optimal profit of the first firm (advantage of the first move). Moreover, the difference between the optimal profit of the two firms is a monotonically increasing function with respect to the 'entangling parameter' γ.

In paper [305] the above results were extended to the case of several firms. The arguments and results are mostly analogous (the calculations being of course heavier). Let us notice only that the operator J of (13.73) or (13.82) is generalized to the operator

$$J = \exp\{-\sum_{i \neq j} \gamma_j (\hat{a}_1^+ \hat{a}_2^+ - \hat{a}_1^- \hat{a}_2^-)\}.$$

13.9 Finite-dimensional quantum mechanics of open systems

The transformations of open quantum systems may be performed by more general operators than unitary. Namely, one defines *operations* between the state spaces $\mathcal{T}_s(H_A)$ and $\mathcal{T}_s(H_B)$ as positive linear maps $\mathcal{T}_s(H_A) \to \mathcal{T}_s(H_B)$ (that take positive linear operators to positive linear operators), which are contractions in the trace norm:

$$0 \leq \text{tr}[T(\rho)] \leq \text{tr}(\rho) \qquad (13.83)$$

for any $\rho \in \mathcal{T}^+(H_A)$. Since $|T(\rho)| = T(|\rho|)$ for a positivity preserving T inequality (13.83) is equivalent to the inequality

$$0 \leq \text{tr}|T(\rho)| \leq \text{tr}|\rho| \qquad (13.84)$$

for any $\rho \in \mathcal{T}_s(H_A)$.

Remark 13.10. Some authors define operations as CP-maps introduced below.

Applying duality (13.3), for any $T \in \mathcal{L}(\mathcal{T}_s(H_A), \mathcal{T}_s(H_B))$ one can define the dual map $T^* : \mathcal{L}(\mathcal{L}_s(H_B), \mathcal{L}_s(H_A))$ via the equation

$$\text{tr}[T(\rho)\sigma] = \text{tr}[\rho T^*(\sigma)]. \qquad (13.85)$$

If T is positive, then T^* is also positive (as follows from (13.85)). Contraction property (13.83) is equivalent to $T^* \mathbf{1} \leq \mathbf{1}$, and the preservation of the trace by T is equivalent to the preservation of unity by T^*: $T^*(\mathbf{1}) = \mathbf{1}$.

The following simple result is crucial for the theory of games.

Lemma 13.1. *Positive contractions preserving or not increasing trace (or preserving or not increasing the unity operator) form a convex compact set in $\mathcal{L}(\mathcal{T}_s(H_A), \mathcal{T}_s(H_B))$.*

Proof. It is straightforward to see that any of the 4 sets mentioned are convex and closed in $\mathcal{L}(\mathcal{T}_s(H_A), \mathcal{T}_s(H_B))$. The only thing to check for compactness is thus the boundedness, and it follows from (13.84). $\square$

This duality allows for the most straightforward method to introduce the important notion of the partial trace. Namely, by duality (13.85), if ρ is a state on $\tilde{H}$, the positive linear map $\sigma \mapsto \sigma \otimes \rho$ from $\mathcal{T}(H) \to \mathcal{T}(H \otimes \tilde{H})$ has the adjoint positive linear map $E_\rho : \mathcal{L}(H \otimes \tilde{H}) \to \mathcal{L}(H)$, called the *partial trace*. This mapping satisfies the equation

$$\mathrm{tr}[E_\rho(A)\sigma] = \mathrm{tr}[A(\sigma \otimes \rho)], \tag{13.86}$$

with $A \in \mathcal{L}(H \otimes \tilde{H}), \sigma \in \mathcal{T}(H)$. Moreover,

$$E_\rho(B \otimes D) = B\,\mathrm{tr}(D\rho), \tag{13.87}$$

because

$$\mathrm{tr}[E_\rho(B \otimes D)\sigma] = \mathrm{tr}[B \otimes D(\sigma \otimes \rho)] = \mathrm{tr}[B\sigma]\mathrm{tr}[D\rho].$$

Since any operator in $\mathcal{L}(H \otimes \tilde{H})$ is a linear combinations of the product operators of type $B \otimes D$, formula (13.87) can be taken as an equivalent definition of the partial trace.

In particular, if $\rho = \mathbf{1}$, formula (13.87) reduces to

$$\mathrm{tr}_{\tilde{H}}(B \otimes D) = E_\mathbf{1}(B \otimes D) = B\,\mathrm{tr}D, \tag{13.88}$$

the new left notation being seemingly the most commonly used one.

For example, if $\psi = (|00\rangle + |11\rangle)/\sqrt{2} \in H_1 \otimes H_2 = \mathbf{C}^2 \otimes \mathbf{C}^2$, the corresponding density matrix is

$$\rho = |\psi\rangle\langle\psi| = \frac{1}{2}\sum_{j,k=0}^{1} |j\rangle\langle k| \otimes |j\rangle\langle k|$$

and its partial trace is

$$\mathrm{tr}_{H_2}\rho = \mathrm{tr}_{H_1}\rho = \frac{1}{2}\sum_{j=0}^{1} |j\rangle\langle j| = \begin{pmatrix} 1/2 & 0 \\ 0 & 1/2 \end{pmatrix}.$$

Important fact is that any state ρ can be written as a partial trace of a pure state, called a *purification* of ρ. In fact, for any state ρ in H, in the basis $|\xi_j\rangle$, where ρ is diagonal, it can be written as $\rho = \sum_j \rho_j |\xi_j\rangle\langle\xi_j|$, and a possible choice of pure state is $|\psi\rangle\langle\psi|$ with

$$|\psi\rangle = \sum_j \sqrt{\rho_j} |\xi_j\rangle \otimes |\eta_j\rangle$$

in $H \otimes H$, where η_j is any orthonormal basis (for instance, $\eta_j = \xi_j$, or $\eta_j = \bar{\xi}_j$). More precisely, $|\psi\rangle$ can be chosen to lie in $H \otimes \tilde{H}$, where the dimension of $\tilde{H}$ equals the rank of ρ (the number of non-vanishing ρ_j). Then

$$|\psi\rangle\langle\psi| = \sum_{j,k} \sqrt{\rho_j \rho_k}(|\xi_j\rangle \otimes |\eta_j\rangle)(\langle\xi_k| \otimes \langle\eta_k|) = \sum_{j,k} \sqrt{\rho_j \rho_k}|\xi_j\rangle\langle\xi_k| \otimes |\eta_j\rangle\langle\eta_k|.$$

Taking partial trace only terms with $j = k$ survive, because $\mathrm{tr}|\eta_j\rangle\langle\eta_k| = \delta_j^k$ yielding

$$\mathrm{tr}_{\tilde{H}}|\psi\rangle\langle\psi| = \sum_j \rho_j |\xi_j\rangle\langle\xi_j| = \rho.$$

The possibility of purification gives rise to the important measures of distances between the states. Namely, one defines the *fidelity* and the *fidelity distance* between two states ρ and γ respectively as

$$F(\rho,\gamma) = \max\{|\langle\xi|\eta\rangle| : \mathrm{tr}_{\tilde{H}}(|\xi\rangle\rangle\xi|) = \rho, \quad \mathrm{tr}_{\tilde{H}}(|\eta\rangle\rangle\eta|) = \gamma\},$$

$$d_F(\rho,\gamma) = \min\{\| \, |\xi\rangle - |\eta\rangle\| : \mathrm{tr}_{\tilde{H}}(|\xi\rangle\rangle\xi|) = \rho, \quad \mathrm{tr}_{\tilde{H}}(|\eta\rangle\rangle\eta|) = \gamma\}.$$

Employing the bases $|e_i^A\rangle\langle e_i^A|$ in $\mathcal{T}_s(H_A)$ and $|e_i^B\rangle\langle e_i^B|$ in $\mathcal{T}_s(H_B)$ one can describe an operator $T : \mathcal{T}_s(H_A) \to \mathcal{T}_s(H_B)$ via its matrix

$$T_{(j,l),(i,k)} = \mathrm{tr}[|e_l^B\rangle\langle e_k^B|T(|e_i^A\rangle\langle e_j^A|)] = \langle e_k^B|T(|e_i^A\rangle\langle e_j^A|)e_l^B\rangle, \qquad (13.89)$$

so that

$$T(|e_i^A\rangle\langle e_j^A|) = \sum_{l,k} \mathrm{tr}[|e_l^B\rangle\langle e_k^B|T(|e_i^A\rangle\langle e_j^A|)]|e_l^B\rangle\langle e_k^B| = \sum_{l,k} T_{(j,l),(i,k)}|e_l^B\rangle\langle e_k^B|.$$

$$\qquad (13.90)$$

This matrix provides another representation for T as an operator in $H_A \otimes H_B$ acting as

$$T(e_k^A \otimes e_l^B) = \sum_{i,j} T_{(ij),(kl)} e_i^A \otimes e_j^B. \qquad (13.91)$$

Of interest are the invertible operations and the operations that preserve pure states. As an example let us see how they look like for the qubits (for the extension to arbitrary Hilbert spaces (see [222])).

Proposition 13.2. *Let* $T : \mathcal{T}_s(\mathbf{C}^2) \to \mathcal{T}_s(\mathbf{C}^2)$ *be an operation (a positive linear contraction).*

(i) If T *is invertible and* T^{-1} *is also an operation, then*

$$T(A) = UAU^{-1} \tag{13.92}$$

with U *a unitary or anti-unitary operator in* $\mathbf{C}^2$.

(ii) If T *preserves pure states, then either* T *is given by (13.92) or*

$$T(\rho) = \mathrm{tr}[\rho B]|\psi\rangle\langle\psi| \tag{13.93}$$

with some $B \in \mathcal{L}(C^2)$ *and a unit vector* ψ.

Proof. (i) Any element A of $\mathcal{T}_s(\mathbf{C}^2)$ can be written as (13.17) with real x_j. The operator A is seen to be positive if and only if

$$x_0 \geq 0 \text{ and } x_0^2 - x_1^2 - x_2^2 - x_3^2 \geq 0.$$

Let T be given by the matrix T_{jk}, $j, k \in \{0, 1, 2, 3\}$, in the coordinates $\{x_j\}$. First of all, since both T and T^{-1} do not increase trace, it follows that they both preserve trace, and hence $T_{00} = 1$ and $T_{01} = T_{02} = T_{03} = 0$. Therefore T can be described by the vector $a = (T_{10}, T_{20}, T_{30})$ and the 3×3-matrix $B = T_{ij}$ with $i, j \geq 1$.

Let us denote by x the vectors in $\mathbf{R}^3$ with coordinates x_1, x_2, x_3. The condition of the preservation of positivity implies that

$$\|x\| \leq 1 \implies \|a + Bx\| \leq 1.$$

Hence B does not increase volume and therefore $|\det B| \leq 1$. Since the same is true for B^{-1} it follows that $\det(B) = \pm 1$. Consequently B maps the unit ball $\|x\| \leq 1$ bijectively and onto the unit ball centered at $-a$. But by linearity the image of B should be a symmetric set (with each y it should also contain $-y$), and consequently $a = 0$. Hence B maps $\|x\| \leq 1$ bijectively onto itself and hence it is a linear isometry and thus an orthogonal matrix. If $B \in SO(3)$, then T is obtained via the dressing with $u \in SU(2)$. If $B \in O(3)$ with $\det(O) = -1$, then T is obtained by dressing with an anti-unitary operator (see (13.18)).

(ii) Let us prove it under the additional simplifying assumption that T preserves the trace (general case just a bit more lengthy). Then, as in (i), we can conclude that $T_{00} = 1$ and $T_{01} = T_{02} = T_{03} = 0$, and T can be described by the vector $a = (T_{10}, T_{20}, T_{30})$ and the 3×3-matrix $B = T_{ij}$ with $i, j \geq 1$. By the requirement of the preservation of pure states it then follows that T transform the ball $\|x\| \leq 1$ into the ball $\|a + Bx\| \leq 1$ in such a way that the boundary is also transformed into the boundary. It is then seen by simple topological considerations that this is possible either when $a = 0$ and B is invertible or when $B = 0$. In the first case we are back to (i), and in the second case

$$T : x_0 I + x_1\sigma_1 + x_2\sigma_2 + x_3\sigma_3 \mapsto x_0(I + a_1\sigma_1 + a_2\sigma_2 + a_3\sigma_3).$$

By the requirement of the conservation of purity, a is a vector of unit norm, and we get (13.93).

Remark 13.11. Much more complicated argument allows one to fully characterise all positive linear contractions $T : \mathcal{T}_s(\mathbf{C}^2) \to \mathcal{T}_s(\mathbf{C}^2)$. It turns out that any such operation is a finite sum of the operators of two types (referred to as completely positive and completely copositive operators): $\rho \mapsto V\rho V^*$ and $\rho \mapsto V\rho^T V^*$, where $\rho^T = \bar{\rho}$ is the transpose matrix to ρ (that equals to its complex conjugate by self-adjointness) and V some linear operators in $\mathbf{C}^2$, see [304], or other arguments leading to the same conclusion in [299] and [276]. This result does not extend to higher dimensions.

The most important class of operations constitute the so called *completely positive (CP)* operations. To define them let us notice that for any Hilbert space H the tensor product $H \otimes \mathbf{C}^n = H^n$ can be represented by n-dimensional vectors with elements from H:

$$h = (h_1, \ldots, h_n) = h_1 \otimes e_1 + \cdots + h_n \otimes e_n.$$

Moreover, the state space

$$\mathcal{T}(H \otimes \mathbf{C}^n) = \mathcal{T}(H) \otimes \mathcal{T}(\mathbf{C}^n)$$

can be identified with the space of $n \times n$-matrices with elements from $\mathcal{T}(H)$. In fact, if $\rho \in \mathcal{T}(H)$ and $A = (A_{jk}) \in \mathcal{T}(\mathbf{C}^n)$, then

$$(\rho \otimes A)(h) = \sum_j (\rho \otimes A)(h_j \otimes e_j) = \sum_{j,k} \rho(h_j)A_{kj}e_k,$$

so that

$$[(\rho \otimes A)(h)]_k = \sum_j A_{kj}\rho(h_j).$$

Moreover, for $A = (A_{jk}) \in \mathcal{T}(H \otimes \mathbf{C}^n)$, we can write

$$A = \sum_{j,k} A_{jk} \otimes E_{jk}$$

with the matrix E^{jk} with the elements $[E^{jk}]_{lm} = \delta_j^l \delta_m^k$, and therefore

$$\mathrm{tr}_{\mathbf{C}^n} A = \sum_j A_{jj}. \tag{13.94}$$

Any linear operator $T : \mathcal{T}(H_A) \to \mathcal{T}(H_B)$ can be lifted to $T_n : \mathcal{T}(H_A \otimes \mathbf{C}^n) \to \mathcal{T}(H_B \otimes \mathbf{C}^n)$ as

$$T_n(X \otimes Y) = (T(X) \otimes Y).$$

This definition means that T_n acts on matrices with elements from $\mathcal{T}(H_A)$ by transforming each elements by means of T.

Physics arguments suggest that realizable transformations of quantum state spaces $\mathcal{T}(H)$ should be positive and remain positive after lifting to $H \otimes \mathbf{C}^n$. Thus a linear map $T : \mathcal{T}(H_A) \to \mathcal{T}(H_B)$ is called *n-positive*, if $T_n : \mathcal{T}(H_A^n) \to \mathcal{T}(H_B^n)$ is positive, that is

$$(x, T_n(a)x) = \sum_{i,j}(x_i, T(a_{ij})x_j) \geq 0,$$

for any $x = (x_1, \ldots, x_n) \in H_B^n$, $a \in \mathcal{T}(H_A^n)$ positive. The map T is *completely positive (CP)* if T_n are positive for all $n \in N$.

From the definition of duality (13.85), it is straightforward to see that T is CP if and only if T^* is CP.

One is mostly interested in trace preserving CP map, referred to as $TP-CP$ operations. Such operations are also called *quantum communication channels* or *quantum-quantum channels* (in quantum communications) or *physically realizable operations* (in quantum computing).

Since any positive operator $a \in \mathcal{T}(H_A^n)$ can be written as $a = b^*b$ with some $b \in \mathcal{T}(H_A^n)$ (for instance one can choose $b = b^* = \sqrt{a}$), so that $a_{ij} = \sum_k b_{ki}^* b_{kj}$ with some $b_{ik} \in \mathcal{L}(H)$, and the operator b^*b is positive for any $b \in \mathcal{T}(H_A^n)$, it follows that T_n is positive if and only if

$$\sum_{i,j,k}(x_i, T(b_{ki}^* b_{kj})x_j) = (x_i, T[(b^*b)_{ij}]x_j) \geq 0 \qquad (13.95)$$

for any $x = (x_1, \ldots, x_n) \in H_B^n$.

Choosing matrices b such that $b_{ki} = \delta_m^k b_i$ with some fixed m and some b_i (that is a matrix with only one non-vanishing row) it follows from (13.95) that

$$\sum_{i,j}(x_i, T(b_i^* b_j)x_j) \geq 0 \qquad (13.96)$$

for any $x = (x_1, \ldots, x_n) \in H_B^n$ and $b_i \in \mathcal{L}(H)$. In other words, the matrix $T[b_i^* b_j]$ is positive definite for any $b_1, \ldots, b_n \in \mathcal{T}(H_A)$. Mappings T for which this holds for any n are often referred to as the mappings of *positive type* or *positive definite*. By linearity, (13.96) also implies (13.95). Thus we arrive at important conclusion that T being of positive type is an equivalent property to being completely positive.

Remark 13.12. A mapping $E : \Omega \times \Omega \to \mathcal{T}(H)$ for any set Ω is called positive definite or of positive type if the matrix $E(a_i, a_j)$ is positive definite in H^n for any n and $a = (a_1, \ldots, a_n) \in \Omega^n$. Thus positive type used above refers to the mapping $\mathcal{T}(H_A) \times \mathcal{T}(H_A) \to \mathcal{T}(H_B)$ given by $(a, b) \mapsto a^* b$.

The following result from [298] gives the fundamental *Stinespring representation* for CP maps.

Theorem 13.1. *The linear mapping $T : \mathcal{T}(H_A) \to \mathcal{T}(H_B)$ is completely positive, if and only if*

$$T(X) = V^* \rho(X) V \tag{13.97}$$

with some $*$-representation ρ of $\mathcal{T}(H_A)$ in some finite-dimensional Hilbert space H (that is ρ is a linear mapping $\mathcal{T}(H_A) \to \mathcal{T}(H)$ such that $\rho(a^*) = [\rho(a)]^*$, $\rho(1) = 1$, and $\rho(ab) = \rho(a)\rho(b)$) and a linear $V : H_B \to H$. If $T(1) = 1$, then the operator V is a (possibly partial) isometry: $V^* V = 1$.

Proof. If (13.97) holds, then

$$\sum_{i,j} (x_i, T(b_i^* b_j) x_j) = \sum_{i,j} (V x_i, \rho(b_i^*) \rho(b_j) V x_j)$$

$$= \sum_{i,j} (\rho(b_i) V x_i, \rho(b_j) V x_j) = \left\| \sum_i \rho(b_i) V x_i \right\|^2 \geq 0.$$

Conversely, let T be CP. On the tensor product $\mathcal{T}(H_A) \otimes H_B$ we can define the Hermitian form (linear with respect to the second variable and conjugate linear with respect to the first one) as follows:

$$(\phi, \psi)_T = \sum_{i,j} (x_i, T(b_i^* a_j) y_j)$$

for $\phi = \sum_i b_i \otimes x_i$, $\psi = \sum_j a_j \otimes y_j$. By (13.95) this form is positive definite, that is $(\phi, \phi)_T \geq 0$ for any ϕ. Hence it satisfies the Cauchy-Schwarz inequality $|(\phi, \psi)_T|^2 \leq (\phi, \phi)_T (\psi, \psi)_T$. Therefore the null-space N of this form, $N = \{\phi : (\phi, \phi)_T = 0\}$, is a closed subspace and the quotient space $H = \mathcal{T}(H_A) \otimes H_B / N$ is a Hilbert space.

The natural representation $\rho : \mathcal{T}(H_A) \to \mathcal{T}(H)$ is obtained by projecting the mapping $\rho' : \mathcal{T}(H_A) \to \mathcal{T}(\mathcal{T}(H_A) \otimes H_B)$ defined by

$$\rho'(X) : \sum_i b_i \otimes x_i \mapsto \sum_i X b_i \otimes x_i,$$

to H. The mapping $x \to 1 \otimes \mathbf{x}$ induces the linear operator $V : H_B \to H$. Equation

$$(\rho(X) V x, V x)_T = (X \otimes x, 1 \otimes \mathbf{x})_{\mathbf{T}} = (\mathbf{T(X)x, x})$$

implies (13.97).

Theorem 13.1 implies the following corollary. The linear mapping T : $\mathcal{T}(H_A) \to \mathcal{T}(H_B)$ is CP if and only it is $\min(n,m)$-positive, where n and m are the dimensions of H_A and H_B. In fact, in the proof above only n-positivity was used. If $n > m$, we can turn to the adjoint mapping T^*, where m-positivity would suffice to get the Stinespring representation.

Using the theory of representations, one can make formula (13.97) even more concrete. Namely, it is known (see e.g. Section 22 of [290]) that any representation ρ of $\mathcal{T}(H_A)$ in some finite-dimensional Hilbert space H is equivalent (up to a trivial representation) to the direct sum of a finite number of identical representations. That is, $H = H_0 + \sum_{k=1}^{K} H_k$ (orthogonal sum of subspaces) and

$$\rho(X) = \sum_{k=1}^{K} U_k^* X U_k$$

with U_k isometric bijections $H_k \to H$. Substituting this formula in (13.97) one obtains the following fundamental *Kraus* or *Choi-Kraus representation* for CP maps.

Theorem 13.2. *The linear mapping* $T : \mathcal{T}(H_A) \to \mathcal{T}(H_B)$ *is completely positive, if and only if*

$$T(X) = \sum_{k=1}^{K} V_k X V_k^* \tag{13.98}$$

with $K \le nm$ *(n and m are the dimensions of* H_A *and* H_B*) and some linear operators* V_k *in* H_A. *If* T *is trace preserving, then* $\sum_k V_k^* V_k = \mathbf{1}$.

Notice that the last statement is obtained by using the fact that trace preservation of T means that T^* preserves the identity operator.

As the composition of the operators of type (13.98) is clearly of the same type, it follows that the composition of any two CP maps is again CP.

Yet another representation of CP maps in terms of partial traces (also referred to sometimes as the *Stinespring representation*) is of great importance for physical interpretation and realization of these maps (see e.g. [240] or [268]). Let us present it for coinciding H_A and H_B (see more general versions in [240] or [268]).

Theorem 13.3. *For any TP-CP map* $T : \mathcal{T}(H) \to \mathcal{T}(H)$ *there exists a Hilbert space* $\tilde{H}$, *and (i) a partial isometry* $F : H \to H \otimes \tilde{H}$ *such that*

$$T(\rho) = \mathrm{tr}_{\tilde{H}}[F \rho F^*], \tag{13.99}$$

and (ii) a unitary map $U : H \otimes \tilde{H} \to H \otimes \tilde{H}$ and a state $\omega \in \mathcal{T}(\tilde{H})$ such that

$$T(\rho) = \mathrm{tr}_{\tilde{H}}[U(\rho \otimes \omega)U^*]. \tag{13.100}$$

Proof. (i) Let $\tilde{H} = \mathbf{C}^K$ with K the number of terms in (13.98), and let $F : H \to H \otimes \tilde{H}$ is defined by the formula

$$F = \sum_k V_k \otimes |k\rangle : |\xi\rangle \mapsto \sum_k V_k|\xi\rangle \otimes |k\rangle,$$

where $|k\rangle$ are the basis vectors in $\tilde{H}$. It follows that

$$F^*(x \otimes y) = \sum_k V_k^* x \langle k|y\rangle,$$

which can be denoted $\sum_k V_k^* \otimes \langle k|$ by identifying H with $H \otimes \mathbf{C}$. Hence

$$F^*F = \sum_{k,j}(V_k^* \otimes \langle k|)(V_j \otimes |j\rangle) = \sum_{k,j} V_k^*V_j\langle k|j\rangle = \sum_k V_k^*V_k = \mathbf{1},$$

that is, F is a partial isometry. Finally, by (13.88),

$$\sum_k V_k\rho V_k^* = \sum_{k,j}\mathrm{tr}_{\tilde{H}}(V_k\rho V_j^* \otimes |k\rangle\langle j|) = \mathrm{tr}_{\tilde{H}}(F\rho F^*).$$

(ii) Let $\omega = |1\rangle\langle 1|$. Then $(\rho \otimes \omega)_{jk} = \rho\delta_k^j$ for any $\rho \in \mathcal{T}(H)$. Hence

$$[U(\rho \otimes \omega)U^*]_{jk} = U_{k1}\rho U_{j1}^*$$

for any operator $U = (U_{jk}) \in \mathcal{T}(H \otimes \tilde{H})$. Therefore, by (13.94), in order to get (13.100) we need

$$\sum_k U_{k1}\rho U_{k1}^* = \sum_k V_k X V_k^*.$$

Thus it is sufficient to have $U_{k1} = V_k$ for all k. Clearly a unitary operator U exists satisfying this condition, because $\sum_k V_k^*V_k = \mathbf{1}$.

The physical meaning of (13.99) is as follows. It means that the $TP - CP$ transformations are exactly the transformations obtained from pure, that is unitary, transformations performed on a given system combined with another ancillary system, referred to as a reservoir, or environment, or just *ancilla*, and projected on the states of a given system.

Yet another characterization of CP-TP map can be given in terms of its matrix (13.89).

Theorem 13.4. *T is completely positive iff its matrix (13.89) is positive, as the matrix of an operator in $H_A \otimes H_B$, that is,*

$$\sum_{j,l,i,k} \bar{x}_{jl}T_{(j,l),(i,k)}x_{ik} \geq 0 \tag{13.101}$$

for any vector $(x_{jl}) \in H_A \otimes H_B$.

Proof. (i) Let T is completely positive. By Theorem 13.2 and convexity, it is sufficient to show that for $T(\rho) = V\rho V^*$, the matrix

$$T_{(j,l),(i,k)} = \langle e_k^B | V(|e_i^A\rangle\langle e_j^A|)V^* e_l^B \rangle = \langle e_k^B | V e_i^A \rangle \langle V e_j^A | e_l^B \rangle = y_{ik}\bar{y}_{jl}$$

$$(13.102)$$

is positive, where $y_{ik} = \langle e_k^B | V e_i^A \rangle$. But

$$\sum_{j,l,i,k} \bar{x}_{jl} T_{(j,l),(i,k)} x_{ik} = \sum_{ik} x_{ik} y_{ik} \sum_{jl} \bar{x}_{jl}\bar{y}_{jl} = \left| \sum_{ik} x_{ik} y_{ik} \right|^2 \geq 0.$$

(ii) Again by convexity, to show that any positive matrix corresponds to CP map, it is sufficient to show this for the extreme points of the set of positive matrices that have the form $T = |Y\rangle\langle Y|$ with some $Y = \sum_{ij} y_{ij} e_i^A \otimes e_j^B \in H_A \otimes H_B$. These operators act in $H_A \otimes H_B$ as

$$T(e_k^A \otimes e_l^B) = \bar{y}_{ij} y_{kl} e_i^A \otimes e_j^B,$$

and thus have the matrix $T_{ij,kl} = \bar{y}_{ij} y_{kl}$. By (13.102), this is the matrix of the operator $T(\rho) = V\rho V^*$ with $y_{ik} = \langle e_k^B | V e_i^A \rangle$.

As shown in Proposition 13.2, positive TP maps in $\mathcal{T}(\mathbf{C}^2)$ can be described by a vector a and an operator B in $\mathbf{R}^3$. Complete positivity can be fully characterised in terms of certain inequalities involving a and singular values of T. This characterization is however rather nontrivial, see [294]. An interesting point to note is that the operations given by (13.92) with anti-unitary U are positive, but not CP.

Let us point out some basic examples of CP maps. Pinching map (13.28) is a CP map, because it is explicitly defined in the Kraus representation. Given ρ, a state on $\tilde{H}$, the mapping $\sigma \mapsto \sigma \otimes \rho$ from $\mathcal{T}(H) \to \mathcal{T}(\tilde{H})$ is clearly CP, and consequently, the partial trace (13.86) is also CP by the duality. In qubits the mappings

$$T(\rho) = p_0\rho + \sum_{j=1}^{3} \sigma_j\rho\sigma_j, \qquad (13.103)$$

with a probability distribution p_0, p_1, p_2, p_3, are called the *Pauli channels*. They are CP due to the Kraus representation. In the d-dimensional case $\mathbf{C}^d$ with the basis $e_0, \ldots, e_{d-1}$ one introduces the operators X and Z by their actions $Xe_j = e_{j-1}$ for $j > 0$ and $Xe_0 = e_{d-1}$, and $Ze_j = e^{-2\pi i/d}e_j$ for all j. The *generalized Pauli channel*, defined by the formula

$$T(\rho) = \sum_{k,j=0}^{d-1} p_{kj}(X^k Z^j)^* \rho(X^k Z^j), \qquad (13.104)$$

where $\{p_{ij}\}$ is a probability law on $\{0, \ldots, d-1\}^2$, is also a CP map.

13.10 Elements of the general theory of quantum games

A general static (simultaneous) game of N players is a triple $(N, S = S_1 \times \cdots \times S_N, \Pi = (\Pi_1, \ldots, \Pi_N))$, where S_j is the strategy space of jth player and $P_j : S \to \mathbf{R}$ is the payoff of the jth player. General quantum games can be fit into this scheme. Namely, a static simultaneous *quantum game* of N players with finite-dimensional strategies (and separated actions) can be described by N finite-dimensional Hilbert spaces $H_1, \ldots, H_N$, an initial state ρ on $H = H_1 \otimes \cdots \otimes H_N$, a POVM $\{M_\omega\}$, $\omega \in \Omega$, on H (see (13.39), (13.40)) with the set of outcomes Ω, the payoff functions $f = (f_1, \ldots, f_N)$, $f_j : \Omega \to \mathbf{R}$, with $f_j(\omega)$ being the payoff of jth player for the outcome ω, and the choice of strategic spaces S_j for each player, where S_j is a closed subset of the set of all CP-TC mappings in $\mathcal{T}(H_j)$. For a choice (or profile) of strategies $(s_1, \ldots, s_N) \in S$, the final state of the game is assumed to be $(s_1 \otimes \cdots \otimes s_N)\rho$ with the possible outcome ω measured by the POVM $\{M_\omega\}$ being

$$\mathrm{tr}[M_\omega (s_1 \otimes \cdots \otimes s_N)\rho],$$

so that the final payoffs can be calculated by the formula

$$\Pi_j(s_1, \ldots, s_N) = \sum_{\omega \in \Omega} f_j(\omega) \mathrm{tr}[M_\omega (s_1 \otimes \cdots \otimes s_N)\rho]. \tag{13.105}$$

Notice that the introduction of POVM generalises both MW and EWL protocols.

Let us say that the quantum game is played with the *full strategic spaces* if each S_j coincide with the whole set CP-TC mappings in $\mathcal{T}(H_j)$ and the quantum game is played with *full unitary strategic space* (sometimes referred to in this context as pure strategies) if each S_j arises from the set of all unitary operators in H_j.

The following quantum version of the Nash theorem is a straightforward extension of its classical counterpart.

Theorem 13.5. *Any quantum game played with the full strategic spaces has at least one Nash equilibrium.*

Proof. Since the strategic spaces S_j are compact convex sets (as closed subsets of trace preserving positive maps, see Lemma 13.1) and the payoff function (13.105) is linear on each S_j, the proof is exactly the same as the classical version, or otherwise stated, the claim is a particular case of the general Glicksberg theorem (see Section 11.2).

Remark 13.13. Possibly the first precise formulation of this general result appeared in [270], but it was mentioned in particular forms in previous publications.

13.11   Further links and examples

Concluding our introduction to quantum games let us note that the literature on this subject is already quite immense. Further general insights and extensive bibliography can be obtained from various review papers that include [239], [250], [237]. There one can find also references to the big chunk of work devoted to building various quantized versions of all standard examples of classical games (various social dilemmas, etc., like the quantum versions of Monty Hall problem in [233] and [221], and of the Trucker Game in [226], see also "clever Alice" and "stupid Alice" from [60] that was used as a motivation for quantum games in our Chapter 7).

Let us indicate some trends of research which were not even touched upon in our presentation. These trends include the repeated or iterated quantum games initiated in [249], the analysis of the links of quantum games with the Bayesian games of incomplete information (see [220]), and an interesting activity on the expressing (interpreting) in game-theoretic term the fundamental properties of quantum nonlocality and its optimal quantitative characteristics, see e.g. [234] and [243], linking the theory of games with the fundamental problems of quantum communication and teleportation.

Finally, the initial paper [288] being motivated by problems in quantum computation and cryptology, this link is of great importance, see e.g. [237]. As the simplest example illustrating this link let us describe briefly the well known 'Guess the number' game.

Recall that n-qubit systems can be described by the Hilbert space $H^{\otimes n}$, which is the tensor product of n two-dimensional spaces $H = \mathbf{C}^2$. The space $H^{\otimes n}$ has dimension 2^n and its natural basis can be represented by 2^n vectors of the form $|x\rangle$, where

$$x = |x_{n-1} \cdots x_0\rangle = |x_{n-1}\rangle \otimes \cdots \otimes |x_0\rangle$$

is a string of n symbols x_j with values 0 or 1 representing the binary expansion of the corresponding integer x. Let $x \cdot y$ denotes the modulo 2 scalar product of these expansions:

$$x \cdot y = (x_{n-1}y_{n-1} + \cdots + x_0 y_0)(mod\ 2).$$

The *Hadamard operator* or *Hadamard-Walsh operator* on a qubit is the transformation of $\mathbf{C}^2$ given by the matrix

$$W = \frac{1}{\sqrt{2}} \begin{pmatrix} 1 & 1 \\ 1 & -1 \end{pmatrix},$$

or equivalently by its action on the standard basis:

$$W|0\rangle = \frac{1}{\sqrt{2}}(|0\rangle + |1\rangle), \quad W|1\rangle = \frac{1}{\sqrt{2}}(|0\rangle - |1\rangle).$$

The *Hadamard-Walsh operator* on n-qubit system is the tensor product $W^{\otimes n}$ acting as

$$W^{\otimes n}\left(|x_{n-1}\rangle \otimes \cdots \otimes |x_0\rangle\right) = W|x_{n-1}\rangle \otimes \cdots \otimes W|x_0\rangle.$$

Clearly W is a unitary operator such that $[W^{\otimes n}]^2 = \mathbf{1}$. Direct computation shows that

$$W^{\otimes n}|0\cdots 0\rangle = 2^{-n/2} \sum_{x=0}^{2^{n-1}-1} |x\rangle, \tag{13.106}$$

the r.h.s. being the uniform mixtures of all basis states of $H^{\otimes n}$, and generally

$$W^{\otimes n}|y\rangle = \sum_{x=0}^{2^{n-1}-1} (-1)^{x\cdot y}|x\rangle. \tag{13.107}$$

The *Bernstein-Vazirani oracle* with a parameter $a \in H$ is the transformation of H defined by the following action on the basic vectors

$$T_a|x\rangle = (-1)^{a\cdot x}|x\rangle.$$

The *'Guess the number' game* we are talking here is the game between Alice and Bob, where Alice chooses a number a and Bob has to guess it by asking the result of the action of the oracle on some vectors. How many question Bob has to ask to guess a? Classically, when he can use only the basis vectors $|x\rangle$, he needs effectively to get the results for all 2^n vectors thus asking 2^n questions. Remarkably enough, using the full space $H^{\otimes n}$ he can find the answer just with one question. Namely, Bob prepares the initial state

$$|\psi\rangle = 2^{-n/2} \sum_{x=0}^{2^{n-1}-1} |x\rangle,$$

and asks Alice to give him the result of $T_a|\psi\rangle$, that is,

$$T_a|\psi\rangle = 2^{-n/2} \sum_{x=0}^{2^{n-1}-1} (-1)^{a\cdot x}|x\rangle = W^{\otimes n}|a\rangle.$$

It remains for Bob to apply another $W^{\otimes n}$ to get the required number

$$a = W^{\otimes n}T_a|\psi\rangle = W^{\otimes n}W^{\otimes n}|a\rangle.$$

Bibliography

[1] M. Akian, S. Gaubert, V. Kolokoltsov. Invertability of functional Galois connections. C.R. Acad. Sci. Paris, Ser. I **335** (2002), 1-6.

[2] S. Alpern, Sh. Gal. The theory of search games and rendezvous. International Series in Operations Research and Management Science, 55. Kluwer Academic Publishers, Boston, MA, 2003.

[3] W.J. Anderson. Continuous-Time Markov Chains. Probability and its Applications. Springer Series in Statistics. Springer 1991.

[4] D. Applebaum. Probability and Information. Cambridge University Press, 1996.

[5] D. Applebaum. Levy processes and Stochastic Calculus. Cambridge University Press, 2004.

[6] R.J. Aumann, S. Hart (Editors) Handbook of game theory with economic applications. Amsterdam, Elsevier: vol. 1, (1992), vol. 2 (1994), vol. 3 (2002).

[7] R. Avenhaus, D.M. Kilgour. Efficient distributions of arms-control inspection effort. Naval Res. Logist. **51:1** (2004), 1-27.

[8] R. Avenhaus. Applications of inspection games. Math. Model. Anal. **9:3** (2004).

[9] R. Avenhaus, M. J. Canty. Playing for time: a sequential inspection game. European J. Oper. Res. **167:2** (2005), 475-492.

[10] R. Axelrod. The Evolution of Cooperation. New York: Basic Books, 1984.

[11] F. Baccielli et al. Synchronization and Linearity: an Algebra for Discrete Event Systems. New York, John Wiley, 1992.

[12] M. Balinski, H. Peyton Young. Fair representation. Meeting the ideal of one man, one vote. Yale University Press, New Haven, 1982.

[13] E.A. Barbashin. To the theory of dynamic systems (in Russian). Uchebnie zapiski MGU **135** (1949), 110-133.

[14] T. Basar, G.J. Olsder. Dynamic Noncoopeative Game Theory. Sec. Edition, SIAM, 1999.

[15] T. Bass. Road to ruin. Discover **13** (1992), 56-61.

[16] M. Beckmann, C.B. McGuire, C.B. Winsten. Studies in the Economics of Transportation. Yale University Press, 1956.

[17] V.P. Belavkin. Measurement, Filtering and Control in Quantum Open Dynamical Systems. Reports on Mathematical Physics **43:3** (1999), 405-425.

[18] V.P. Belavkin, V. Kolokoltsov. On general kinetic equation for many particle systems with interaction, fragmentation and coagulation. Proc. Royal Soc. Lond. A **459** (2003), 727-748.

[19] R.E. Bellman. Dynamic programming. Princeton Univ. Press and Oxford Univ. Press, 1957.

[20] J. Bentham. Deontology or, the Science of Morality, 2 volumes (ed. by John Bouring) London, 1834.

[21] E.R. Berlenkamp, J.H. Conway, R.K. Guy. Winning Ways for your Mathematical Plays. Academic Press, London and New York, 1982.

[22] E.R. Berlenkamp, J.H. Conway, R.K. Guy. Winning Ways for your Mathematical Plays. vol. 2. Games in Particular. Academic Press, London and New York, 1982.

[23] P. Bernard. Robust control approach to option pricing, including transaction costs. Ann. Intern. Soc. Dynam. Games **7**, Birkhauser Boston, MA, 391-416.

[24] H.S. Bierman, L. Fernandez. Game Theory with Economic Applications. Addison-Wesley, 1998.

[25] K. Binmore. Fun and Games. A text on game Theory. D.C. Heath and Company. Lexington, Massachusetts, Toronto, 1992.

[26] J. M. Binner, L.R. Fletcher, L. Khodarinova, V. Kolokoltsov. Optimal Strategic Investment in a Duopoly. Aston Business School Working Paper Series RP0412, 2006.

[27] J.C. de Borda. Memoire sur les elections au Scrutin. Histoire de l'Academie Royale des Sciences, Paris, 1781.

[28] D. Braess. Uber ein paradoxon der verkehrsplunung. Unternehmensforschung **12** (1968), 258-268.

[29] S.J. Brams. The Presidential Election Game. New Haven: Yale University Press, 1978.

[30] S.J. Brams. Biblical Games. A Strategic Analysis of Stories in the Old Testament. Cambridge, Mass: The MIT Press, 1980.

[31] S.J. Brams. Superpower Games. Yale University Press, 1985.

[32] S.J. Brams. Negotiaton Games. Routledge 2003.

[33] S.J. Brams, A.D. Taylor. The Win-Win Solution: Guaranteeing Fair Shares to everybody. W.W. Norton and Company, 1999.

[34] V.M. Bure, O.A. Malafeyev. Coordinated strategies in repeated finite n-person games. Vestnik St. Petersburg Univ. Math. **28:1** (1995), 59-61.

[35] V.M. Bure, O.A. Malafeyev. Some game-theoretical models of conflict in finance. Nova J. Math. Game Theory Algebra **6:1** (1996), 7-14.

[36] D. Challet, M. Marsili. Minority Games. Interacting agents in financial markets. Damien Challet, Matteo Marsili. Oxford University Press 2004.

[37] F.A. Chimenti. Pascal's wager: a decision-theoretic approach. Mathematics Magazine **63:5** (1990), 321-325.

[38] J.E. Cohen, P. Horowitz. Paradoxical behavior of mechanical and electrical networks. Nature **325** (1991), 699-701.

[39] A.M. Colman. Game Theory and Its Applications in the Social and Biological Sciences. Oxford, Butterworth-Heinemann Press, 1995.

[40] A.M. Colman. Game Theory and Experimental Games. Pergamon Press, 1982.

[41] Condorcet, Marquis de. Essai sur l'application de l'analyse à la probabilité des decisions rendues à la pluralité des voix. Paris 1785.

[42] J.H. Conway. On Numbers and Games. London Math. Soc. Monograph 6, Academic Press, London and Ney York, 1956.

[43] A.C.C. Coolen. The Mathematical Theory of Minority Games. Statistical mechanics of interacting agents. Oxford University Press 2004.

[44] V. Corradi, R. Sarin. Continuous Approximations of Stochastic Evolutionary Game Dynamics. J. of Economic Theory **94** (2000), 163-191.

[45] M.G. Crandoll, P.-L. Lions. Viscosity solutions of Hamilton-Jacobi equations. Trans. Amer. Math. Soc. **277:1** (1983), 1-42.

[46] R. Dawkins. The selfish gene. Oxford University Press, 1989.

[47] M.A. Dimand, R.W. Dimand. A History of Game Theory, v. 1. Routledge, 1996.

[48] L.S. Dodgson (Lewis Carroll). The Principles of parliamentary Repeesntation. 1st edn. London, 1884.

[49] J. Eisert, M. Wilkens, M. Lewenstein. Quantum Games and Quantum Strategies. Phys. Rev. Lett. **83:15** (1999), 3077-3081.

[50] P. Espinosa and Mariel. A model of optimal advertising expenditures in a dynamic duopoly. Atlantic Economic Journal **29** (2001), 135-161.

[51] R. Farquharson. Theory of Voting. Oxford: Basil Blackwell, 1969.

[52] T. Ferguson, C. Melolidakis. On the inspection game. Naval Res. Logist. **45:3** (1998), 327-334.

[53] P.C. Fishburn. The theory of social choice. Princeton, Princeton University Press, 1973.

[54] W. Fleming. The convergence problem for differential games. J. Math. Anal. Appl. **1** (1961), 102-116.

[55] A. T. Fomenko. Mathematical Impressions. AMS, Providence, 1990.

[56] D. Fudenberg, D.K. Levine. The theory of learning in games. MIT Press Series on Economic Learning and Social Evolution, 2. MIT Press, Cambridge, MA, 1998.

[57] D. Fudenberg, J. Tirole. Game theory. MIT Press, Cambridge, MA, 1991.

[58] H. Gintis. Game Theory Evolving. Princeton University Press, 2000.

[59] J. Glaze, C.-T. Ma. Efficient Allocation of a Prize: King Solomon's dilemma. Games and Economics Behavior **1** (1989), 222-233.

[60] A.A. Grib, G.N. Parfenov. Can a game be quantum? (moshet li igra bit kvantovoi?) Zap. Nauchn. Sem. S.-Peterburg. Otdel. Mat. Inst. Steklov. (POMI) 291 (2002), Vopr. Kvant. Teor. Polya i Stat. Fiz. 17, 131–154, 280; translation in J. Math. Sci. (N.Y.) 125:2 (2005), 173-184.

[61] E. Grigorieva et al. The private value single item bisection auction. Economic Theory **30** (2007), 107-118.

[62] P.M. Grundy. Mathematics and Games. Eureka **2** (1939), 6-8.

[63] Yu. V. Gurin, D.V. Shakov. Big book of games and entertainments. Petersburg, Kristal, 2000 (in Russian).

[64] R.K. Guy (Editor). Conbinatorial Games. Proceedings of Simposia in Applied Mathematics, v. 43, AMS, 1991.

[65] D.M. Hanssens, L.J. Parsons and R.L. Schultz. Market Response Models: Econometric and Time Series Analysis. Kluwer Academic Publishers, Norwell, 1990.

[66] S.P. Hargreaves Heap, Y. Varoufakis. Game Theory: a Critical Introduction. Routledge, London.

[67] A. Haurie, P. Macotte. On the relationship between Nash-Cournot and Wardrop equilibria. Networks **15** (1985), 295-308.

[68] O. Hernandez Lerma. Adaptive Markov Control Processes. Springer-Verlag, N.Y. 1989.

[69] O. Hernandez Lerma. Discrete Time Markov Control Processes. Springer-Verlag, N.Y. 1996.

[70] M.W. Hirsh. Differential topology. Corrected reprint of the 1976 original. Graduate Texts in Mathematics, 33. Springer-Verlag, New York, 1994.

[71] W. Hodges. Building Models by Games. Cambridge University Press.

[72] J. Hofbauer, K. Sigmund. Evolutionary Games and Population Dynamics. Cambridge University Press, 1998.

[73] J. Hofbauer, K. Sigmund. Evolutionary Game Dynamics. Bulletin of the AMS **40:4** (2003), 479-519.

[74] Z. Hucki, V. Kolokoltsov. Pricing of rainbow options: game theoretic approach. To appear in International Game Theory Review **9:2** (2007).

[75] A. Iqbal, A.H. Toor. Darwinism in quantum systems? Phys. Letters A **294** (2002), 261-270.

[76] R. Isaacs. Differential games. A mathematical theory with applications to warfare and pursuit, control and optimization. John Wiley and Sons, 1965.

[77] H.J. Jensen. Self-organized critiacality: emergent complex behavior in physical and biological systems. Cambridge lecture notes in physics **10**, Cambridge Univ. Press, 1998.

[78] S. Jorgensen. A survey of some differential games in advertising. Journal of Economic Dynamics and Control **4** (1982), 341-369.

[79] S. Jorgensen, G. Zaccour. Differential Games in Marketing. Kluwer Academic, 2004.

[80] H. Kahn. On Thermonuclear War. Princeton University press, 1960.

[81] M. Kandori. Introduction to repeated games with private monitoring. J. of Economic Theory **102** (2001), 1-15.

[82] S. Karlin. Mathematical Methods and Theory in Games, Programming and Economics, v. 1. Pergamon Press, 1959.

[83] L.A. Khodarinova, J.M. Binner, L.A. Fletcher, V.N. Kolokoltsov, P. Wysall. Controlling alliances through executing pressure. Nottigham Trent University Research Report 5/04, 2004.

[84] J.M. Binner, L.A. Fletcher, L.A. Khodarinova, V.N. Kolokoltsov. External pressure on alliances. Aston working paper series RP0635 (2006) ISBN 1854496395.

[85] A. Yu. Kitaev, A.H. Shen, M.N. Vyalyi. Classical and Quantum Computation. Graduate Studies in Mathematics, v. 47, AMS, 2002.

[86] V.N. Kolokoltsov. Localization and analytic properties of the simplest quantum filtering equation. Rev. Math. Phys. **10:6** (1998), 801-828.

[87] V.N. Kolokoltsov. Nonexpansive maps and option pricing theory. Kibernetica **34:6** (1998), 713-724.

[88] V.N. Kolokoltsov. Games for all. Short lecture course on game theory. Mathematics and Statistics Research Report 6/04, Nottingham Trent University, 2004.

[89] V.N. Kolokoltsov. On linear, Additive, and Homogeneous Operators in Idempotent Analysis. In: Advances in Soviet Mathematics **13** (1992), Idempotent Analysis, (Eds. V.P. Maslov and S.N. Samborski), 87-101.

[90] V.N. Kolokoltsov. Idempotent Structures in Optimisation. Proc. Intern. Conf. devoted to the 90-th anniversary of L.S. Pontryagin, v. 4, VINITI, Moscow (1999), 118-174 (in Russian). Engl. transl. J. Math. Sci., NY **104:1** (2001), 847-880.

[91] V.N. Kolokoltsov. Introduction of a new Maslov-type currency (coupons) as a means of solving a market game under non-equilibrium prices. Dokl. Acad. Nauk **320:6** (1991). Engl Transl. in Sov. Math. Dokl. **44:2** (1992), 624-629.

[92] V.N. Kolokoltsov. Turnpikes and Infinite Extremals in Markov processes of decision making. Matem. Zametki (in Russian) **46:4** (1989), 118-120.

[93] V. N. Kolokoltsov. Measure-valued limits of interacting particle systems with k-nary interactions I. Probab. Theory Relat. Fields **126** (2003), 364-394.

[94] V. N. Kolokoltsov. Measure-valued limits of interacting particle systems with k-nary interactions II. Stochastics and Stochastics Reports **76:1** (2004), 45-58.

[95] V. N. Kolokoltsov. On Markov processes with decomposable pseudo-differential generators. Stochastics and Stochastics Reports **76:1** (2004), 1-44.

[96] V. N. Kolokoltsov. Kinetic equations for the pure jump models of k-nary interacting particle systems. Markov processes and Related Fields **12** (2006), 95-138.

[97] V. N. Kolokoltsov. Nonlinear Markov Semigroups and Interacting Lévy Type Processes. Journal Stat. Phys. (2007).

[98] V. N. Kolokoltsov. Nonlinear Markov processes and kinetic equations. Monograph. Cambridge University Press, 2010.

[99] V.N. Kolokoltsov, O.A. Malafeyev. Introduction to the analysis of many agent systems of competition and cooperation (in Russian). Sankt Petersburg University Press, 2007.

[100] V.N. Kolokoltsov, V.P. Maslov. Idempotent Analysis as a tool in control theory. Part 1 Funkts. Anal. i Prilosh. **23:1** (1989), 1-14. Engl. Transl. Funct. Anal. Appl.

[101] V.N. Kolokoltsov, V.P. Maslov. Bellman's differential equation and Pontryagin's maximum principle for multicriteria optimization problems. Dokl.

Acad. Nauk SSSR **324:1** (1992), 29-34.

[102] V.N. Kolokoltsov, V.P. Maslov. Idempotent Analysis and its Application to Optimal Control. Moscow, Nauka, 1994 (in Russian).

[103] V.N. Kolokoltsov, V.P. Maslov. Idempotent Analysis and its Applications. Kluwer, 1997.

[104] V.Yu. Korolev, M. Rey. Statistical Analysis of volatility of financial time series and turbulent plasmas by the method of moving separation of mixtures. In: V.Yu. Korolev, N.N. Skvortsova (Eds.) Stochastic Models of Structural Plasma Turbulence, VSP Leiden, Boston, 2006, p. 245-344.

[105] N.N. Krasovskii, A.I. Subbotin. Positional-differential games (Pozicionnye differencial'nye igry). Moskow: Nauka, 1974 (in Russian). French Translation: Jeux différentiels. Moscow: Editions Mir, 1977.

[106] N.N. Krasovskii, A.I. Subbotin. Game Theoretical Contrl Problems. Springer, New York, 1988.

[107] K. Kuratowski. Topology, v. 2. Academic Press 1968.

[108] Y.F. Lim, K. Chen, C. Jayaprakash. Scale invariant behavior in a spatial game of prisoners' dilemma. Physical Review E, **65:2** (2002).

[109] G.L. Litvinov, V.P. Maslov (Eds.). Idempotent Mathematics and Mathematical Physics. Contemporary Mathematics v. 377, AMS 2005.

[110] O.A. Malafeyev. On the existence of the generalized value of pursuit games. Vestnik LGU (In Russian) **19** (1972), 41-47.

[111] O.A. Malafeyev. Dynamic games with dependent motions (in Russian). Dokl. Akad. Nauk SSSR **213** (1973), 783–786.

[112] O.A. Malafeyev. Equilibrium points in dynamical games. Cybernetics **3** (1974), 111-118.

[113] O.A. Malafeyev. Stationary strategies in differential games. J. of Comput. Math. and Mathemat. Physics **17:1** (1977), 221–225.

[114] O.A. Malafeyev. Stable non-cooperative N-person games. Vestnik of Leningrad State University **4:8** (1978), Seria 1, 32–35.

[115] O.A. Malafeyev. Internal metrics and equilibrium points in non-cooperative games. Vestnik of Leningrad State University **4:3** (1979), Seria 1, 46–48.

[116] O.A. Malafeyev. Existence of equilibrium points in non-cooperative differential two-person games. Vestnik of Leningrad State University **4:7** (1980), Seria 1, 12–16.

[117] O.A. Malafeyev. Solutions in differential N-person games. Cybernetics **1** (1983), 68-72.

[118] O.A. Malafeyev. Stability of solutions for mixed extentions in non-cooperative games. Vestnik of Leningrad State University **2:1** (1984), Seria 1, 17–22.

[119] O.A. Malafeyev. A differential game with a finite number of players. (Spanish) Investigación Oper. **7:2** (1986), 23–32.

[120] O.A. Malafeyev. Stability of the problems of multicriteria optimization and conflict control of dynamical systems. Leningrad State Univesity, 1990 (In Russian).

[121] O.A. Malafeyev. Dynamical competitive one-sector model of growth. Proceedings of the All-Union summer School on Stability, Development and

Growth, Irkutsk, 4-11 July, 1994. vol. 2, 131-137 (In Russian).

[122] O.A. Malafeyev. Control Conflict Systems. Sankt-Peterburg State University, 2000 (In Russian).

[123] O.A. Malafeyev, V.N. Kiselev. Game-theoretic search with several participants (Russian). In: Investigations in the geometry of simple pursuit. Yakutsk. State Univ., Yakutsk, 1991, 46-55.

[124] O.A. Malafeyev, A.I. Murav'ev (Eds.) Modelling of conflict situation in social and economical systems. Sankt-Peteburg State University of Economics and Finances, 1998 (In Russian).

[125] O.A. Malafeyev, S.A. Nemnyugin. A generalized dynamical model of the motion of a system in multicomponent external field with stochastic components. Teoret. Mat. Fiz. **107:3** (1996), 433–438. Engl. transl. Theoret. and Math. Phys. **107:3** (1997), 770–774.

[126] O.A. Malafeyev, M.S. Troeva. A game-theoretical model for a controlled process of heat transfer. System modelling and optimization (Prague, 1995), 243-250, Chapman and Hall, London, 1996.

[127] O.A. Malafeyev, V.B. Vilkov. A property of von Neumann-Morgenstern solutions. (Russian) Vestnik Leningrad. Univ. Mat. Mekh. Astronom. **4** (1988), 24-26. Engl. transl. Vestnik Leningrad Univ. Math. **21:4** (1988), 28-30.

[128] O.A. Malafeyev, A.F. Zubova. Mathematical and computer modeling of social economic systems with many agent interaction (in Russian). Sankt-Petersburg State University, 2006.

[129] O.A. Malafeyev, Zd. Wyderka. On the existence of Nash equilibrium in a non-cooperative, n-person, linear differential games with measures as coefficients. Appl. Math. Comput. Sci. **5:4** (1995), 689-701.

[130] V.P. Maslov. Méthodes opé ratorielles (in French). Mir, Moscow 1987.

[131] V.P. Maslov. Quantum Economics (in Russian). Nauka, Moskow 2006.

[132] V.P. Maslov, K.A. Volosov (Eds.). Mathematical Aspects of Computer Engineering. Mir, Moscow, 1988.

[133] L. Marinato, T. Weber. A quantum approach to static games of complete information. Phys. Letters A **272** (2000), 291-303.

[134] J. Maynard Smith. Evolution and the Theory of Games. Cambridge University Press, 1982.

[135] J. Maynard Smith and G.R. Price. The Logic of Animal Conflict. Nature 246 (1973), 15-18.

[136] W.M. McEneaney. A robust control framework for option pricing. Math. Oper. Research **22** (1997), 201-221.

[137] W.M. McEneaney. Distributed Dynamic Programming for Discrete-Time Stochastic Control and Idempotent Algorithms. Automatica **47:3** (2011), 443-451.

[138] A. Mehlmann. The Game's Theory Afoot! Game Theory in Myth and Paradox. AMS, 2000. Translated from German.

[139] A. Mehlmann. Applied Differential Games. Plenum Press, New York, 1988.

[140] D.A. Meher. Quantum Strategies. Phys. Rev. Lett. **82:5** (1999), 1052-1055.

[141] M. Mesterton-Gibbons. An introduction to game theoretic modelling. Second Edition, AMS, 2001.

382   Understanding Game Theory

[142] H. Moulin. Axioms of cooperative decision making. Cambridge, Cambridge
       University Press, 1988. Russian Translation Moscow, Mir, 1991.
[143] H. Moulin. Fair division and collective welfare. The MIT press, Cambridge,
       2003.
[144] R.B. Myerson. Game theory. Analysis of conflict. Harvard University Press,
       Cambridge, MA, 1991.
[145] J. von Neuman, O. Morgenstern. Theory of Games and Economic Behavior.
       Princeton University Press, 1944.
[146] J. Norris. Cluster Coagulation. Comm. Math. Phys. **209** (2000), 407-435.
[147] M.A. Nowak. Evolutionary Dynamics (Exploring the Equations of Life).
       Harvard University Press, 2006.
[148] M.A. Nowak, R.M. May. Evolutionary games and spatial chaos. Nature
       259 (1992), 826-829.
[149] M. Nowak, K. Sigmund. A strategy of win-stay, lose-shift that outperforms
       tit-for-tat in the prosoner's dilemma game. Nature **264** (1993), 56-58.
[150] G.J. Olsder. Differential Game-Theoretic Thoughts on Option Pricing and
       Transaction Costs. International Game Theory Review **2:2,3** (2000), 209-
       228.
[151] L.A. Petrosian. Differential Games of Persuit (in Russian), LGU Press,
       1977.
[152] L.A. Petrosyan, O.A. Malafeyev. Differential multicriterial n-person games.
       (Russian) Vestnik Leningrad. Univ. Mat. Mekh. Astronom. **3** (1989), 27-31,
       Engl. transl. Vestnik Leningrad Univ. Math. **22:3** (1989), 33-38.
[153] L.A. Petrosyan, V.V. Mazalov (Eds.) Game theory and applications, 1.
       Nova Science Publishers, Inc., Hauppauge, NY, 1996.
[154] L.A. Petrosyan, V.V. Mazalov (Eds.) Game theory and applications, 2.
       Nova Science Publishers, Inc., Hauppauge, NY, 1996.
[155] L.A. Petrosyan, D.W. Yeung (Eds.) ICM Millennium Lectures on Games.
       Springer, 2003.
[156] C. Piga. Competition in a duopoly with sticky price and advertising. Int.
       Journal Industrial Organisation **18** (2000), 595-614.
[157] B.P. Pshenichnii, V.V. Ostapenko. Differential Games (in Russian).
       Naukova Dumka, Kiev, 1992.
[158] E.W. Piotrowski, J. Stadkowski. Quantum market games. Physica A **312**
       (2002), 208-216.
[159] W. Poundstone. Prisonner's Dilemma: John von Neuman, Game Theory,
       and the Puzzle of the Bomb. Doubleday, New York, 1992.
[160] P.J. Proudon. Essais de Philosophie pritique. Paris, 1861. Liberalism
       Against Populism. San Francisco, Freeman, 1982.
[161] A. Rapoport, A.M. Chammah. Prisoner's Dilemma: A Study in Conflict
       and Cooperation. University of Michigan Press, Ann arbor, Sec. Edition
       1970.
[162] J. Rawls. A theory of Justice. Cambridge, MA: Belknap.
[163] W. Riker. Liberalism Against Populism. San Francisco, Freeman, 1982.
[164] I.V. Romanovski. Turnpile theorems for semi-Markov decision processes (in
       Russian). Trudy Mat. Inst. Steklova **111** (1970), 208-223.

[165] Yu. A. Rosanov. Probability, Random Processes and Statistics. Moscow, Nauka 1985 (in Russian).

[166] A.E. Roth et al. Bargaining and Market Behavior in Jerusalem, Ljubljana, Pittsburgh, and Tokyo: An Experimental Study. American Economic Review **81:5** (1991), 1068-1095.

[167] T. Roughgarden, E. Tardos. How Bad is selfish Routing? Journal of the ACM, **49:2** (2002), 236-259.

[168] J.J. Rousseau. A Discourse on Inequality. English Translation in London, Penguin, 1984.

[169] M. Rubinstein. Somewhere over the Rainbow and Return to OZ, RISK 4, (1995), 63-66 and RISK 7 (1995), 67-71.

[170] B. Russel. Common Sense and Nuclear Warfare. New York, Simon and Schuster, 1959.

[171] S.N. Samborski, A.A. Tarashan. On semirings appearing in multicriteria optimization problems. Dokl. Acad. Nauk SSSR **308:6** (1989), 1309-1312.

[172] R. Selten. The Chain-Store Paradox. Theory and Decision **9** (1978), 127-159.

[173] S.P. Sethi. Deterministic and stochastic optimisation of a dynamic advertising model. Optimal Control Applications and Methods **4** (1983), 179-184.

[174] G. Shafer, V. Vovk. Probability and Finance. It is only a game! Wiley, 2001.

[175] A.N. Shiryaev. Essentials of Stochastic Finances: Facts, Models, Theory. World Scientific, 1999.

[176] R.P. Sprague. Über mathematishe Kampfspiele. Tohoku Math. J. **41** (1935-1936), 438-444.

[177] S. Stahl. A gentle introduction to game theory. AMS, Providence, Rhode Island, 1991.

[178] R. Straffin. Topics in the theory of voting. The UMAP Expository Monograph Series. Boston, Birkhaüser, 1980.

[179] R. Straffin. Game Theory and Strategy. Washington, 1993.

[180] A.I. Subbotin. Generalized solutions of first-order PDEs. The dynamical optimization perspective. Basel: Birkhäuser, 1994.

[181] A. Sudbery. Quantum Mechanics and the Particles of Nature. An outline for mathematicians. Cambridge University Press, 1986.

[182] Y. Tauman, N. Watanabe. The Shapley value of a patent lisencing game: the asymptotyic equivalence to non-cooperative results. Economic Theory **30** (2007), 135-149.

[183] S. Tijs. Introduction to Games Theory. Hindustan Book Agency, 2003.

[184] N.N. Vorobiev. Game Theory. New York, Springer Verlag 1977.

[185] N.N. Vorobiev. Extremal Algebra of positive matrices (in Russian). Electron. Informationsverarb. und Kybernet. **3:1** (1967), 39-71.

[186] J.Y. Wakano, N. Yamamura. A simple learning strategy that realizes robust cooperation better than Pavlov in iterated prisoner's dilemma. J. of Ethology **19:1** (2001), 1-8.

[187] Q. Wang and Z. Wu. A duopolistic model of dynamic competitive advertising. European Journal of Operational Research **128** (2001), 213-226.

[188] J.N. Webb. Game Theory. Decisions, Interaction and Evolution. Springer Undergraduate Mathematics Series. Springer 2007.

[189] J.W. Weibull. Evolutionary Game Theory. The MIT Press 1995.

[190] R. J. Williams. Sufficient Conditions for Nash Equilibria in N-Person Games over Reflexive Banach Spaces. J. of Optimization Theory and Applications **30:3** (1980), 383-394.

[191] W.-t. Wu and J.-h. Jiang. essential equilibrium points of N-person noncooperative games. Scientia sinica **10:11** (1962), 1307-1322.

[192] F. Yang, C. Wu, Q. He. Applications of Ky Fan's inequality on σ-compact set to variational inclusion and n-person game theory. J. Math. Anal. Appl. **319** (2006), 177-186.

[193] D.W. Yeung, L.A. Petrosyan. Cooperative stochastic differential games. Springer Series in Operations Research and Financial Engineering. Springer, New York, 2006.

[194] J. Yu. On Nash Equilibria in N-Person Games over Reflexive Banach Spaces. J. of Optimization Theory and Applications **73:1** (1992), 211-214.

[195] A. Ziegler. A Game Theory Analysis of Options. Springer 2004.

Literature added to the Second Edition:

[196] T. S. Aidt. Economic Analysis of corruption: a survey. The Economic Journal **113: 491** (2009), F632-F652.

[197] R. Alicki and K. Lendi. Quantum Dynamical Semigroups and Applications. Springer-Verlag, Berlin Heidelberg 2007.

[198] D. Arce and T. Sandler. Counterterrorism: A Game-Theoretic Analysis. Journal of Conflict Resolution **49** (2005), 183-200.

[199] Ph. Artzner et al. Coherent Measures of Risk. Mathematical Finance **9** (1999), 203-228.

[200] J. P. Aubin, D. Pujal and P. Saint-Pierre. Dynamic management of portfolios with transaction costs under tychastic uncertainty. In: M. Breton and H. Ben-Ameur (Eds.). Numerical Methods in Finance, Springer 2005, 59-89.

[201] R. Avenhaus and T. Krieger. Inspection games over time - Fundamental models and approaches. Monograph to be published in 2019.

[202] Averboukh Yu. Extremal shift rule for continuous-time zero-sum Markov games. Dynamical games and application 7:1 (2017) 1-20.

[203] G. S. Becker and G. J. Stigler. Law enforcement, Malfeasance, and Compensation of Enforces. The Journal of Legal Studies **3:1** (1974), 1-18.

[204] H. Beecher Stowe. Uncle Tom's cabin. Blackie, 1963.

[205] M. Benaim and J.-Y. Le Boudec, A class of mean field interaction models for computer and communication systems. Performance Evaluation **65** (2008), 823-838.

[206] S.C. Benjamin. Comment on "A quantum approach to static games of complete information" ArXiv: quant-ph/000812 (2000).

[207] S. C. Benjamin and P. M. Hayden. Comments on "Quantum Games and Quantum Strategies". Phys. Rev. Lett 87:6 (2001), 069801.

[208] S.C. Benjamin and P. M. Hayden. Multi-Player Quantum Games. ArXiv: quant-ph/0007038v2 (2001).

[209] A. Bensoussan. On the theory of option pricing. Acta Appl. Math. **2** (1984), 139-158.

[210] A. Bensoussan, J. Frehse, P. Yam. Mean Field Games and Mean Field Type Control Theory, Springer, 2013.

[211] A. Bensoussan, S. Hoe, M. Kantarcioglu, A Game-Theoretical Approach for Finding Optimal Strategies in a Botnet Defense Model. Decision and Game Theory for Security First International Conference, GameSec 2010, Berlin, Germany, November 22-23, 2010. Proceeding, T. Alpcan, L. Buttyan, and J. Baras (Eds.), Vol. 6442 pp. 135-148.

[212] A. Bensoussan, J. D. Diltz, H. Sing Ru. Real options games in complete and incomplete markets with several decision makers. SIAM J. Financial Math. **1** (2010), 666-728.

[213] P. Bernard, N. El Farouq and S. Thiery. Robust control approach to option pricing and interval models: a representation theorem and fast algorithm. SIAM Journal on Control and Optimization, **46:6** (2007), 2280-2302.

[214] P. Bernhard, J. Engwerda, B. Roorda, J.M. Schumacher, V. N. Kolokoltsov, P. Saint-Pierre and J.-P. Aubin. The Interval Market Model in Mathematical Finance: Game-Theoretic Methods. Birkhäuser, 2012.

[215] N. H. Bingham and R. Kiesel. Risk-Neutral Valuation. Sec. Edition, Springer 2004.

[216] J. M. Binner, F. Ciardiello, L. R. Fletcher, V. N. Kolokoltsov. Nash equilibria in a Duopolistic Market Share Model. International Game Theory Review (IGTR), Special issue devoted to the SING13 Conference, v. 21, No. 2 (2019).

[217] P. E. Caines, "Mean Field Games", *Encyclopedia of Systems and Control*, Eds. T. Samad and J. Ballieul. Springer Reference 364780; DOI 10.1007/978-1-4471-5102-9 30-1, Springer-Verlag, London, 2014.

[218] R. Carmona and F. Delarue. Probabilistic Theory of Mean Field Games with Applications, v. I, II. Probability Theory and Stochastic Modelling v. 83, 84. Springer, 2018.

[219] P. Carr, H. Geman and D. B. Madan. Pricing and Hedging in Incomplete Markets. J. Financial Econ. **32** (2001), 131-167.

[220] T. Cheon and A. Iqbal. Bayesian Nash Equilibria and Bell Inequalities. Journal of the Physical Society of Japan. 77:2 (2008), 024801.

[221] Ch.-F. Li, Y.-Sh. Zhang, Y.-F. Huang and G.-C. Guo. Quantum strategies of quantum measurements. Physics Letters A 280 (2001), 257-260.

[222] E.B. Davies. Quantum Theory of Open Systems, 1976.

[223] J. Du, H. Li, X. Xu, M. Shi, X. Zhou, R. Han. Remark on Quantum Battle of the Sexes Game. ArXiv: quant-ph/0103004 (2001).

[224] J. Du, H. Li, X. Xu, M. Shi, X. Zhou, R. Han. Nash equilibrium in the Quantum Battle of the Sexes Game. ArXive: quant-ph/0010050v3 (2001).

[225] J. Du, H. Li, X. Xu, X. Zhou, R. Han. Phase-transition-like Behavior of Quantum Games. ArXiv: quant-ph/0111138v4 (2003).

[226] J. Du, H. Li, X. Xu, X. Zhou, R. Han. Multi-Player and Multi-Choice

Quantum Game. ArXiv: quant-ph/0010092v4 (2002).

[227] J. Du, H. Li, X. Xu, X. Zhou, R. Han. Entanglement Enhanced Multiplayer Quantum Games. Physics Letters A, 302 (2002), 229-233.

[228] J. Eisert, M. Wilkens and M. Lewenstein. Quantum Games and Quantum Strategies. Phys Rev Lett 83:15 (1999), 3077-3080.

[229] B.-G. Englert, Ch. Kurtsiefer and H. Weinfurter. Universal unitary gate for single-photon two-qubit states. Phys. Rev A 63, 032303.

[230] J. R. Faria and D. Arce. A Vintage Model of Terrorist Organizations. Journal of Conflict Resolution **56:4** (2012), 629-650.

[231] H. Fielding. The history of the life of the late Mr Jonathan Wilde the Great. H. Hamilton, 1947.

[232] W. H. Fleming and H. M. Soner. Controlled Markov processes and viscosity solutions. Springer, 2006.

[233] A.P. Flitney and D. Abbot. Quantum version of the Monty Hall problem. Phys Rev A 65 (2002), 062318.

[234] A.P. Flitney et al. Equivalence between Bell inequalities and quantum minority games. Phys Lett A 373 (2009), 521-524.

[235] P. Frackiewicz. A new model for quantum games based in the Marinatto-Weber approach. J. Phys. A: Math. Theor. 46 (2013), 275301.

[236] D. A. Gomes and J. Saude. Mean field games models – a brief survey. Dyn. Games Appl. 4:2 (2014), 110–154.

[237] J. Orlin Grabbe. An introduction to Quantum Game Theory. ArXiv:quant-ph/0506219 (2005).

[238] O. Guéant, J.-M. Lasry and P.-L. Lions. Mean Field Games and Applications. Paris-Princeton Lectures on Mathematical Finance 2010. Lecture Notes in Math. 2003, Springer, Berlin, p. 205-266.

[239] H. Guo, J. Zhang and G. J. Koehler. A survey of quantum games. Decision Support Systems 46 (2008), 318-332.

[240] M. Hayashi. Quantum Information Theory. Mathematical foundation. Second Edition. Springer Graduate Texts in Physics. Springer, Berlin, 2006, 2017.

[241] M. Huang, R. Malhamé, P. Caines. Large population stochastic dynamic games: closed-loop Mckean-Vlasov systems and the Nash certainty equivalence principle. Communications in information and systems 6 (2006), 221–252.

[242] M. Huang, P. Caines and R. Malhamé. Large-Population Cost-Coupled LQG Problems With Nonuniform Agents: Individual-Mass Behavior and Decentralized ϵ-Nash Equilibria. IEEE Trans Automat Control 52:9 (2007), 1560–1571.

[243] A. Iqball and D. Abbot. Constructing quantum games from a system of Bell's inequalities. Phys Lett A 374 (2010), 3155-3163.

[244] A. Iqbal and A.H. Toor. Equilibria of Replicator Dynamics in Quantum Games. ArXiv: quant-ph/0106135.

[245] A. Iqbal and A.H. Toor. Quantum mechanics gives stability to a Nash equilibrium. Phys. Rev. A 65:2 (2002), 022306.

[246] A. Iqbal and A. H. Toor. Darwinism in quantum systems? Phys. Lett. A

294, 5-6 (2002), 261-270.
[247] A. Iqbal and A. H. Toor. Quantum cooperative games. Phys. Lett. A 293, 3-4 (2002), 103-108.
[248] G. Jumarie. Derivation and solutions of some fractional Black-Scholes equations in coarse-grained space and time. Application to Merton's optimal portfolio. Comput. Math. Appl. **59:3** (2010), 1142-1164.
[249] R. Kay, N. Johnson and S. C. Benjamin. Evolutionary quantum game. J. Phys. A: Math. Gen 34 (2001), L547-L552.
[250] F. S. Khan, N. Solmeyer, R. Balu and T. Humble. Quantum games: a review of the history, current state, and interpretation. Quantum Information Processing (2018), 17:309.
[251] V.N. Kolokoltsov. *The stochastic Bellman equation as a nonlinear equation in Maslov spaces. Perturbation theory.* Dokl. Akad. Nauk **323:2** (1992), 223-228. Engl. transl. in Sov. Math. Dokl. **45:2** (1992), 294-300.
[252] V. N. Kolokoltsov. Markov processes, semigroups and generators. DeGruyter Studies in Mathematics v. 38, DeGruyter, 2011.
[253] V. Kolokoltsov. Generalized Continuous-Time Random Walks (CTRW), Subordination by Hitting Times and Fractional Dynamics. Theory of Probability and its Applications **53:4** (2009), 594-609.
[254] V. N. Kolokoltsov. Game theoretic analysis of incomplete markets: emergence of probabilities, nonlinear and fractional Black-Scholes equations. Risk and Decision Analysis 4 (2013), 131-161.
[255] V. N. Kolokoltsov. Game-theoretic analysis of rainbow options in incomplete markets. In: P. Bernhard, J. Engwerda, B. Roorda, J.M. Schumacher, V. N. Kolokoltsov, P. Saint-Pierre and J.-P. Aubin. The Interval Market Model in Mathematical Finance: Game-Theoretic Methods. Birkhäuser, 2012, pp. 217-290.
[256] V. N. Kolokoltsov. Nonlinear Markov games on a finite state space (mean-field and binary interactions). International Journal of Statistics and Probability **1:1** (2012), 77-91.
[257] V. N. Kolokoltsov. The evolutionary game of pressure (or interference), resistance and collaboration (2014). http://arxiv.org/abs/1412.1269. MOR (Mathematics of Operations Research), 42:4 (2017), 915-944.
[258] V. Kolokoltsov. The probabilistic point of view on the generalized fractional PDEs. Fractional Calculus and Applied Analysis 22:3 (2019), 543-600.
[259] V. Kolokoltsov. Differential Equations on Measures and Functional Spaces. Monograph, Birkhauser, 2019.
[260] V. N. Kolokoltsov and A. Bensoussan. Mean-field-game model of botnet defence in cyber-security (2015). http://arxiv.org/abs/1511.06642 AMO (Applied Mathematics and Optimization) 74(3), 669-692,
[261] V. N. Kolokoltsov and O. A. Malafeyev. Mean field game model of corruption (2015). Dynamics Games and Applications. 7:1 (2017), 34-47. Open Access.
[262] V. Kolokoltsov and O. Malafeyev. Corruption and botnet defense: a mean field game approach International Journal of Game Theory. Online First: http://link.springer.com/article/10.1007/s00182-018-0614-1.

[263] V. N. Kolokoltsov and O. A. Malafeyev. Many Agent Games in Socio-economic Systems: Corruption, Inspection, Coalition Building, Network Growth, Security. Mongraph, Springer Nature, 2019.

[264] V. Kolokoltsov, V. Korolev, V. Uchaikin. Fractional Stable Distributions. J. Math. Sci. (N.Y.) **105:6** (2001), 2570-2577.

[265] V. Kolokoltsov and W. Yang. The turnpike theorems for Markov games. Dynamic Games and Applications **2: 3** (2012), 294-312.

[266] A. N. Kolmogorov. Selected Works in Mathematics and Mechanics, Moscow 1985 (in Russian), 136-137.

[267] P. L. Krapivsky and S. Redner. Organization of growing random networks. Phys. Rev. E **63** (2001), 066123.

[268] K. Kraus. States, Effects, and Operations. Springer Lecture Notes in Physics, v. 190. Springer, Berlin, 1983.

[269] J.-M. Lasry, P.-L. Lions. Jeux à champ moyen, I. Le cas stationnaire. Comptes Rendus Mathematique Acad. Sci. Paris 343:9 (2006), 619-625.

[270] Ch. F. Lee and N. F. Johnson. Efficiency and formalism of quantum games. Phys Rev A 67 (2003), 022311.

[271] H. Li, J. Du and S. Massar. Continuous variable quantum games. Physics Letters A 306 (2002), 73-78.

[272] J. E. Littleword. A mathematician's miscellany. London, 1957.

[273] J. Liu, Y. Tang and Z. R. Yang. The spread of disease with birth and death on networks. arXiv:q-bio/0402042v3. Published in Journal of Statistical Mechanics: Theory and Experiment, 2004.

[274] C. F. Lo and D. Kiang. Quantum Stackelberg duopoly. Physics Letters A 318 (2003), 333-336.

[275] T. J. Lyons. Uncertain volatility and the risk-free synthesis of derivatives. Applied Mathematical Finance **2** (1995), 117-133.

[276] W. Majewski. On the structure of positive maps: Finite-dimensional case. J. Math. Phys. 53 (2012), 023515.

[277] O. A. Malafeyev (Eds.) Introduction to modeling corruption systems and processes, (in Russian), v. 1 and 2. Stavropol, 2016.

[278] O. A. Malafeyev, N. D. Redinskikh and G. V. Alferov. Electric circuits analogies in economics modeling: Corruption networks. Proceedings of ICEE-2014 (2nd International Conference on Emission Electronics), DOI: 10.1109/Emission.2014.6893965, Publisher: IEEE.

[279] O. A. Malafeyev, S. A. Nemnyugin, E. P. Kolpak and A. Awasthi. Corruption dynamics model, AIP Conference Proceedings 1863, 170013 (2017); https://doi.org/10.1063/1.4992358.

[280] O. A. Malafeyev, D. Saifullina, G. Ivaniukovich, V. Marakhov and I. Zaytseva. The model of multi-agent interaction in a transportation problem with a corruption component. AIP Conference Proceedings 1863, 170015 (2017); https://doi.org/10.1063/1.4992360.

[281] V. P. Maslov. Nonlinear Averages in Economics. Mathematical Notes **78:3** (2005), 347-363.

[282] V. V. Mazalov. Mathematical game theory and Applications. John Wiley, 2014.

[283] V. V. Mazalov and A. N. Rettieva. Fish war and cooperation maintenance. Ecological Modelling 221 (2010), 1545-1553.

[284] M. M. Meerschaert and E. Scala. Coupled continuous time random walks in finance. Physica A **370** (2006), 114-118.

[285] M. M. Meerschaert, E. Nane and Y. Xiao. Correlated continuous time random walks. Statistics and Probabilt Letters **79:9** (2009), 1194-1202.

[286] R. V. Mendes. The Quantum Ultimatum Game. Quantum Infomation Processsing 4:1 (2005), 1-12.

[287] B. B. De Mesquita. The Predictioneer's Game. Random House 2010.

[288] D. A. Meyer. Quantum strategies. Phys Rev Lett 82:5 (1999), 1052-1055.

[289] P.-A. Meyer. Quantum Probability for Probabilists. Springer, 1993.

[290] M. A. Naimark. Normed Algebra. Wolters-Noordhoff, Groningen, 1972. Translation from Russian, Moscow, Nauka, 1968.

[291] M. E. J. Newman. Power Laws, Pareto distribtions and Zipf's law. Arxiv:cond-mat/0412004v3, 2006.

[292] B. Roorda, J.M. Schumacher and J. Engwcrda. Coherent Acceptability Measures in Multiperiod Models. Mathematical Finance **15:4** (2005), 589-612.

[293] B. P. Rosendorff and T. Sandler. Too Much of a Good Thing?: The Proactive Response Dilemma. Journal of Conflict Resolution **48** (2005), 657-671.

[294] M. B. Ruskai, S. Szarek and E. Werner. An analysis of completely positive trace-preserving maps on $\mathcal{M}_2$. Linear Algebra and its Applications 347 (2002), 159-187.

[295] F. Salten. Bambi, A life in the Woods. Engl. Transl. Simon and Schuster, 1928.

[296] F. Selleri. Quantum Paradoxes and Physical Reality. Kluwer Academic, 1990.

[297] S. N. Smirnov. Guaranteed deterministic approach to superhedging (in Russian). Mathematical theory of Games and Applications 10:4 (2018), 59-99.

[298] W. F. Stinespring. Positive functions on C^* -algebras. Proc. Amer. Math. Soc. 6 (1955), 211-216.

[299] E. Stormer. Positive linear maps of operator algebras. Acta Math. 110 (1963), 233-278.

[300] V. E. Tarasov. Fractional Dynamics: Applications of Fractional Calculus to Dynamics of Particles, Fields and Media. Springer 2011.

[301] M. Tugan-Baranovsky. Modern Socialism in its Historical Development. Saint Petersburg, 1906.

[302] V.V. Uchaikin, V.M. Zolotarev. Chance and Stability: Stable Distributions and their Applications. VSP, 1999.

[303] X.-T. Wang. Scaling and long-range dependence in option pricing I: pricing European option with transaction costs under the fractional Black-Scholes model. Phys. A. **389:3** (2010), 438-444.

[304] S.L. Woronowicz. Positive maps of low dimensional matrix algebras. Rep. Math. Phys. 10:2 (1976), 165-183.

[305] J. Zhou, L. Ma and Y. Li. Mulytiplayer quantum games with continous-variable strategies.

Index

ϵ-Nash equilibrium, 305

alleles, 146
ancilla, 369
anonymity, 99
anti-linear map, 322, 329
anti-unitary maps, 329
auction
  Dutch, 32
  flower, 32
  increasing bid, 34
  Vickrey, 32
average income, 269
axiom of participation, 101

backward induction, 53
  paradoxes, 46
Baire theorem, 243
basic postulate of quantum
  measurement, 331
battle of the sexes, 19
beam splitter, 334
Bell basis, 330
Bellman equation, 274, 316
  Duhamel formula, 275
Bellman operator, 217, 267, 283
Bernstein-Vazirani oracle, 373
Bertrand's duopoly, 131
best response principal, 304
binary interaction, 303
Bloch sphere, 327

Borda rule, 98
Borda score, 98
bottle imp, 41
Braess paradox, 36
budget set, 91

centipede game, 45
chicken game, 22
Choi-Kraus representation, 368
coalition form, 170
combinatorial game, 115
  fuzzy, 119
  impartial, 118
  negation, 120
  negative, 119
  positive, 119
  subtraction, 120
  sum, 119
  zero, 119
common knowledge of rationality, 14
compensation, reward and fitness
  principles, 85
completely positive maps, 366
compromise set, 89, 248
  ultimate, 89
conditional expectation in quantum
  systems, 332
Condorcet consistent rules, 97
Condorcet paradox, 96
Condorcet winner, 96
Coplend rule, 98

Coplend score, 98
core of the game, 82, 84, 170
cost-sharing game, 82, 84
coupons, 92
Cournot's duopoly, 129

de-quantization, 272
deadlock, 24
density matrix, 325
derivative securities, 25
dictator, 104
  benevolent, 85
dictator rule, 104
diet problem, 93
differential game, 50
Dirac's notations, 324
directed graph, 274
discounting, 54
dollar auction game, 34
dominance
  strict, 8
  weak, 15
dominated strategy
  elimination, 8
  iterated elimination, 14, 31
  strictly, 8, 29
  weakly, 15, 29
domineering, 117
duality, 162
  theorem, 162
dynamic game, 48
dynamic programming, 163

egalitarianism, 88
Einstein-Podolskii-Rosen (EPR)
  paradox, 335
entangled states, 324
  maximally, 324
entropy, 148, 326
Equality of payoffs lemma, 75, 134,
  136
equilibrium prices, 92
evolutionary stable strategy (ESS),
  65, 77, 142
excess demand, 92
exogenous rights, 85

extensive form, 46

fidelity, 363
fidelity distance, 363
function
  closed, 243
  semicontinuous, 243
function of individual demand, 92

game tree, 42
game with transferable utility, 84
general will versus will of all, 30
generalized dynamic system, 255
  accessibility function, 255
  trajectory, 255
generic property, 155
genetic locus, 146
Grundy's theorem, 123
guaranteed payoff, 25, 62
Guess the number game, 373

Hadamard-Walsh operator, 373
half wave plates, 334
Hamilton-Jacobi-Bellman, 279
Hausdorf distance, 167, 243
Hawk and Dove game, 67
Hawk-and-Dove game, 313
  nonlinear, 314
Head and Tail game, 4
hedge, 217
Hilbert-Schmidt norm, 323

idempotent algebra, 271
idempotent convolution, 283
idempotent semigroup, 272
infinitely repeated games, 53
inspection game, 165, 188
intensities of jumps, 300
Isaacs equation, 281

Königsberg bridges, 90
Kant's imperative, 20
kinetic equation, 290, 302
Kolmogorov's matrix, 300

Lagrange function, 163

Lanchester's model, 51
lattice, 111
  distributive, 112
law of diminishing returns, 184, 198
law of large numbers, 69
liars and truth tellers, 40
lifeboat story, 87
limit of a sequence, 62
Lion and Man game, 50
logic, 111
  quantum, 112
lower and upper games, 258
Lyapunov stability, 145, 149

Mach-Zender interferometer, 334
MAD strategy, 10
market wars, 51
market-entry game, 44
Markov chain, 300
Markov control problem, 316
Markov models of interaction, 288
master equation, 319
Matching Pennies, 4
max-plus algebra, 271
mean-field interacting particle
  system, 301
MFG consistency problem, 317
  with major player, 320
minimax strategies, 25
minimax theorem, 139
minority game, 312
minority principle, 102
mirror operator, 334
mixed states of a quantum system,
  325
  entangled, 326
  separable, 326
mixed strategy, 69, 133
monomorphic population, 77, 141
monotonicity, 100
Morra, 174

Nash equilibrium, 17, 29
Nash theorem, 73
natural selection, 144
neutrality, 99

Nim, 117
Nim number, 121
non-cooperative game, 29
non-selective measurement, 332
nonlinear Markov games, 51
normal form, 4, 29

optimality principle, 279
option, 215
  maturity date, 216
  multiple-strike, 215
  portfolio, 216
  rainbow or colored, 215
  spread, 216
order, 110

Pareto optimality, 8, 100, 246
Pareto set, 283
partial trace, 362
Pascal's wager, 28
passive optical devices, 334
Pauli channel, 370
Pauli matrices, 327
phase shifter, 334
plurality rule, 96
plurality with runoff, 97
polymorphic population, 77, 141
population profile, 141
positive operator valued measure
  (POVM), 336
postman problem, 90
pressure and resistance game, 306
principle of optimality, 316
prisoner's dilemma, 7, 9
  three-player, 350
  three-player, quantum, 351
profile, 133
projection valued measure (PVM),
  336
public goods game, 13
pure coordination game, 19
pure states of a quantum system, 322
purity, 326
pursuit game, 49

quantum game, 371

full strategic spaces, 371
    pure or unitary strategic space, 371
quarter wave plates, 334
qubit, 107, 112
qubits, 322
qunits, 322
qutrits, 322

reinforcement, 101
replicator dynamics, 313
replicator dynamics (RD), 143, 294
    stochastic, 296
risk, 74
robust control, 217
Rock-Paper-Scissors, 5
    n-person extension, 148
    in nature, 66

sacrifice game, 18
Schmidt decomposition, 324
Schrödinger cat, 110
selfish routing, 36
sequential majority comparisons, 98
sex ratio, 80
sex-ratio game, 313
Shapley vector, 170
Sherlock Holmes, 174
Solomon's wisdom, 42
spin vector, 331
    component, 331
St. Petersburg game, 74
Stackelberg duopoly, 132
stag hunt, 24
stand-alone principle, 82, 84, 170
Stern-Gerlach experiment, 331
Stinespring representation, 367, 368
stochastic games, 63
Stokes parameters, 327
strategic equivalence, 246
structural stability, 153, 244
subgame perfection, 49
subset
    first category, 243
    nowhere dense, 243

residual, 243
    second category, 243
Sun Tsu, 28
symmetric games, 6

tac-tix, 117
tensor product, 324
Tit-for-Tat, 58
trace norm, 323
transition matrix, 300
transition probabilities, 300
transition rates, 300
transportation problem, 93
transversality, 250
trigger strategy, 55
tropical algebra, 272
turnpike, 266
    on strategies, 270
    on the state space, 270
two-player game, 4

ultimatum game, 52
Uta stansburiana, 66
utilitarianism, 85, 88
utility function, 26, 91
    collective, 88

value of the game, 189
viscosity solutions, 262
von Neumann-Morgenstern solution,
    264
voting, 95
voting by successive veto, 102

Walras law, 92
War and Peace, 51
Wardrop equilibrium, 37
watch dog effect, 110
Wigner theorem, 329

Zeno paradox, 110
    quantum, 110
zero sum games, 5

CPSIA information can be obtained
at www.ICGtesting.com
Printed in the USA
LVHW050923220820
663883LV00008B/16

9 789811 214851